AF691171

15

Topics in Organometallic Chemistry

Editorial Board:

J. M. Brown · P. H. Dixneuf · A. Fürstner · L. S. Hegedus
P. Hofmann · P. Knochel · G. van Koten · S. Murai · M. Reetz

Topics in Organometallic Chemistry

Recently Published and Forthcoming Volumes

Surface and Interfacial Organometallic Chemistry and Catalysis
Volume Editors: C. Copéret, B. Chaudret
Vol. 16, 2005

Chiral Diazaligands for Asymmetric Synthesis
Volume Editors: M. Lemaire, P. Mangeney
Vol. 15, 2005

Palladium in Organic Synthesis
Volume Editor: J. Tsuji
Vol. 14, 2005

Metal Carbenes in Organic Synthesis
Volume Editor: K. H. Dötz
Vol. 13, 2004

Theoretical Aspects of Transition Metal Catalysis
Volume Editor: G. Frenking
Vol. 12, 2005

Ruthenium Catalysts and Fine Chemistry
Volume Editors: C. Bruneau, P. H. Dixneuf
Vol. 11, 2004

New Aspects of Zirconium Containing Organic Compounds
Volume Editor: I. Marek
Vol. 10, 2004

Precursor Chemistry of Advanced Materials
CVD, ALD and Nanoparticles
Volume Editor: R. Fischer
Vol. 9, 2005

Metallocenes in Stereoselective Synthesis
Volume Editor: T. Takahashi
Vol. 8, 2004

Transition Metal Arene π-Complexes in Organic Synthesis and Catalysis
Volume Editor: E. P. Kündig
Vol. 7, 2004

Organometallics in Process Chemistry
Volume Editor: R. D. Larsen
Vol. 6, 2004

Organolithiums in Enantioselective Synthesis
Volume Editor: D. M. Hodgson
Vol. 5, 2003

Organometallic Bonding and Reactivity: Fundamental Studies
Volume Editor: J. M. Brown, P. Hofmann
Vol. 4, 1999

Activation of Unreactive Bonds and Organic Synthesis
Volume Editor: S. Murai
Vol. 3, 1999

Lanthanides: Chemistry and Use in Organic Synthesis
Volume Editor: S. Kobayashi
Vol. 2, 1999

Alkene Metathesis in Organic Synthesis
Volume Editor: A. Fürstner
Vol. 1, 1998

Chiral Diazaligands for Asymmetric Synthesis

Volume Editors: Marc Lemaire · Pierre Mangeney

With contributions by
J. A. J. Breuzard · M. L. Christ-Tommasino · O. Chuzel · M. Fontecave
J. M. Fraile · J. I. García · O. Hamelin · M. Lemaire · P. Mangeney
J. A. Mayoral · S. Ménage · O. Riant · S. Roland · D. Savoia · E. Schulz

Springer

The series *Topics in Organometallic Chemistry* presents critical reviews of the present and future trends in polymer and biopolymer science including chemistry, physical chemistry, physics and material science. It is adressed to all scientists at universities and in industry who wish to keep abreast of advances in the topics covered.
As a rule, contributions are specially commissioned. The editors and publishers will, however, always be pleased to receive suggestions and supplementary information. Papers are accepted for *Topics in Organometallic Chemistry* in English.
In references *Topics in Organometallic Chemistry* is abbreviated *Top Organomet Chem* and is cited as a journal.

Springer WWW home page: http://www.springeronline.com
Visit the TOMC content at http://www.springerlink.com/

Library of Congress Control Number: 2005933266

ISSN 1436-6002
ISBN-10 3-540-26064-1 Springer Berlin Heidelberg New York
ISBN-13 978-3-540-26064-6 Springer Berlin Heidelberg New York
DOI 10.1007/b105252

This work is subject to copyright. All rights are reserved, whether the whole or part of the material is concerned, specifically the rights of translation, reprinting, reuse of illustrations, recitation, broadcasting, reproduction on microfilm or in any other way, and storage in data banks. Duplication of this publication or parts thereof is permitted only under the provisions of the German Copyright Law of September 9, 1965, in its current version, and permission for use must always be obtained from Springer. Violations are liable for prosecution under the German Copyright Law.

Springer is a part of Springer Science+Business Media

springeronline.com

© Springer-Verlag Berlin Heidelberg 2005
Printed in Germany

The use of registered names, trademarks, etc. in this publication does not imply, even in the absence of a specific statement, that such names are exempt from the relevant protective laws and regulations and therefore free for general use.

Cover design: *Design & Production* GmbH, Heidelberg
Typesetting and Production: LE-TeX Jelonek, Schmidt & Vöckler GbR, Leipzig

Printed on acid-free paper 02/3141 YL – 5 4 3 2 1 0

Volume Editors

Marc Lemaire

Universite Claude Bernard Lyon 1
308 (CPE)
UMR5181, CASO
43 bvd du 11 novembre 1918
69622 Villeurbanne, France
marc.lemaire@univ-lyon1.fr

Pierre Mangeney

Institut de Chimie Moléculaire FR 2769
Université P. et M. Curie
4 place Jussieu, T44-45, 2 ét., bte 183
75252 Paris cedex 05, France
mangeney@ccr.jussieu.fr

Editorial Board

Dr. John M. Brown

Dyson Perrins Laboratory South Parks Road
Oxford OX13QY
john.brown@chem.ox.ac.uk

Prof. Alois Fürstner

Max-Planck-Institut fur Kohlenforschung
Kaiser-Wilhelm-Platz 1
45470 Miihlheim an der Ruhr, Germany
fuerstner@mpi-muelheim.mpg.de

Prof. Peter Hofmann

Organisch-Chemisches Institut
Universität Heidelberg
Im Neuenheimer Feld 270
69120 Heidelberg, Germany
ph@phindigo.oci.uni-heidelberg.de

Prof. Gerard van Koten

Department of Metal-Mediated Synthesis
Debye Research Institute
Utrecht University
Padualaan 8
3584 CA Utrecht, The Netherlands
vankoten@xray.chem.ruu.nl

Prof. Manfred Reetz

Max-Planck-Institut fur Kohlenforschung
Kaiser-Wilhelm-Platz 1
45470 Miilheim an der Ruhr, Germany
reetz@mpi.muelheim.mpg.de

Prof. Pierre H. Dixneuf

Campus de Beaulieu
University de Rennes 1
Av. du Gl Leclerc
35042 Rennes Cedex, France
Pierre.Dixneuf@univ-rennesl.fr

Prof. Louis S. Hegedus

Department of Chemistry Colorado State
University Fort Collins
Colorado 80523-1872
USA
hegedus@lamar. colostate.edu

Prof. Paul Knochel

Fachbereich Chemie
Ludwig-Maximilians-Universität
Butenandstr.5-13
Gebäude F
81377 München, Germany
knoch@cup.uni-muenchen.de

Prof. Shinji Murai

Faculty of Engineering
Department of Applied Chemistry
Osaka University
Yamadaoka 2-1, Suita-shi Osaka 565
Japan
murai@chem.eng.osaka-u.ac.jp

Topics in Organometallic Chemistry Also Available Electronically

For all customers who have a standing order to The Handbook of Environmental Chemistry, we offer the electronic version via SpringerLink free of charge. Please contact your librarian who can receive a password or free access to the full articles by registering at:

springerlink.com

If you do not have a subscription, you can still view the tables of contents of the volumes and the abstract of each article by going to the SpringerLink Homepage, clicking on "Browse by Online Libraries", then "Chemical Sciences", and finally choose The Handbook of Environmental Chemistry.

You will find information about the

- Editorial Board
- Aims and Scope
- Instructions for Authors
- Sample Contribution

at springeronline.com using the search function.

Preface

Although much more recent than the use of chiral diphosphine, the use of chiral amine and chiral diamine as a ligand for metal mediated catalysis plays an increasing role in organic synthesis and catalysis. Compared to diphosphine ligands, nitrogen analogs present several advantages, mainly in relation to the chemistry of nitrogen organic compounds. Indeed, the chemistry of nitrogen-containing compounds appears to be much better documented than that of phosphorous derivatives. The higher stability of the amine compared to phosphine with regard to oxidation is also an advantage, which may explain the fast development of such reagents and ligands. Moreover, and in contrast to phosphine, amines and more generally, other nitrogen-containing molecules are ubiquitously present in the biological catalytic system. This fact gives us a specific area of research and acts as an inspiring model. Even if we limit our research to diamine and diamine derivatives, the use of such structures are numerous in different areas such as natural product and drug synthesis, multistep diastereoselective synthesis, organometallic chemistry and asymmetric catalysis. It is therefore not possible to deal with all such domains and we have focused on some recent results that we believe to have widespread potential applications, e.g. asymmetric catalysis.

The first chapter by Savoia deals with the synthesis of chiral diamines. From the original synthesis of Corey's diamine, a tremendous amount of research has been published in order to give access to this particularly interesting class of compounds that are often the basic framework of diazaligands. The second chapter, written by O. Chuzel and O. Riant, describes the use of a particular diamine, i.e. sparteine and isosparteine. These molecules were one of the first chiral diamines used for asymmetric catalysis thanks to the stable chirality bearing of the nitrogen atom and to the complexing properties of such molecules. As early as 1973, J.P. Guetté et al. demonstrated the possibility of an enantioselective Reformatsky reaction using a similar catalyst. During the last ten years, many uses of sparteine and isosparteine as ligands have been evaluated.

The third chapter presented by E. Schulz deals with the use of dinitrogen-containing ligands in three important asymmetric methods of C – C bond formation: asymmetric cyclopropanation, the Dields–Alder reaction and allylic substitution.

In these three cases, ligands containing two nitrogen atoms appear to produce results similar or often superior to most of the other type of ligands. Although a review was published on this area five years ago, the recent results are of considerable practical interest and an updated review seems perfectly justified.

The difficult separation and recycling of the expansive (and often toxic) catalyst is one of the main limitations of practical applications of asymmetric catalysis by transition metal complexes. Diamines also present several advantages in this area. J.P. Fraile, J.M. Garcia, and J.A. Mayoral have submitted as the fourth chapter a review article describing the possibility of using non-covalent bonding between chiral nitrogen-containing complexes and either solid or liquid supports in order to facilitate separation and recycling of the catalyst. The recent interest in chiral diamine is also related to this chemistry of similar functional groups. Diaza carben is obviously one of the more promising new ligands for metal transition catalysis. The review article written by P. Mangeney and S. Roland shows the versatility of chiral diamines as building blocks in the synthesis of diazacarbene. Diamine could also be very easily transformed into diurea and dithiourea, and these functional groups appear to be efficient alternative ligands for both catalysis by transition metal complexes and organo-catalysis. This latter process gave spectacular results in important areas such as the asymmetric Strecker reaction, as described in the sixth chapter. The final chapter may be the most intriguing in terms of recent developments of diamines in organometallic chemistry. Fontecave and Hamelin describe the access of "chiral at metal" complexes: this is of major fundamental interest and the first results with such complexes in asymmetric catalysis are also described. We believe that the potential applications of such fascinating structures could be of major interest in asymmetric catalysis for the next few years.

Paris, Villeurbanne, September 2005 Marc Lemaire and Pierre Mangeney

Contents

Top Organomet Chem (2005) 15: 1–58
DOI 10.1007/b136347
© Springer-Verlag Berlin Heidelberg 2005
Published online: 14 September 2005

Progress in the Asymmetric Synthesis of 1,2-Diamines from Azomethine Compounds

Diego Savoia

Dipartimento di Chimica "G. Ciamician", Università di Bologna, via Selmi 2, 40126 Bologna, Italy
diego.savoia@unibo.it

Abstract Recent advances in the preparation of optically pure 1,2-diamines by stereoselective addition reactions to azomethine compounds (imines, iminium ions, oximes, hydrazones, nitrones) are discussed. The methods scrutinized involve either the addition of carbon nucleophiles (organometallic reagents, carbanions, enolates, silyl enol ethers, carbon radicals and radical anions) to aldehyde derivatives, e.g., aldimines, and the complementary addition of hydrogen and hydride reagents to ketone derivatives, e.g., ketimines. Focus is given to the scope, advantages and limitations of the diverse methodologies and some possible developments are envisaged.

Keywords Azomethine compounds · 1,2-Diamines · Organometallic compounds · Reduction · Stereoselectivity

Abbreviations

Boc	*tert*-Butyloxycarbonyl
Cbz	Carboxybenzyl
CSA	(+)-Camphorsulfonic acid
de	Diastereomeric excess
DMF	*N,N*-Dimethylformamide
dr	Diastereomeric ratio
ee	Enantiomeric excess
HMPTA	Hexamethylphosphoric triamide
PMP	*para*-Methoxyphenyl
THF	Tetrahydrofuran
TMEDA	Tetramethylethylenediamine
TMS	Trimethylsilyl

1 Introduction

The 1,2-diamino moiety is present in many chiral natural products with valuable biological properties and in drugs where the pharmacological activity is related to their absolute configuration. Moreover, enantiopure 1,2-diamines are widely used as chiral auxiliaries and ligands in asymmetric synthesis and catalysis, especially the C_2-symmetric *trans*-1,2-diaminocyclohexane and *syn*-1,2-diphenyl-1,2-diaminoethane and their *N,N′*-disubstituted derivatives. The synthesis and chemistry of 1,2-diamines [1, 2] was reviewed a few years ago, and the many possible synthetic routes to these compounds from different functional groups have been thoroughly described. The "chiral pool" provides a lot of optically pure compounds that can be used as the starting material en route to properly designed, optically pure 1,2-diamines, either maintaining (part of) the existing stereocenter(s), or creating a new stereocenter(s) exploiting the asymmetric induction of the present one(s). Such asymmetric transformations have been increasingly employed in the last few decades, as an alternative to optical resolution [1, 3, 4] and enzyme-catalyzed kinetic resolution [5–7], which are principally based on the use of optically pure monocarboxylic and dicarboxylic acids.

In this chapter, the field of investigation is restricted to syntheses of 1,2-diamines with general structure **1** from precursors containing one or two prochiral azomethine functions C = N – Z (Scheme 1), involving in the first step the stereoselective formation of one or two stereocentres in the ethylene tether linking the two nitrogen atoms. To obtain the primary 1,2-diamines, the nitrogen substituent *Z* has to be removed in a subsequent step, by methodologies which must not affect the stereochemical integrity of the intermediate compounds. Specifically, route A involves the stereoselective formation of the C1 – C2 bond by the reductive coupling of monofunctional compounds **2**. This occurs by a preliminary single electron transfer from an electrode or a chemical reductant species, and is generally suited to the preparation of symmetrically 1,2-disubstituted 1,2-diamines **1** ($R^1 = R^2$). Also, the addition of an α-aminoalkyl metal reagent to monoazomethine compounds **2** leads to the same target **1** (route B). On the other hand, routes C–F exploit transformations of unsaturated compounds already possessing the N – C – C – N skeleton. In fact, in routes C and D the chiral α-amino azomethine compounds **3** and **4**, arbitrarily depicted with the *R* configuration, undergo addition of an organometallic reagent and a reducing agent (hydrogen, hydride), respectively. Finally, symmetrically or unsymmetrically 1,2-disubstituted 1,2-diamines **1** can be obtained by the analogous reactions of 1,2-bis(azomethine) compounds, i.e., 1,4-diazadienes **5** and **6** (routes E and F). The reactivity of the azomethine compound and consequently the applicability of a given methodology and/or the choice of the reagent are dependent on the nature of both the C and the N substituents (imine, oxime, hydrazone, nitrone). These methods are complementary to other ones which exploit the addition of nitrogen compounds to C = C bonds [1, 8]. The reduction of enantiopure α-aminoamides and α-aminonitriles as well as the building of an achiral two-carbon tether between the nitrogen atoms are not included in this survey, as they do not involve stereoselective steps. Moreover,

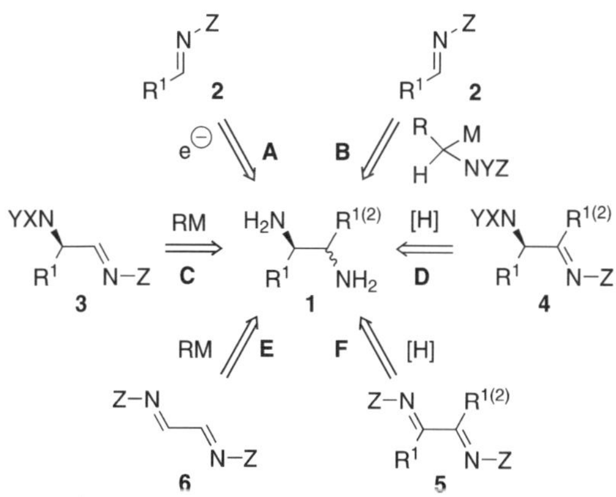

Scheme 1 Routes to 1,2-diamines from azomethine compounds

axially chiral 2,2′-bis(azaheterocycles) and 2-(1-aminoalkyl)azaheterocycles are not considered as 1,2-diamines.

It should be underlined that three diastereomers exist for symmetrically 1,2-disubstituted 1,2-diamines **1** ($R^1 = R^2$), i.e., the (*R*,*R*) and (*S*,*S*) enantiomers of *syn*-**1** and the *meso* form *anti*-(*R*,*S*)-**1**. On the other hand, in the case of different R^1 and R^2 substituents there are four diastereomers for **1**, i.e., two enantiomers for both the *syn* and *anti* diastereomers. As concerns the way by which the control of the absolute configuration of the novel stereocenter(s) can be achieved during the C – C or C – H bond(s) formation, one takes advantage of a preexisting stereocenter(s) in the substrates **3** and **4** (substrate-induced diastereoselectivity), or in the removable nitrogen substituent *Z* (chiral auxiliary) of **2**, **5** and **6** (auxiliary-induced diastereoselectivity). A third option is the use of a chiral coreagent, commonly a ligand coordinating the metal center of the organometallic reagent, metal hydride or Lewis acid salt (reagent-induced enantioselectivity). In the case where two adjacent sterocenters are formed (routes A, E and F) one is also faced with the problem of controlling the simple (*syn*/*anti*) diastereoselectivity.

2 Synthesis of 1,2-Diamines from Monoimines

2.1 Reductive Coupling of Imines

2.1.1 Introduction

The reductive coupling of imines (imino pinacol coupling) is a direct route to symmetrically 1,2-disubstituted 1,2-diamines, but both the efficiency and the stereochemical control are problematic. The relevance of these drawbacks is dependent on the nature of the azomethine compound, i.e., the nature of both the carbon and the nitrogen substituents, and the one-electron reduction method (metal, low-valent metal salt, electrochemistry). First of all, almost exclusively aromatic imines, i.e., those derived from aromatic aldehydes, are reactive enough to give the desired 1,2-diamines, which can be accompanied by the monoamine coming from a mere reduction step. On the other hand, aliphatic imines are more resistant to reduction, and aromatic ketimines can only be reduced to monoamines, owing to steric hindrance in the coupling step. Oximes and hydrazones are not useful substrates, owing to the possible cleavage of the N – O and N – N bonds in the reducing conditions.

In order to achieve a satisfactory preparation of C_2-symmetric (*R*,*R*)- or (*S*,*S*)-diamines, it is necessary to avoid or minimize the formation of the *meso* (*R*,*S*)-diastereomer. It is noteworthy that the photochemically promoted coupling prevalently leads to the *meso* 1,2-diamines. As a consequence of the poor reactivity of aliphatic imines, the desirable couplings of achiral aliphatic imines $R^1R^2CH^* - CH = NR^3$ and $R^1XCH^* - CH = NR^3$ (X is a heterosubstituent) and the intramolecular coupling of chiral 1,n-diimines, e.g., $R^*N = C(CH_2)_{n-2}C = NR^*$, have never been reported. Actually, the reductive coupling of aliphatic imines could be achieved using an excess of samarium diiodide [9] or the bimetallic redox system $Al/PbBr_2$ in the presence of $AlBr_3$ [10] in refluxing tetrahydrofuran (THF). However, it is expected that satisfactory stereocontrol should not be obtained from chiral aliphatic imines in these hard conditions.

2.1.2 Mechanism

The reductive coupling of imines can follow different pathways, depending on the nature of the one-electron reducing agent (cathode, metal, low-valent metal salt), the presence of a protic or electrophilic reagent, and the experimental conditions (Scheme 2). Starting from the imine **7**, the one-electron reduction is facilitated by the preliminary formation of the iminium ion **8** by protonation or reaction with an electrophile, e.g., trimethylsilyl (TMS) chloride. Alternatively, the radical anion **9** is first formed by direct reduction of the imine **7**, followed by protonation or reaction with the electrophile, so giving the same intermediate α-amino radical **10**. The 1,2-diamine **11** can be formed from the radical **10** by dimerization (and subsequent removal of the electrophile) or addition to the iminium ion **8**, followed by one-electron reduction of the so formed aminyl radical. In certain cases/conditions the radical **9** can be further reduced to the carbanion **12**, which then attacks the

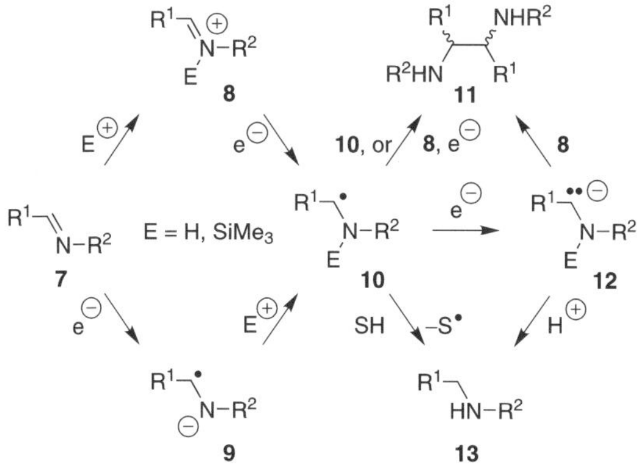

Scheme 2 Mechanism of the electrochemical or metal-promoted reductive coupling of imines in an acidic medium

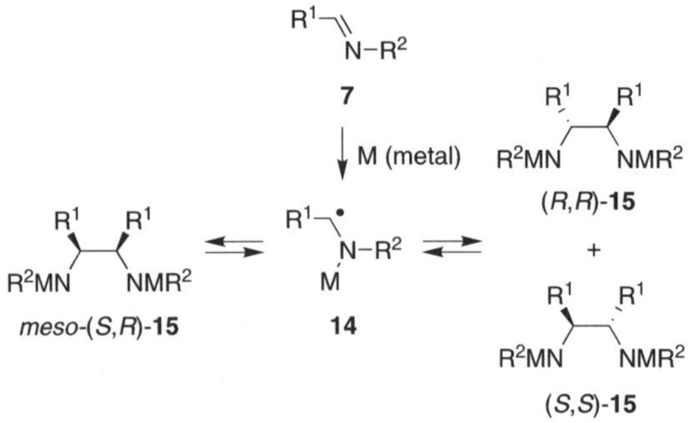

Scheme 3 Equilibration of *meso*- and *d,l*-1,2-diamines in the metal-promoted reductive coupling of imines

iminium ion **8** (this is described in Sect. 2.2.1). The amine **13** can be formed as a by-product from the radical **9** by H radical abstraction in the reaction medium, or by proton quenching of the carbanion **12**.

If a highly reducing metal is used, e.g., alkali metal, in the absence of the proton source or electrophile, the aromatic imine **7** (R^1 is aryl) is reduced to the α-amido radical **14** (Scheme 3). It must be underlined that the radical coupling leading to the diamide **15** is reversible at room temperature or even at 0 °C, depending on the nature of the metal [11]. As a consequence of the equilibrium, the ratio *meso*-to-*dl*-**15** is under thermodynamic control. Taking advantage of this, we can accomplish the isomerization of *meso* 1,2-diamines to the *dl* mixtures by treatment with 2 equiv of butyllithium in THF at room temperature or lithium/isobutylene at 35–40 °C, but the conversion should be monitored to avoid substantial or complete formation of the corresponding monoamine [12].

2.1.3
Synthesis of C_2-Symmetric Acyclic 1,2-Diaryl-Substituted 1,2-Diamines

Aromatic imines are the preferred substrates for the reductive coupling, e.g., by electrochemical reduction, owing to the lower reduction potentials with respect to aliphatic imines. Moreover, the success of the electrochemical coupling crucially depends on the nature of the electrodes used [13]. Generally, the control of the absolute stereochemistry of the two newly formed stereocentres is accomplished by the use of a chiral N substituent (chiral auxiliary). As a fact, the electrochemical reduction of a ketimine derived from 1-phenylethylamine gave poor results in terms of both yield and diastereoselectivity, and very low asymmetric induction was observed using enantiopure ammonium salts as coelectrolytes [14]. On the other hand, the electroreduction of imines derived from chiral nonracemic α-amino esters in the presence of chlorotrimethylsilane and triethylamine mainly gave ketals of 2,4-disubstituted azetidin-3-ones with high stereocontrol [15]. Conversely, a satisfactory conversion of the imine **16** to diamine **17** was accomplished by reduction with zinc and methanesulfonic acid, although the diastereoselectivity was only moderate [16]. The best results were obtained by introducing a provisional three-carbon tether between the two valine moieties of the reactive aromatic imines (see the coupling of the diimines **18** and **19** to give the 1,4-diazaheterocycles **20**). Particularly, complete stereocontrol was obtained for **20c**, which is derived from 2-hydroxybenzaldehyde, but the yield was low. The linear 1,2-diamines were then obtained by hydrolysis of the ester groups and oxidation of the carboxylates with lead tetraacetate, and were isolated as *N*,*N*′-di(Cbz) derivatives, where CbZ is carboxybenzyl, e.g., **21** (Scheme 4). Very low diastereoselectivies were observed working on the corresponding diimines derived from valine amides; this can be due to the use of *N*,*N*-dimethylformamide (DMF) as the solvent, instead of THF, in which the substrates are insoluble [16].

The coupling of N-substituted benzaldimines, mediated by the zinc–copper couple in the presence of (+)-camphorsulfonic acid (CSA) in DMF, was investigated. The best results were obtained for the imine **22**, and the optimal balance of yield, diastereoselectivity and enantioselectivity for the diamine **23** was obtained using 3 equiv of (+)-CSA [17] (Scheme 5). How-

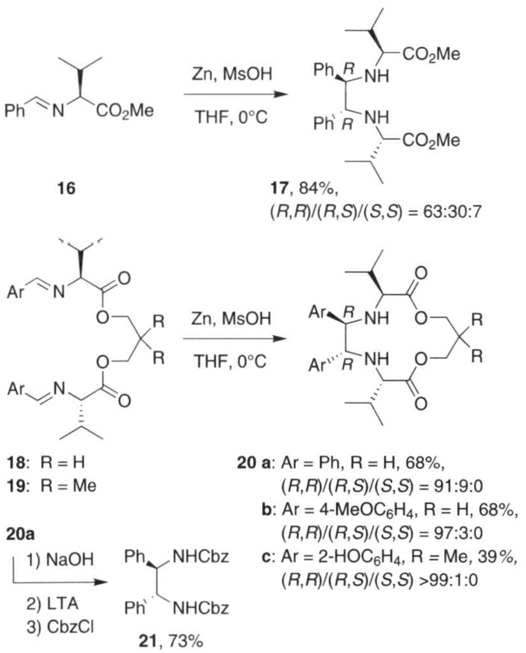

Scheme 4 Intramolecular Zn–Me_3SiCl-promoted reductive coupling of O,O′-tethered bis(*N*-arylidene valinates)

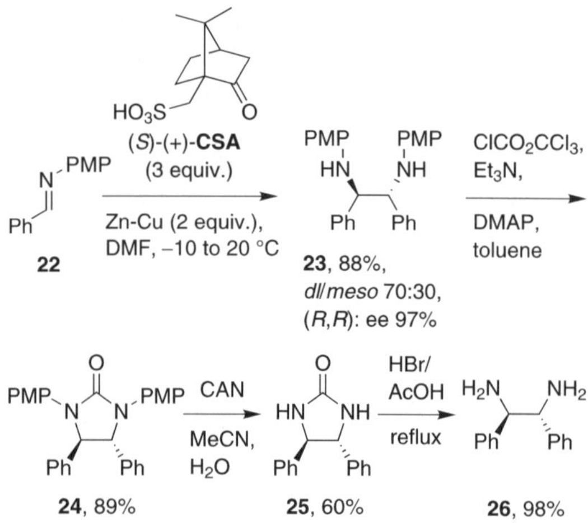

Scheme 5 Enantioselective zinc-promoted reductive coupling of benzaldimines in the presence of (+)-camphorsulfonic acid

ever, the amount of the *meso* dimer **23** was relevant, and the separation of the *dl* mixture required conversion of the crude product to the (–)-myrtenal aminal, followed by chromatography, or 1,3-imidazolidin-2-one **24**. The latter was obtained enantiomerically pure by recrystallization, then converted to the imidazolidinone **25** by oxidative removal of the *N*-(4-methoxyphenyl) substituent and finally to the free primary diamine (*R*,*R*)-**26** by ring cleavage in acidic medium. Unfortunately, the scope of the method was not fully defined.

Samarium diiodide promotes the coupling of *N*-benzylidene 1-phenylethylamine in the presence of ytterbium triflate giving the corresponding 1,2-diamine with high yield and a *syn*-to-*syn*-to-*anti* ratio of 46 : 16 : 38 in optimized experimental conditions [18]. Hence, the choice of the N substituent (chiral auxiliary) is crucial to achieve high stereoselectivity and at the same time to avoid the simple reduction pathway. As a fact, in the coupling of chiral benzaldimines with metallic samarium in the presence of a catalytic amount of iodine, *O*-methyl-(*S*)-valinol and especially *O*-methyl-(*R*)-phenylglycinol gave good performances. For example, the diamine **28** was exclusively formed from **27**, while reduction to the amine was almost suppressed (Scheme 6). On the other hand, in the same conditions *N*-benzylidene (*S*)-valinol was quantitatively reduced to *N*-benzyl (*S*)-valinol, and methyl (*S*)-valinate did not react [19]. It is noteworthy that the method does not work satisfactorily on benzaldimines bearing *ortho*-methoxy and *ortho*-bromo substituents, e.g., imine **29**, from which the by-product **31** was formed in addition to the expected product **30**. Protection of the alcoholic function in the imine **32** derived from (*R*)-phenylglycinol can be avoided by treatment with samarium and trimethylaluminum, which caused evolution of methane, then addition

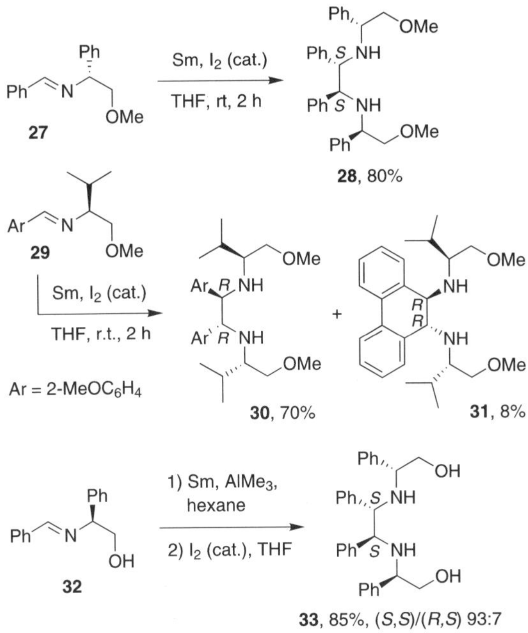

Scheme 6 Sm-promoted reductive coupling of chiral *C*-aryl imines

of a catalytic amount of iodine mainly led to the 1,2-diamine (*S*,*S*)-**33** with very high diastereoselectivity [20] (Scheme 6). Almost the same result was obtained using (*S*)-valinol as the chiral auxiliary. It is expected that primary (*S*,*S*)- and (*R*,*R*)-1,2-diamines can be prepared by both methods, removing the N substituents by oxidative protocols, e.g., with periodic acid–methyl amine, although this final step has not been reported.

Enantiomerically pure *ortho*-substituted $Cr(CO)_3$ benzaldimine complexes **34**, prepared from the corresponding resolved enantiomerically pure aldehyde complexes, underwent reductive coupling by treatment with samarium diiodide to stereoselectively give the $Cr(CO)_3$ diamine complexes **35** with good yields, together with the reduction products **36** [21]. The metal-free diamines **37** were then obtained by exposure to sunlight and air. It is noteworthy that the method is tolerant of *ortho* functional groups such as halogens and methoxy. Moreover, the planar chiral mono-$Cr(CO)_3$-complexed 2,2′-diiminodiphenyl **38** gave the *trans*-1,2-diamine **39** in optically pure form by SmI_2-mediated intramolecular coupling [22]. By an analogous route, the optically pure diferrocenyldiimine **40** was converted to the diamine **41** with high diastereoselectivity, then the optically pure diamine was obtained by crystallization [23].

The reductive coupling of the chiral planar di(imine) **42**, prepared from the corresponding chiral (*R*)-dialdehyde, in turn available as a single enantiomer by standard reactions, can lead to four stereoisomers. The use of

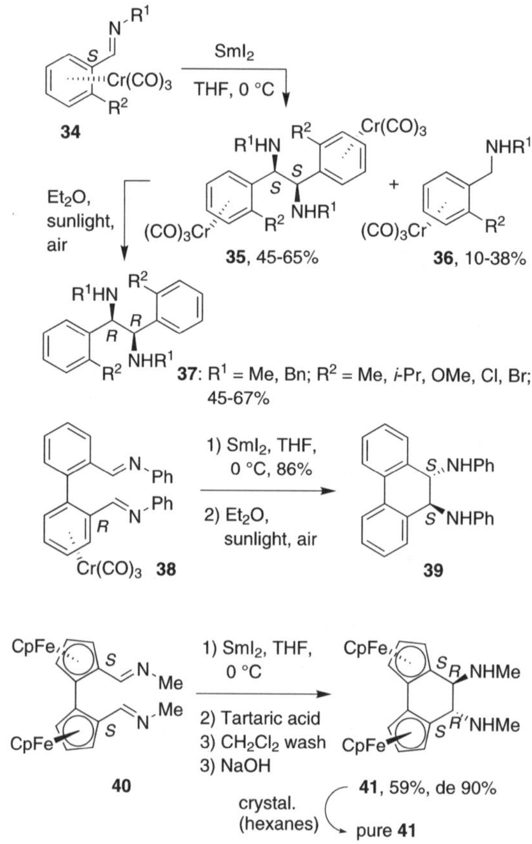

Scheme 7 SmI_2-promoted reductive coupling of optically pure tricarbonylchromium-complexed benzaldimines and ferrocenyl diimines

zinc–methanesulfonic acid gave a better yield of the product **43** than samarium diodide. In both cases, a single C_2-symmetric *trans* diastereomer and a *trans*-to-*cis* ratio of more than 98 : 2 were obtained [18] (Scheme 8). Addition of ytterbium triflate or other additives was detrimental for both the yield and diastereoselectivity. The configuration of the *trans* diastereomer could not be unambiguously determined, but the (*R*,*S*,*S*) configuration was assumed because the other *trans* diamine, (*R*,*R*,*R*)-**43**, is stabler than (*R*,*S*,*S*)-**43**, and the two *cis*-diamines have comparable stability, as resulted from MM3* calculations. Only one *cis*-diamine was obtained; hence, the reaction should be under kinetic control, in agreement with the prevalent formation of (*R*,*S*,*S*)-**43**.

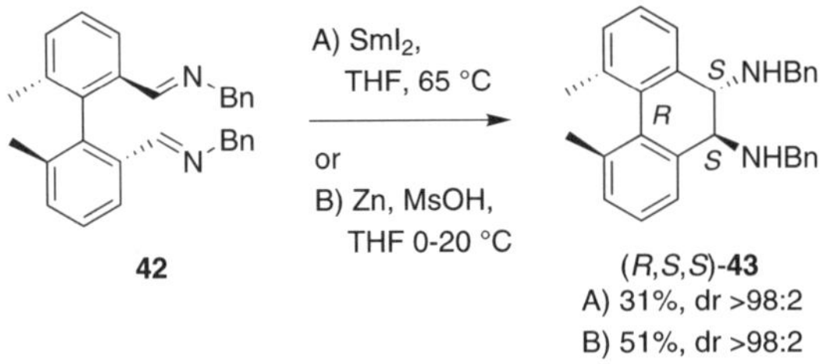

Scheme 8 Intramolecular reductive coupling of a planar chiral diimine

2.1.4
Synthesis of Trisubstituted and Tetrasubstituted Piperazines

A chiral piperazine can be constructed by the intramolecular reductive coupling of a diimine derived from the condensation of a chiral 1,2-diamine with 2 equiv of an aldehyde. For example, the diimines **44** have been converted to 2,3,6-trisubstituted piperazines **45**. The electrochemical reduction method, using a lead cathode in DMF in the presence of methanesulfonic acid, can be useful, being only slightly affected by the nature of the Ar substituent (the starting aromatic aldehyde); as a matter of fact, the piperazines **45** were obtained as single diastereomers with good to excellent yields [24]. Similarly, the diimines **46** derived from enantiopure 1,2-diaminocyclohexane underwent cyclization to the 1,4-diazadecalines **47**. In this case, the reaction was also performed using samarium diiodide in the presence of boron trifluo-

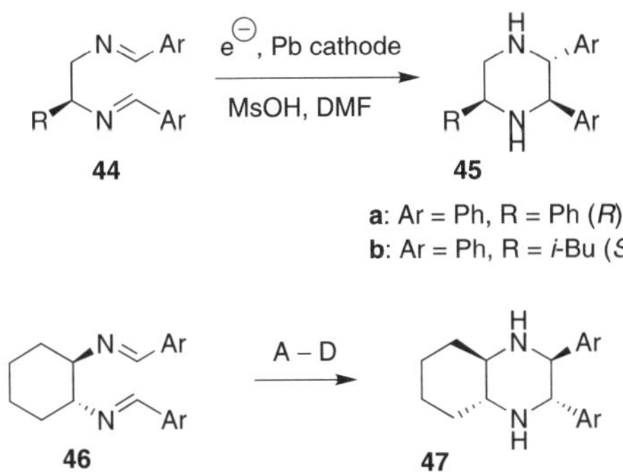

a: Ar = Ph, R = Ph (*R*), 92%
b: Ar = Ph, R = *i*-Bu (*S*), 87%

A) e^{-}, Pb cathode, MsOH, DMF
B) SmI_2, BF_3, THF, 65 °C
C) Zn, MsOH, THF, 0-20 °C
D) Mn, TFA, MeCN/toluene (9:1)

a: Ar = Ph; **A** 59%, **B** 60%, **C** 82%, **D** 99%
b: Ar = 4-FC_6H_4; **B** 42%, **C** 84%
c: Ar = 4-$MeOC_6H_4$; **B** 44%, **C** 41%
d: Ar = 2-HOC_6H_4; **A** 78%
e: Ar = 2-$MeOC_6H_4$; **A** 72%

Scheme 9 Synthesis of enantiopure piperazines by reductive coupling of chiral 2,5-diaza-1,5-dienes

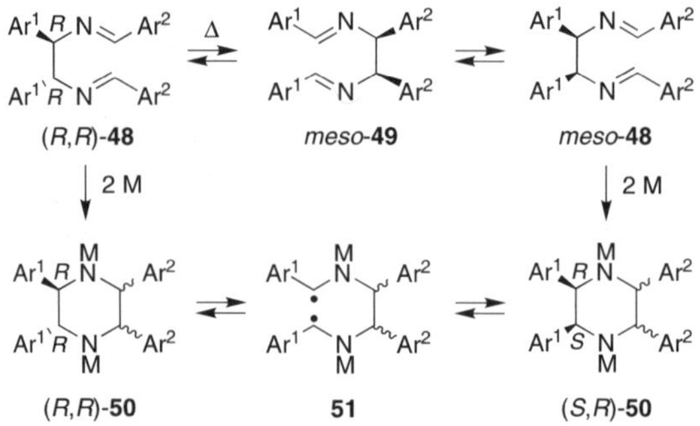

Scheme 10 Diaza-Cope rearrangement of chiral 2,5-diaza-1,5-dienes and homolytic C – C bond cleavage of dimetal piperazides cause loss of stereochemical purity

ride in refluxing THF [18], zinc–methanesulfonic acid [18], and manganese-trifluoroacetic acid [25] (Scheme 9). Remarkably, all the reactions were completely stereoselective.

It is noteworthy that piperazines have not been prepared by reduction of diimines derived from the condensation of 1,2-diarylethylenediamine and aromatic aldehydes. In this regard, it should be observed that problems may arise in the preparation of these diimines, because they can undergo a diaza-Cope rearrangement [26, 27] above room temperature, even during their preparation in refluxing benzene, causing loss of stereochemical integrity (see the equilibration of (*R*,*R*)-**48** to *meso*-**48** through *meso*-**49**, Scheme 10). Moreover, the piperazine diamide (*R*,*R*)-**50** formed by reduction of (*R*,*R*)-**48** with a metal can undergo homolysis of the C2 – C3 bond (and analogously the C5 – C6 bond) even at room temperature, as previously described in Scheme 3, so giving the diamido diradical species **51** and ultimately an equilibrium mixture of diastereomers, e.g., (*R*,*R*)-**50** and (*S*,*R*)-**50**.

2.1.5 Perspectives

In the future, further studies should be addressed to improve the chemoselectivity and diastereoselectivity of the reductive coupling process, especially searching for novel reagents and milder experimental conditions. As a matter of fact, a few novel reductive coupling procedures which showed improved efficiency and/or stereoselectivity have not been further applied to optically active imines. For example, a new electrochemical procedure which makes use of the "spatially addressable electrolysis platform" with a stainless steel cathode and a sacrificial aluminum anode has been developed for imines derived from aromatic aldehydes, and the use of the *N*-benzhydryl substituent allowed 1,2-diamines to be obtained with good yields and *dl*-to-*meso* ratios of more than 99 : 1 [28]. So, there is room to investigate protocols for enantioselective reactions.

Moreover, a dramatic increase of the reaction rate was observed when the coupling of aromatic imines mediated by samarium diiodide was carried out in the presence of both water and a tertiary amine or tetramethylethylenediamine (TMEDA) [29], causing the almost instantaneous formation of the 1,2-diamine, although with undetermined diastereoselectivity. Similarly, the samarium diiodide promoted reductive coupling of iminium ions formed in situ by reacting aliphatic aldehydes with secondary amines and benzotriazole occurred at temperatures as low as – 70 °C [30]. Even in this case a mixture of diastereomers with undetermined ratio was obtained; nevertheless, the item of diastereoselectivity induced by a chiral amine (auxiliary) is worthy of investigation.

An approach to the preparation of asymmetrically 1,2-disubstituted 1,2-diamines has been reported: the zinc–copper-promoted reductive coupling of two different *N*-(4-substituted)phenyl aromatic imines, one bearing a 4-methoxy and the other a 4-chloro substituent, in the presence of either boron trifluoride or methyltrichlorosilane, gave a mixture of the three possible 1,2-diamines, where the mixed one predominated [31]. Low degrees of asymmetric induction were observed using 1-phenylethylamine, phenylglycinol and its *O*-methyl ether, and several α-amino acid esters as the chiral auxiliaries; meanwhile the homocoupling process was not avoided (M.Shimizu, personal communication).

2.2 Addition of α-Amino Carbon Nucleophiles to Imines

2.2.1 α-Amino Organometallic Reagents

(η^2-Imine)metal complexes, better described as azametallacyclopropanes, are formed under proper experimental conditions, e.g., at high temperature or in the presence of hexamethylphosphoric triamide (HMPTA), by the reactions of aromatic aldimines and ketimines with lanthanide and early transition metals or their salts, and some of them could be isolated [32–38]. They can add to aldehydes or imines, so acting as α-amino organometallic reagents. For example, the diastereoselective coupling of a preformed (η^2-benzaldimine)titanium complex with aromatic and aliphatic imines has been described in the racemic series (more than 90%, diastereomeric ratio, dr, more than 97 : 3) [37]. Moreover, the (*N*-TMS benzaldimine)zirconocene complex, prepared from *N*-TMS benzylamine by a dehydrogenative route, was coupled with *N*-TMS benzaldimine to give the symmetrical zirconium diamide with good yield and high *syn* diastereoselectivity [39]. Most importantly, it is possible to prepare complexes from a chiral imine and use them in reactions with a different imine, possibly activated by an electron-withdrawing substituent. In fact, the zirconium-mediated asymmetric coupling of the chiral imine **52** with aldehydes, occurring through the preliminary preparation of the diastereomeric complexes **53**, which behave as chiral α-amino carbanions, has been reported [40] (Scheme 11). The opposite sense of asymmetric induction was observed by preparing the complex at different temperatures, demonstrating that an equilibrium between (*S*,*R*)-**53** and (*R*,*R*)-**53** occurred at high temperature. Hence, the same complexes could also be used to prepare symmetrically or unsymmetrically 1,2-disubstituted 1,2-diamines (*S*,*R*)-**54** and (*R*,*R*)-**54** by reaction with an electrophilic imine.

The asymmetric synthesis of unsymmetrical vicinal diamines by samarium diiodide induced reductive coupling of nitrones derived from aliphatic aldehydes with optically pure *N*-*tert*-butanesulfinyl aromatic imines has been recently reported [41]. For example, the reaction between nitrone **55** and

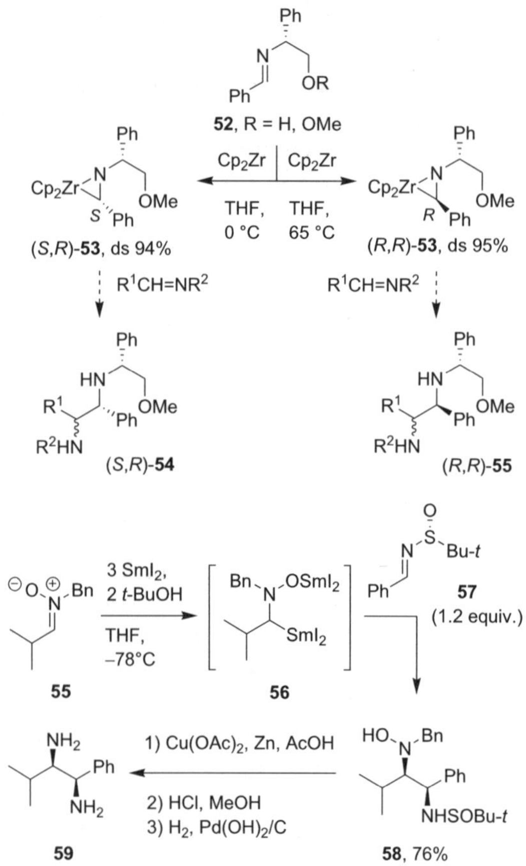

Scheme 11 Asymmetric syntheses of 1,2-diamines with α-amino organometallic reagents

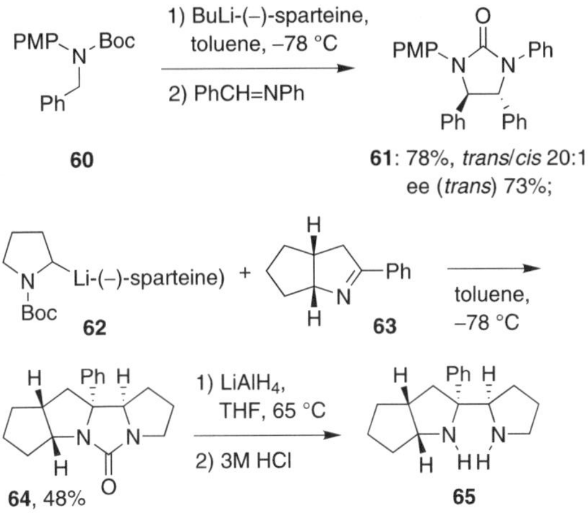

Scheme 12 (–)-Sparteine-mediated addition of α-amino organometallic reagents to imines

sulfinimine **57** gave the 1,2-diamine derivative **58** as a single diastereomer with good yield (Scheme 11). In this case, it was supposed that the dianion **56** is formed from the nitrone and then attacks the imine, because no reaction was observed in conditions where the reductive homocoupling of the nitrone by a radical mechanism occurred. The free diamine **59** was then obtained by routine procedures. It is noteworthy that the presence of *tert*-butanol improved the yield, whereas no reaction was observed when both substrates were aromatic or aliphatic, probably owing to an electronic effect. A range of different aliphatic nitrones were also tested with other aromatic sulfinimines, providing the expected coupling products with good yields, apart from sterically hindered nitrones, and high to complete diastereoselectivities.

The *n*-BuLi-(–)-sparteine complex is generally used to asymmetrically metallate N-substituted carabamates. For example, *N*-benzyl-*N*-Boc-*p*-anisidine **60**, where Boc is *tert*-butyloxycarbonyl, was metallated and the resulting chiral organolithium-(–)-sparteine reagent was added to *N*-benzylidene aniline to give the substituted imidazolidinone **61** with good yield and high stereoselectivity [42] (Scheme 12). Similarly, the lithium reagent **62** was prepared from *N*-Boc pyrrolidine and added to optically pure ketimines, e.g., **63**, to give imidazolidinones, e.g., **64**, as single stereoisomers with low to moderate yields [43]. In this case, (–)-sparteine had no influence on the stereoselectivity, since the use of TMEDA as a bidentate ligand gave **64** with the same stereochemical outcome, although with lower yield. The free diamine **65** was then obtained by cleavage of the imidazolidinone ring.

2.2.2
α-Nitro Carbanions

The condensation of nitro compounds and imines, the so-called aza-Henry or nitro-Mannich reaction, has recently emerged as a powerful tool for the enantioselective synthesis of 1,2-diamines through the intermediate β-amino nitro compounds. The method is based on the addition of a nitronate ion (α-nitro carbanion), generated from nitroalkanes, to an imine. The addition of a nitronate ion to an imine is thermodynamically disfavored, so that the presence of a protic species or a Lewis acid is required, to activate the imine and/or to quench the adduct. The acidic medium is compatible with the existence of the nitronate anion, as acetic acid and nitromethane have comparable acidities. Moreover, the products are often unstable, either for the reversibility of the addition or for the possible β-elimination of the nitro group, and the crude products are generally reduced, avoiding purification to give the desired 1,2-diamines. Hence, the nitronate ion is an equivalent of an α-amino carbanion.

Reactions of nitro compounds with chiral imines have only recently been described. Either chiral 1-phenylethylamine (auxiliary) or the glyceraldehyde acetonide aldehyde was used as the chiral precursors of the imines **66** and **68**, which reacted with 3-mesyloxynitropropane to give the 3-nitropyrrolidines (*dl*)-**67** and **69**, respectively, with good diastereoselectivity. In fact, both products were obtained (almost) exclusively as *trans* diastereomers with high level of asymmetric induction, but the configurations of the newly formed stereocenters were not determined [44] (Scheme 13). *N*-Boc imines can be formed

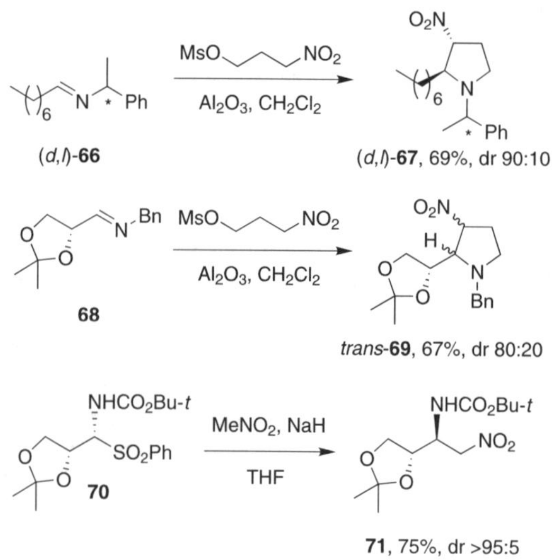

Scheme 13 Diastereoselective addition of α-nitrocarbanions to chiral imines

in situ from α-amido sulfones in the presence of the nitro compound in the basic medium, meanwhile generating the nitronate ion. For example, starting from glyceraldehyde, the α-amido sulfone **70** was prepared, then treated with nitromethane and sodium hydride to give the β-amido nitro compound **71** with good yield and apparently complete diastereoselectivity [45]. The nature of both the carbamoyl group and, especially, the starting chiral aldehyde affects the diastereoselectivity in analogous reactions.

Most efforts have been directed to the development of enantioselective catalytic procedures applied to prochiral substrates. For example, the addition of nitroalkanes to *N*-phosphinoyl aromatic imines **72** is catalyzed by heterobimetallic complexes derived from optically pure binaphthol (Scheme 14). The addition of nitromethane, but not homologous nitroalkane, to **72** in toluene–THF solution at – 40 °C was catalyzed by the basic complex **73** (20 mol %), prepared from ytterbium triisopropoxide, potassium *tert*-butoxide and (*R*)-binaphthol in the optimized ratio 1 : 1 : 3, affording the β-amino nitro compound **74** with high yields and enantiomeric excesses (ees) [46]. Then, treatment with samarium diiodide allowed the simultaneous reduction of the nitro group to primary amine and cleavage of the *N*-phosphinoyl substituent, e.g., leading to (*R*)-1-phenylethylenediamine, isolated as the dihydrochloride **75**. The optimized protocol for reactions of nitroalkanes required the complex **76** as a catalyst, so the β-amino nitro compounds **77** were obtained as mixtures of *anti* and *syn* diastereomers with moderate to good enantioselectivities (ee *anti* up to 83%), then the primary 1,2-diamine, e.g., **78**, was routinely prepared [47].

The highly diastereoselective and enantioselective (ee 95%) synthesis of α,β-diamino acid esters has been obtained by the addition of preformed silyl nitronates, e.g., **80**, to the *N*-*p*-methoxyphenyl glyoxylate imine **79** using copper(II) bisoxazoline complex **81** as the catalyst (Scheme 14). The reaction was carried out at – 100 °C, because the uncatalyzed racemic reaction proceeded to some extent at – 78 °C [48]. The *anti* diastereomer (*R*,*R*)-**82** was obtained with excellent yield, diastereoselectivity and enantioselectivity, then the monoprotected diamine **82** was prepared by reduction with Raney nickel. The *N*-tosyl substituent was unsuitable for the reaction, and many other Lewis acid–chiral ligand complexes, including the copper triflate complex **84**, provided the product (*R*,*R*)-**82** with lower enantioselectivities. Moreover, the reaction of the imine **79** with nitropropane, catalyzed by the complex **84**, could be performed under very user-friendly conditions (room temperature, open air, undistilled CH_2Cl_2 as the solvent) in the presence of a catalytic amount of triethylamine, providing the product (*R*,*R*)-**82** with high yield and excellent diastereoselectivity and enantioselectivity [49]. It is noteworthy that 20 mol % of triethylamine was strictly necessary, since no reaction occurred with 10 mol % and no diastereoselection was observed with 40 mol % loading. It is expected that other nitro compounds can be used with similar results.

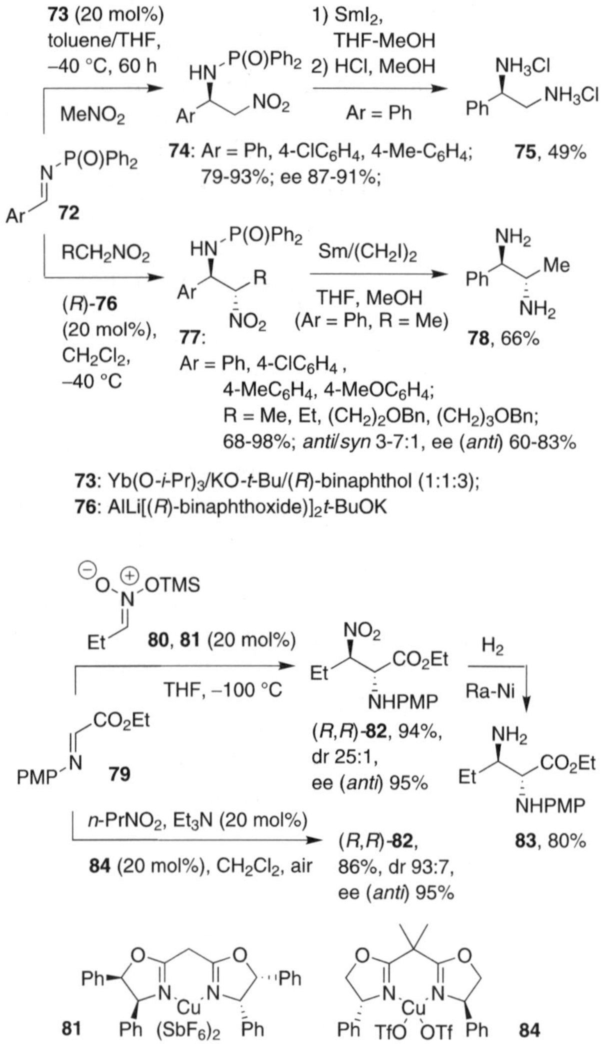

Scheme 14 Catalytic asymmetric aza-Henry (nitro-Mannich-type) reactions

Recently, enantioselective organo-catalytic procedures for the aza-Henry reaction have been disclosed. The presence of either an acidic or a basic function appears to be a requisite of the catalyst. In fact, the condensation of nitromethane with *N*-phosphinoyl arylimines **72** is catalyzed by the chiral urea **85** derived from (*R*,*R*)-1,2-diaminocyclohexane and gives the product (*R*)-**74** with good yield and moderate enantioselectivity (Scheme 15) [50]. The *N*-phosphinoyl substituent is determinant, as the addition of nitromethane to the *N*-phenyl benzaldimine failed and the reaction of the *N*-tosyl benzaldimine gave the expected adduct with quantitative yield but almost no

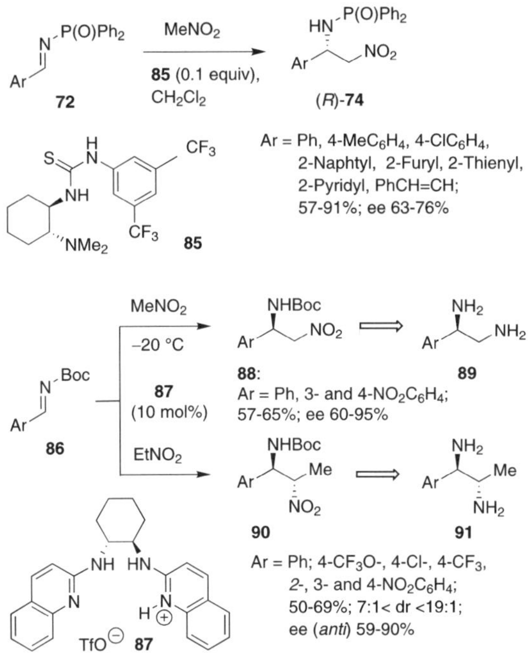

Scheme 15 Organo-catalytic enantioselective aza-Henry reactions

enantioselectivity. It is noteworthy that an effective and highly diastereoselective solvent-free aza-Henry reaction of nitroalkanes and *N*-phosphinoyl arylbutylimines and *tert*-butylimines, catalyzed by 1,1,3,3-tetramethylguanidine, has been described [51].

Similarly, the reaction of nitro compounds with the *N*-Boc aromatic imines **86** occurred in the presence of the enantiopure protic catalyst **87**, which is a white, crystalline bench-stable salt [52] (Scheme 15). The reactions of nitromethane, very slow at – 20 °C, were accelerated in the presence of 10 mol % of **87**, and the β-amino compounds **88** were obtained with moderate yields and moderate to high enantioselectivities. Positive results were also obtained in the corresponding reactions of nitropropane to give the products **90**. Hence, the primary diamines **89** and **91** are available by this route, which is advantageous for the significantly lower cost and toxicity of the catalyst and its easy removal from the reaction mixture simply by a basic wash. These results should stimulate further research on the development of new acid-catalyzed systems.

Like the nitronate ion, the cyanide ion is synthetically equivalent to the aminomethyl carbanion $(CH_2NH_2)^-$, because of the possible reduction of $-CN$ to the $-CH_2NH_2$ group. Consequently, the addition of cyanide ion to imines to give α-aminonitriles (Strecker-type reaction) is a viable route to 1,2-diamines. As a matter of fact, a number of diastereoselective and catalytic enantioselective Strecker-type reactions have been described, but they were generally directed to the synthesis of α-amino acids by hydrolysis of the intermediate α-aminonitriles. Nevertheless, some Strecker-type reactions and another use of α-aminonitriles are described in Sects. 2.3.2 and 2.4.3.

2.2.3 α-Amino Enolates and α-Amino Silyl Enol Ethers

The addition of enolates derived from α-amino ester derivatives is a useful method for the preparation of optically active α,β-diamino acids. The asymmetric induction can be provided by the use of a chiral auxiliary as the N substituent of the imine. The *N*-sulfinyl imine (*S*)-**92** is particularly reactive and undergoes addition of lithium enolates even at low temperature. In fact, the addition of the enolate derived from *N*,*N*-dibenzyl ethyl glycinate occurred to the *re* face of the imine to give the 2,3-diamino ester derivative **93** with *syn* diastereoselectivity, then the free diamino ester **94** was readily obtained by a routine two-step procedure [53]. It is noteworthy that the analogous reaction with the enolate derived from *N*-diphenylmethylidene ethyl glycinate occurred with higher yield and opposite simple diastereoselectivity (1.6 equiv is strictly required), as a consequence of the opposite geometry of the enolate, giving the *anti* adduct **95** [53]. In this case, both the N substituents of **95** could be removed in acidic conditions, so that the free diamino ester **96** was obtained in a single step.

Conversely, lithium enolates of *N*-benzylidene-α-substituted-α-amino acid methyl esters added to the *si* face of aryl sulfinimes (*S*)-**97**, presumably by a concerted mechanism, when allowing the temperature to rise from – 78 to 4 °C, affording the imidazolidines (*R*,*R*)-**98** with high diastereoselectivity [54] (Scheme 16). Next, cleavage of the imidazolidine ring and removal of the sulfinyl N substituent was accomplished by treatment with phosphoric or trifluoroacetic acid, to give the 2,3-diamino esters (2*R*,3*R*)-**99**. Cleavage of the aminal moiety while preserving the sulfinyl group is also possible [55]. On the other hand, the reactions with glycine methyl ester derivative occurred only in the presence of boron trifluoride (3.25 equiv), presumably by forming in situ the sulfinimine complex (*S*)-**100**, then allowing the mixtures to stand in 0.1 M chloroform solutions for at least 24 h. In this case, the imidazolidines (*S*,*R*)-**101** were obtained with moderate to good diastereoselectivities, presumably by a stepwise mechanism [54]. By treatment of the imidazolidines with lithium aluminum hydride, concomitant reduction of the ester and cleavage of the aminal function occurred, e.g., the preparation of the diamino alcohol (2*S*,3*R*)-**102** [56].

The asymmetric synthesis of 2,3-diamino acids can be accomplished by the addition of chiral enolates to prochiral imines. For example, reaction of morpholine-2-one **103**, derived from (*S*)-phenylglycinol, with *N*-benzyl benzaldimine in the presence of pyridine and *para*-toluenesulfonic acid at high

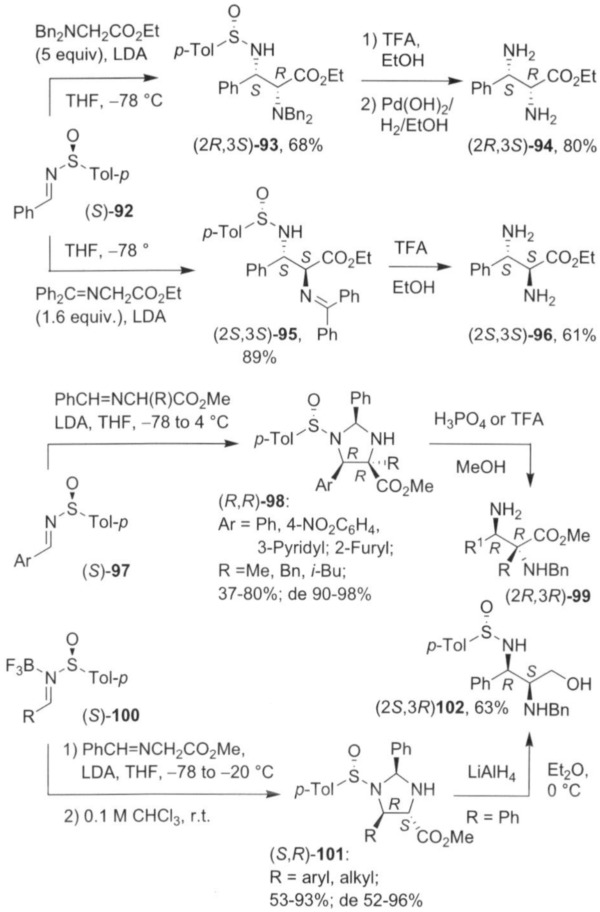

Scheme 16 Diastereoselective synthesis of 2,3-diamino esters and alcohols by the addition of achiral glycine iminoester enolates to chiral *N*-sulfinylimines

temperature generated the chiral azomethine ylid **104**, whose cycloaddition with same imine gave the bicyclic compound **105** as a single diastereomer with moderate yield. Even higher yields were obtained using 4-substituted benzaldimines. Then, hydrogenolysis of **105** afforded the *syn* 2,3-diamino acid **106** [57] (Scheme 17). Moreover, the base-promoted addition of the chiral nonracemic nickel(II) glycine complex **107** with *N*-(4-methoxyphenyl) trifluoromethylimine allowed for an efficient synthesis of the 2-trifluoromethyl-2,3-diamino acid **108** [58]. Both the yield and the dr of the intermediate nickel complexed adducts were affected by the nature of the base and the presence of lithium chloride, and the best diastereoselectivity (dr 99 : 1) was obtained using triethylamine in the presence of lithium chloride.

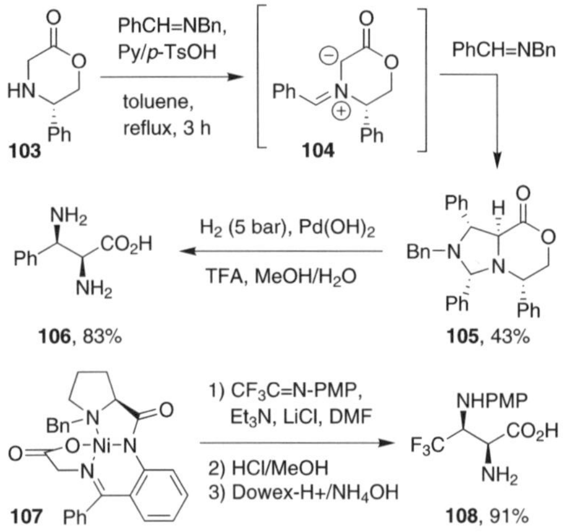

Scheme 17 Asymmetric synthesis of 2,3-diaminoacids by the addition of a chiral iminoester enolate to achiral imines

Finally, the enantioselective condensation of enolizable α-amino ester derivatives with imines can be carried out in the presence of catalytic amounts of a chiral ligand–metal complex and a base. For example, the reaction of methyl isocyanoacetate with the *N*-tosyl arylimines **109** to give the imidazolines **111** was catalyzed by the gold complex formed in situ from gold(I) chloride–dimethyl sulfide in dichloromethane and the chiral ferrocenyldiamine **110**, probably also acting as a base [59] (Scheme 18). From nine phenyl-substituted imines and α-naphthylimine, the imidazolines were obtained as *cis*/*trans* mixtures (ratios more than 90 : 10) and moderate to good ee (46–88%). However, the *cis* isomer was isolated by column chromatography and a single recrystallization allowed enantiomerically pure *cis*-imidazolines **111** to be obtained with moderate yields. Then treatment with concentrated HCl in ethanol gave the monoprotected 2,3-diamino acid esters **112** with quantitative yields. In one case, the free diamino acid was prepared from **112** (Ar is Ph) by reaction with phenol and HBr–AcOH at reflux temperature, followed by addition of water and neutralization with propylene oxide.

The analogous catalytic asymmetric Mannich reaction of *N*-(diphenylmethylidene)glycine methyl ester with aliphatic and aromatic *N*-tosyl imines, e.g., the benzaldimine **109**, took place in THF or toluene at – 20 °C in the presence of 4-Å molecular sieves, triethylamine (10 mol %), and a complex prepared from a copper salt ($CuClO_4$ or $CuPF_6$) and an enantiomerically pure P,N ligand (Scheme 18). Among the several ligands tested in this reaction, optimum performances were given by complex **114**, where the steric hindrance of the *p*-phenyl P substituent in the ligand was determinant. The product **115** was obtained with good diastereoselectivity (*syn*-to-*anti* ratio of 80 : 20)

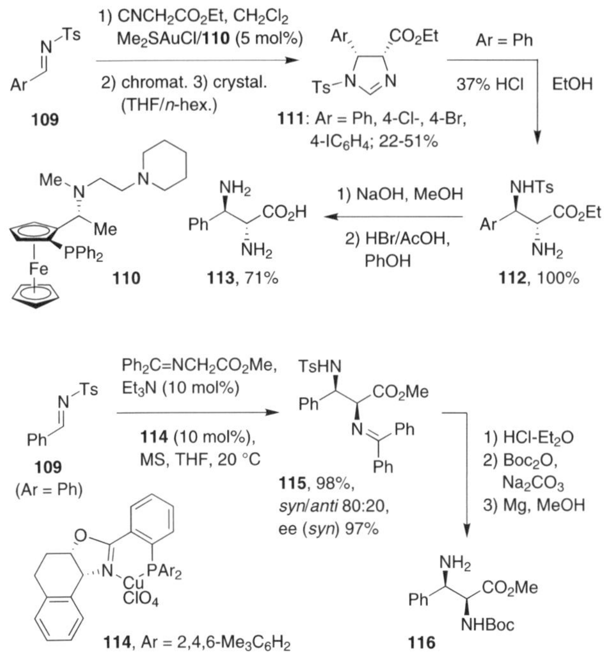

Scheme 18 Reagent-induced enantioselective catalytic synthesis of 2,3-diamino esters by addition of α-amino and α-iminoester enolates to imines

and excellent enantioselectivity for both diastereomers, then *syn*-**115** was converted to the monoprotected diamino acid **116** [60]. Similar levels of diastereoselectivity and enantioselectivity were obtained using diverse aliphatic imines, whereas from *C*-aryl imines the *syn*-to-*anti* ratios were lower, although the enantioselectivity was excellent for both diastereomers. On the other hand, the glyoxylate imine underwent addition with low levels of diastereoselectivity and enantioselectivity.

The addition of silyloxypyrrole **118** to *N*-sulfinyl imines (*S*,*S*)-**117** and (*S*,*R*)-**117**, which differ in the configuration of the sulfur stereocenter, gave the unsaturated γ-lactams (*S*,*S*,*S*)-**119** and (*S*,*R*,*R*)-**119**, respectively, with complete and opposite diastereoselectivity. As a matter of fact, the configuration of the α-stereocenter of the imine C substituent, being the same in both starting imines, did not affect the stereocontrol [61] (Scheme 19). The anti-influenza compound A-315675 **120** could be constructed by this this methodology, which therefore can be more extensively exploited for the enantioselective synthesis of 2-(1-aminoalkyl)pyrrolidines. Compounds possessing the 2,2′-dipyrrolidine skeleton were prepared by the addition of the silyloxypyrrole **118** to pyrrolidinium ions, e.g., the one formed in situ from the pyrrolidine derivative **121**, derived from D-glyceraldehyde acetonide. In this case,

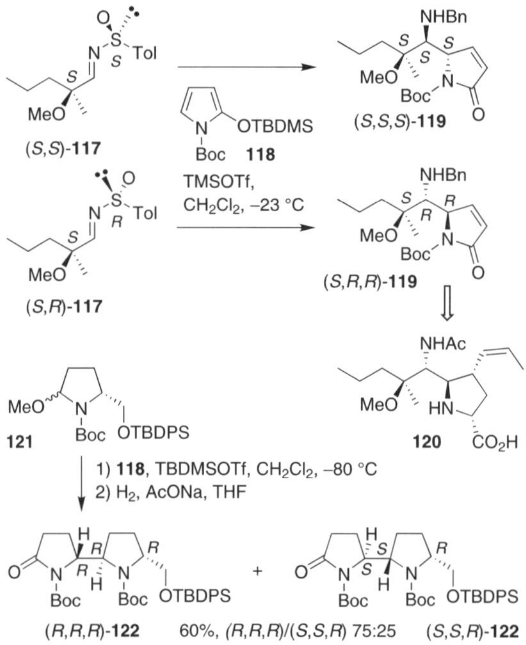

Scheme 19 Asymmetric synthesis of 2-(1-aminoalkyl)pyrrolidines and 2,2′-dipyrrolidines from chiral imines and iminium ions

a mixture of (*R*,*R*,*R*)-**122** and (*S*,*S*,*R*)-**122** was obtained with moderate 1,3-asymmetric induction but complete *syn* diastereoselectivity [62] (Scheme 19).

2-Silyloxypyrroles bearing a chiral N substituent can be prepared from optically pure amines and then added to a prochiral imine or iminium ion. For example, the 2-silyloxypyrrole **123** was prepared from (*R*)-phenylglycinol and employed in a three-component reaction together with benzaldehyde and aniline in propionitrile at low temperature, in the presence of a catalytic amount of ytterbium triflate and molecular sieves. In these conditions, the imine–Lewis acid complex was formed in situ and underwent nucleophilic addition [63] (Scheme 20). The products **124** were obtained as mixtures of two to four diastereomers with prevalent but modest *anti* diastereoselectivity and moderate enantioselectivity, then separation of the two main *anti* diastereomers could be accomplished after hydrogenation and basic hydrolysis of the benzoate function, i.e., at the stage of compounds **125**. Afterwards, reduction with lithium aluminum hydride and cleavage of the chiral auxiliary by hydrogenolysis gave the enantiomers of 2-(1-phenylamino)ethylpyrrolidine **126**. It should be pointed out that the corresponding primary amines would be available by using the same route for *N*-(*para*-methoxyphenyl)imines, owing to the easy oxidative removal of this substituent.

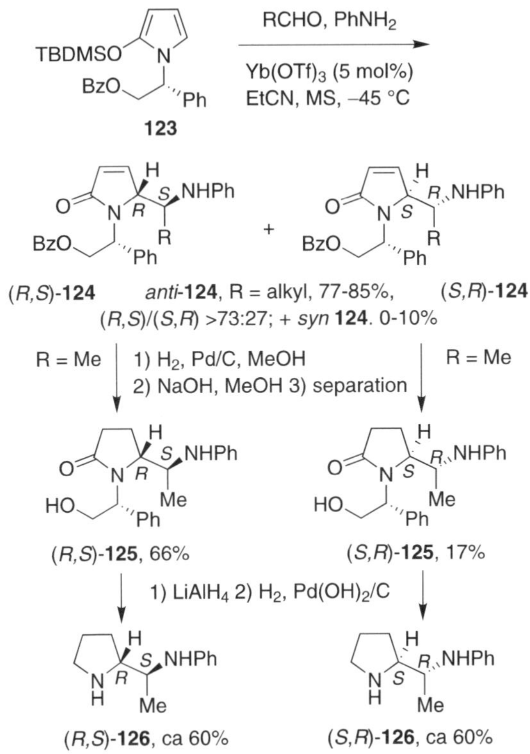

Scheme 20 Asymmetric synthesis of 2-(1-aminoalkyl)pyrrolidines by the addition of a chiral silyloxypyrrole to achiral imines

2.3
Addition of Carbon Nucleophiles to Chiral α-Amino Azomethine Compounds

2.3.1
Addition of Organometallic Reagents

This route has been widely exploited because of the availability of α-amino azomethine compounds from natural (*S*)-α-amino acids, through the corresponding α-amino aldehydes, which are configurationally stable provided that the amino function is suitably protected. Moreover, some α-amino acids are available with the *R* configuration and a number of enzymatic and chemical transformations have been described for the preparation of optically active unnatural α-amino acids. Overall, the route suffers from the additional steps required for protection/deprotection of the amino function and, in the case of hydrazones and nitrones, cleavage of the N – N or N – O bond.

2.3.1.1
α-Amino Imines and Iminium Ions

In this case, the choice of the organometallic reagent is dictated by the nature of the imine N substituent, which also affects the diastereoselectivity. Organolithium compounds in THF and, preferably, organocerium reagents formed by combining equimolar amounts of organolithium compounds, apart from MeLi, and $CeCl_3$, added to *N*-benzyl α-dibenzylamino imines **127** to give the *syn*-diamines **128** with good yields [64] (Scheme 21). In this case, the diastereoselectivity (dr more than 82 : 18) was determined by the preliminary formation of a chelate complex between the imine, acting as a *N,N′* bidentate ligand, and the organometallic reagent, which afterwards attacked the less hindered *si* face of the imine (1,2-asymmetric induction). The pure major diastereomers *syn*-**128** were isolated by column chromatography, then converted to the *N,N′*-disubstituted and unsubstituted *syn*-1,2-diamines **129** and **130**, respectively, by the proper reduction methods. On the other hand, the *N*-tosyl imines **131** underwent nonchelation controlled addition by Grignard reagents, owing to the poor basicity of the imine nitrogen, to give the *anti*-

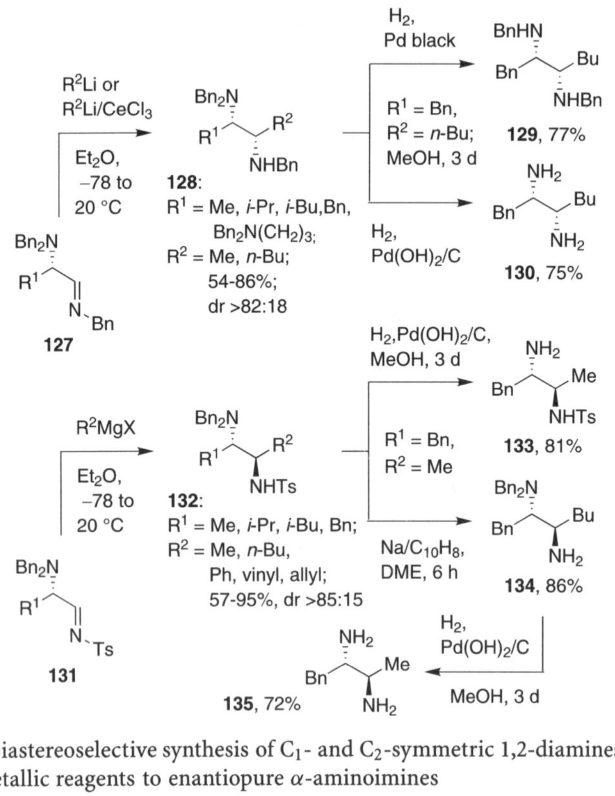

Scheme 21 Diastereoselective synthesis of C_1- and C_2-symmetric 1,2-diamines by addition of organometallic reagents to enantiopure α-aminoimines

diamine derivatives **132**. The monoprotected and free diamines **133–135** were in turn obtained by applying to **132** the proper reducing protocols.

The addition of trimethylsulfonium ylide to the α-amino imine derivative **136** gave 2-(1-aminoalkyl)aziridines **137** with high yields and *anti* diastereoselectivities (*re* face attack) [65] (Scheme 22). In a complementary approach, the boron trifluoride promoted addition of Grignard reagents to the α-aziridino imine **138**, bearing two stereocenters, gave 2-(1-aminoalkyl) aziridines **139** in most cases with (almost) complete *syn* diastereoselectivity [66]. Removal of the N substituents from **137** and **139** was not described, but should be possible by appropriate procedures. Chiral aziridines are potentially useful as auxiliaries or ligands for a number of asymmetric transformations [67]. Moreover, since the aziridine ring can be easily cleaved, diversely functionalized nitrogen compounds can be obtained from them. For example, asymmetrically 1,2-disubstituted *syn*-1,2-diamines **140** might be prepared through the selective cleavage of the aziridine N – C3 bond of **139** by $Pd(OH)_2/C$-catalyzed hydrogenolysis, as described for α-aziridino alcohols [68].

The chiral α-pyrrolidino imine that is formed in situ in the reaction of α-sulfonyl amide **141** with allylic zinc reagents undergoes addition with mod-

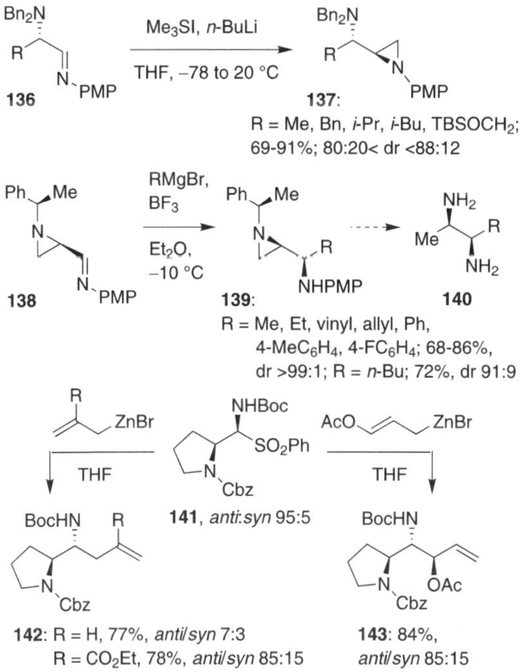

Scheme 22 Stereoselective synthesis of 2-(1-aminoalkyl)aziridines and 2-(1-aminoalkyl) pyrrolidines

erate efficiency and diastereoselectivity to give the 2-(1aminoalkyl)pyrrolidines derivatives **142** and **143** [69] (Scheme 22).

Double asymmetric induction operates when the azomethine compound is derived from a chiral α-amino aldehyde and a chiral amine, e.g., the sulfinimine **144** [70]. In this case, the *R* configuration at the sulfur of the chiral auxiliary, *N*-*tert*-butanesulfinamide, matched with the *S* configuration of the starting α-amino aldehyde, allowing complete stereocontrol to be achieved in the preparation of the diamine derivatives **145** by the addition of trifluoromethyl anion, which was formed from trifluoromethyltrimethylsilane in the presence of tetramethylammonium fluoride (Scheme 23). The substituents at both nitrogen atoms were easily removed by routine procedures; see, for example, the preparation of the free diamine **146**. On the other hand, a lower diastereoselectivity (dr 80 : 20) was observed in one reaction carried out on the imine derived from (*R*)-aldehyde and (*R*)-sulfinamide.

Similarly, the oxazolidine (*R*,*S*)-**147**, obtained as a mixture of epimers at C2 from *N*,*N*-dibenzyl (*R*)-phenylalaninal and *N*-benzyl (*S*)-valinol, reacted with Grignard reagents to form in situ the iminium ion **148**, from which the diamino alcohols **149** were produced as a single diastereomer [71] (Scheme 23). On the other hand, when the oxazolidine derived from the (*S*)-aldehyde was used, the diamino alcohol was obtained as a 70 : 30 mixture of diastereomers. Alhough the preparation of the primary 1,2-diamines was not explored in that paper, compounds **149** would be the precursors of the *syn* 1,2-diamine

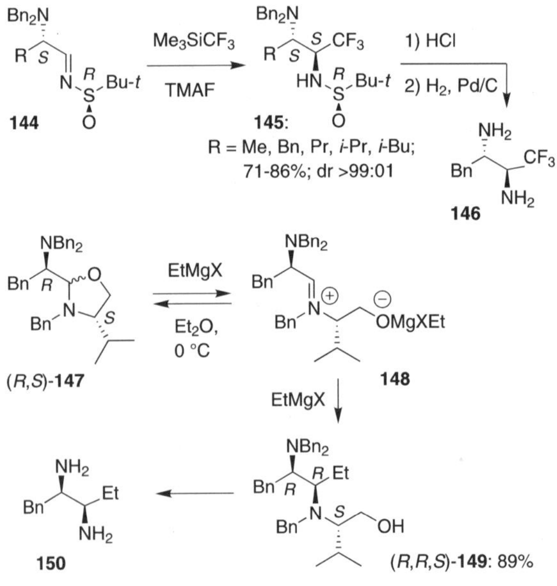

Scheme 23 Double asymmetric induction in the addition of Grignard reagents to chiral α-amino imines and α-amino iminium ions

150, by selective oxidative removal of the chiral auxiliary and subsequent hydrogenolysis of the *N*-benzyl substituents.

The original Kulinkovic hydroxycyclopropanation protocol can be applied to *N*,*N*-dimethylcarboxamides of *N*-allyl α-amino acids **151** to give the diamines **131** with strained bicyclic structures [72] (Scheme 24). The mechanism involves the intramolecular attack of an (η^2-alkene)titanium complex **152** to the carboxamide function, leading to the intermediate titanaoxacyclopentane **153**, followed by formation of the iminium ion **154** and subsequent formation of the cyclopropane ring in **155**. First experiments were carried out on amides of phenylalanine **127a**, tyrosine and tryptophan using cyclopentylmagnesium chloride and chlorotitanium triisopropoxide as the reagents, whereas the serine derivative **127b** required methyltitanium triisopropoxide and cyclohexylmagnesium bromide [73]. Similarly, tricyclic and tetracyclic compounds were prepared starting from proline and 2-indoline carboxylic acid, e.g., the conversion of **156** to **157** [74]. In all cases the products were obtained with good yields but only moderate stereoselectivity. No attempts to set up an enantioselective route to such cyclopropylamines by using chiral ligands on titaniums have been described.

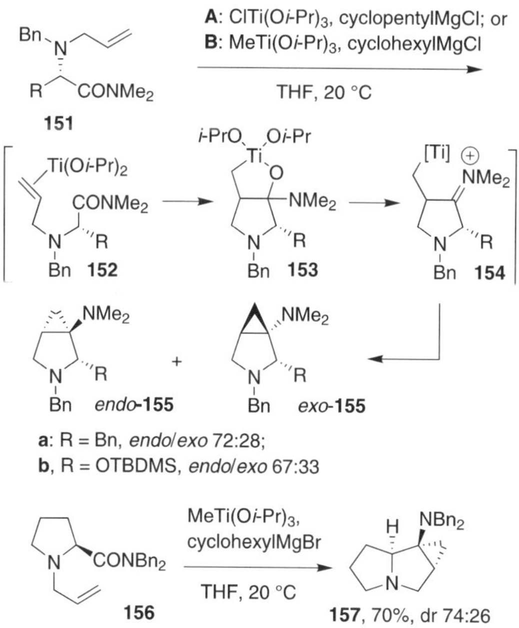

Scheme 24 Titanium-mediated intramolecular reductive cyclopropanation of *N*-allyl-α-amino acid dimethylamides

2.3.1.2
α-Amino Hydrazones

Another route to monosubstituted 1,2-diamines is based on the addition of $RLi/CeCl_3$ or Grignard reagents to chiral hydrazones, in which a stereogenic center is placed in the nitrogen substituent, e.g., **158** [75] (Scheme 23). By quenching the reaction mixtures with acetyl chloride, the *N*-acetyl hydrazines **159** were obtained with good to excellent yields and diastereoselectivities. A drawback of the method is the need to use the reagents in excess. After chromatographic separation of the diastereomers, the N – N bond was cleaved by treatment with lithium/ammonia to give the protected diamines **158**.

Alternatively, chiral hydrazones can be prepared from chiral aldehydes, e.g., **161**, which underwent addition of allyltrichlorosilane in DMF to give the homoallylic hydrazine **162** with high yield and excellent diastereoselectivity [76] (Scheme 25).

2.3.1.3
α-Amino Nitrones

N-substituted and *N*,*N*-disubstituted chiral α-amino nitrones are also available from the corresponding α-amino aldehydes. In a first study, it was shown that the addition of phenylmagnesium bromide to *N*-Boc derivatives exclusively gave the *syn* *N*-Boc α-aminohydroxylamines with good yields, but

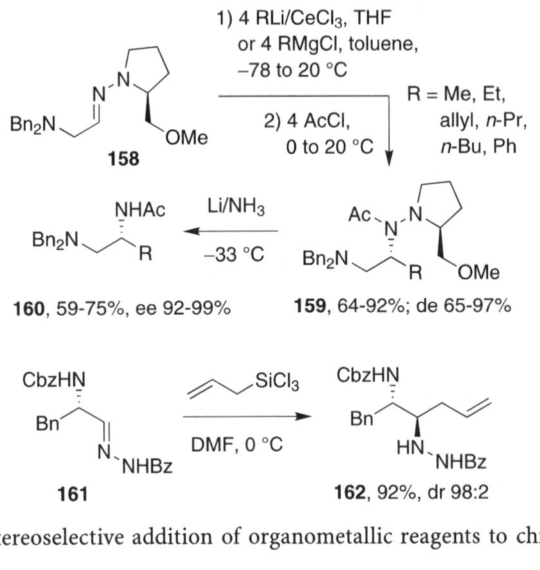

Scheme 25 Diastereoselective addition of organometallic reagents to chiral α-amino hydrazones

methylmagnesium bromide added in a stereorandom fashion [77]; moreover, the adduct of methylmagnesium bromide to an *N*-benzyl-*N*-Boc α-amino nitrone was obtained with only moderate yield, although with complete *syn* selectivity. No other Grignard reagents were examined. More recently, a tunable diastereoselectivity was accomplished during the addition of 2-lithiothiazole, TMS acetylide and Grignard reagents to the L-serine derived nitrones featuring different protections of the amino and hydroxyl functions, so constituting a useful approach to C2 epimers of 2,3-diamino-4-hydroxybutanal and 2,3-diamino-4-hydroxybutanoic acid [78, 79] (Scheme 26). Actually, the addition of the organometallic reagents to the nitrone **163** takes place to give the α-amino hydroxylamine derivatives **164** with (almost) complete *syn* diastereoselectivity (more than 95% by ^{1}H NMR), which was interpreted in terms of steric effects of the substituents. By a sequence of steps involving also the transformation of the hydroxylamine function, the R substituent introduced is then converted to either a carboxylic group (R is acetylide in **164**), so giv-

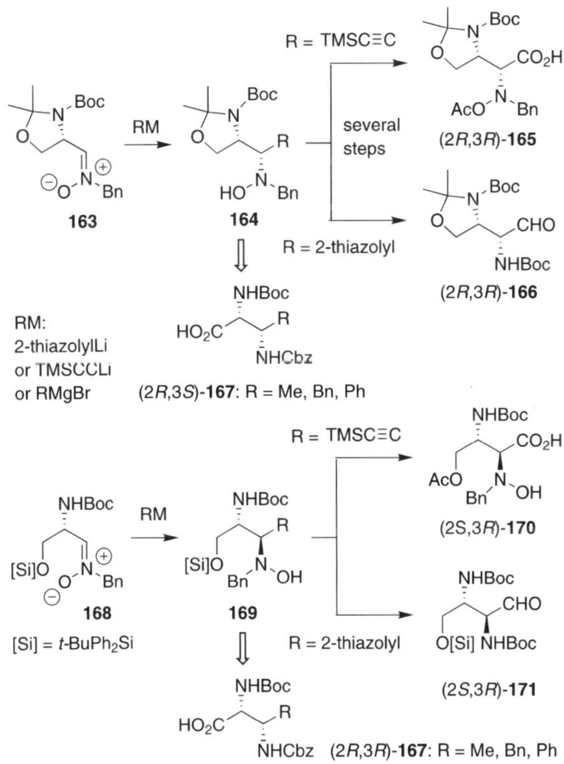

Scheme 26 Diastereoselective synthesis of C2-epimers of 2,3-diamino-4-hydroxybutanal and 2,3-diamino-4-hydroxybutanoic acid

ing the protected diamino acid (2*R*,3*R*)-**165**, or the corresponding aldehyde (2*R*,3*R*)-**166** (R is thiazolyl in **164**) in 50–52% overall yield.

On the other hand, following the same sequences from the differently protected serine-derived nitrone **168**, through the formation of hydroxylamines **169**, C2 epimers of carboxylic acid and aldehydes are obtained, i.e., (2*S*,3*R*)-**170** and (2*S*,3*R*)-**171**. Moreover, the *syn* adducts **164** were exclusively obtained in the addition of Grignard reagents to the nitrone **163**, whereas the same reactions on nitrone **168** occurred with a partial loss of diastereoselectivity [80]. α,β-Diamino acids (2*R*,3*S*)- and (2*R*,3*R*)-**167** can also be prepared from the α-amino hydroxylamines **164** and **169** by reduction, deprotection and oxidation steps. The diastereoselective addition of acetylide anion to *N*,*N*-dibenzyl L-serine phenylimine has been also described [81].

2.3.2 Addition of Nitronates, Enolates, Silyl Ketene Acetals and Cyanide Ion

As previously described, in basic conditions the proline-derived α-sulfonyl amide **141** generates the imine function, which afterwards undergoes addition by a nucleophile, e.g., a nitronate ion; see the diastereoselective synthesis of the diamino nitroalkane derivative **172**, which is the precursor of the piperazine-2-carboxylic acid **173**, through a Nef reaction [45]. Similarly, the addition of the lithium enolate of ethyl acetateto the α-sulfonyl amide **174** gave the diamino ester derivative **175**, wich was then converted to (–)-1-aminopyrrolizidine **176** (Scheme 27).

The *N*-acyliminium methodology has been exploited to prepare pharmacologically active molecules, e.g., GW 311616A [82] and inhibitors of the

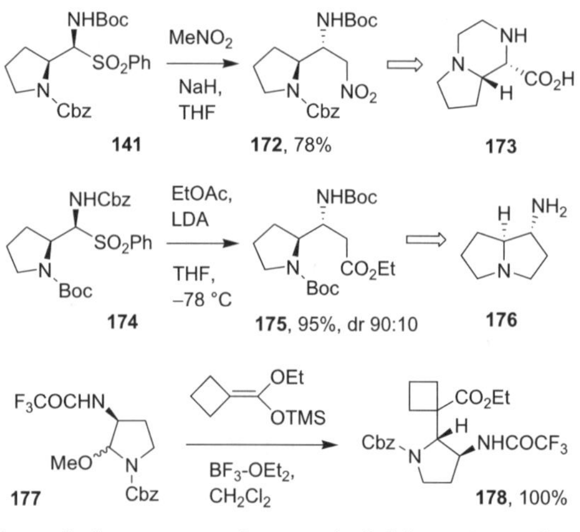

Scheme 27 Addition of nitronates, enolates and silyl ketene acetals to chiral α-amino imines and iminium ions

Hepatite Virus NS3/4A having the pyrrolidine-5,5-*trans*-lactam skeleton [83], starting from (*R*)- and (*S*)-methionine, respectively. The key step is the addition of the proper silyl ketene acetal to an iminium ion, e.g., that generated by treatment of the intermediate **177** with boron trifluoride, which provided the adduct **178** with better diastereoselectivity than other Lewis acids. Inhibitors of hepatitis C virus NS3/4A were efficiently prepared by a similar route from (*S*)-methionine [83]. The addition of indole to a chiral α-amino iminium ion was a completely diastereoselective step in a reported synthesis of tilivalline, a natural molecule which displays strong cytotoxicity towards mouse leukemia L 1210 [84].

The first reports on the Lewis acid promoted addition of TMS cyanide to α-dibenzylamino imines derived from naturally occurring (*S*)-α-amino acids demonstrated that the yield of the 2,3-diaminonitrile derivatives and the diastereoselectivity were affected by the nature of the substrate, most importantly the N substituent, and the Lewis acid. The best results were generally obtained using the imines **179** derived from phenylalaninal, and the preferred Lewis acid was determined for three differently N substituted imines [85] (Scheme 28). In the case of the *N*-benzyl imine, titanium tetrachloride gave the best results in the preparation of **180** (method A), whereas boron trifluoride caused epimerization of the product at room temperature. On the other hand, in the reaction of the *N*-sulfonyl imine no epimerization of the product **180** was observed with the same Lewis acid (method B). Diethylaluminum chloride was the Lewis acid of choice for the addition to the *N*-TMS imine (method C), in this case the TMS protection was lost during the workup. The cyanide group is susceptible to further transformation, e.g., the reduction of **180** with lithium aluminum hydride gave the triamine derivative **181** with moderate yield.

The Strecker reaction has been performed on the aldehyde **182** prepared from L-cysteine [86] (Scheme 28). The imine was formed in situ by treatment with benzylamine, then TMS cyanide was added to afford prevalently in almost quantitative yield the *syn*-diamine **183**, which is the precursor of (+)-biotin **184**. The *syn* selectivity was largely affected by the solvent, toluene being the solvent of choice. Since the aldehyde **182** is chemically and configurationally unstable, a preferred protocol for the synthesis of **183** involved the preliminary formation of the water-soluble bisulfite adduct **185** and the subsequent treatment with sodium cyanide. Although in this case the *syn* selectivity was lower, both diastereomers could be transformed to (+)-biotin.

More recently, the addition of cyanide ion, generated from TMS cyanide and cesium fluoride, to α-aziridino *N*-sulfinyl imines, being chiral either at the α position or at sulfur, has been examined [87] (Scheme 28). The configuration of the newly formed stereocenter was determined only by the chiral (*S*)-sulfinyl group. In fact, the *R* configuration (diastereomeric excess, de, 98%) was obtained from either the α-(*R*)-imine **186** or the α-(*S*)-imine **188**, giving **187** and **189**, respectively. Acyclic 2,3-diaminonitriles can be obtained

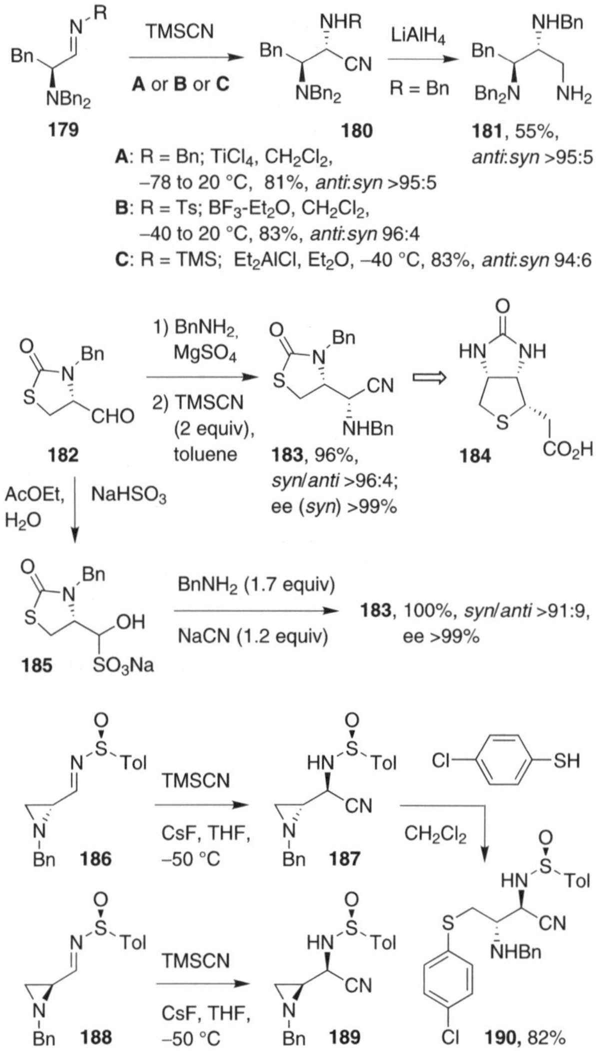

Scheme 28 Diastereoselective synthesis of 2,3-diaminonitriles

by ring-opening reactions of the α-aziridino amines (see the preparation of the sulfide **190** from the aziridine **187**).

Instead, the reaction of TMS cyanide with the chiral imine prepared from racemic 2-benzoylaminocyclohexanone and (*R*)-1-phenylethylamine or (*S*)-1-phenylethylamine in the presence of zinc chloride occurred with low stereocontrol, and the two enantiomers of *trans*-1,2-diaminocyclohexane carboxylic acid were isolated with low yields after several steps [88].

All four isomers of 4-amino-4-carboxyproline have been prepared starting from *trans*-4-hydroxy-L-proline by a sequence of steps involving Bucherer–

Berg reaction of N-protected 4-oxoproline esters (ammonium carbonate, potassium cyanide, EtOH-H_2O or DMF–H_2O 55–60 °C), leading to spirohydantoins through the intermediate aminonitriles [89]. The diastereoselectivity was highly dependent on the bulkiness of the ester function (96 : 4 with *tert*-butylester).

2.3.3 Perspectives

The stereoselective intramolecular addition of an organometallic reagent or a nucleophilic radical to an α-amino azomethine function has not been investigated. For this purpose, N-(ω-functionalized)-α-amino azomethine compounds **191** should be prepared starting from optically pure α-amino aldehydes (Scheme 29). The ω-functionality X,Y in **191** can be the halide, alkene, alkyne or carbonyl groups. All of them allow the formation in situ of organometallic and radical (or ketyl radical anion) species **192** and **193**, respectively, capable of intramolecularly attacking the electrophilic C = N bond to give the 3-aminoazacycloalkanes **194**. A few reactions described in the literature indicate the feasibility of these routes. They include transition-metal- or SmI_2-promoted intramolecular reductive coupling of bifunctional compounds, such as aldehyde/oxime [90], aldehyde/hydrazone [91] and alkene–alkyne/imine–oxime–hydrazone [92–94]. Alternatively, one can exploit the

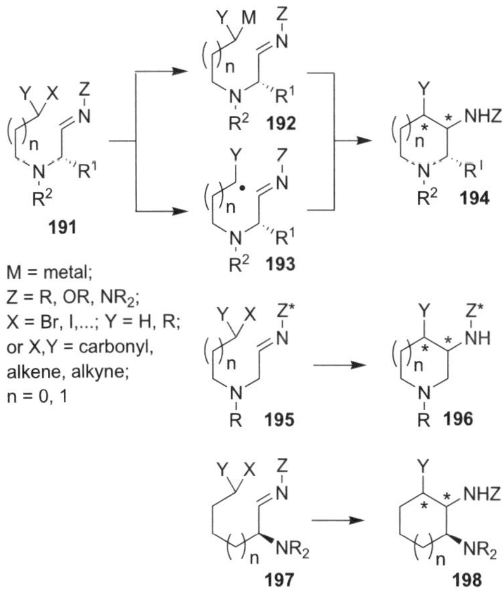

Scheme 29 Possible stereoselective routes to 3-aminoazacycloalkanes and substituted 1,2-diaminocycloalkanes

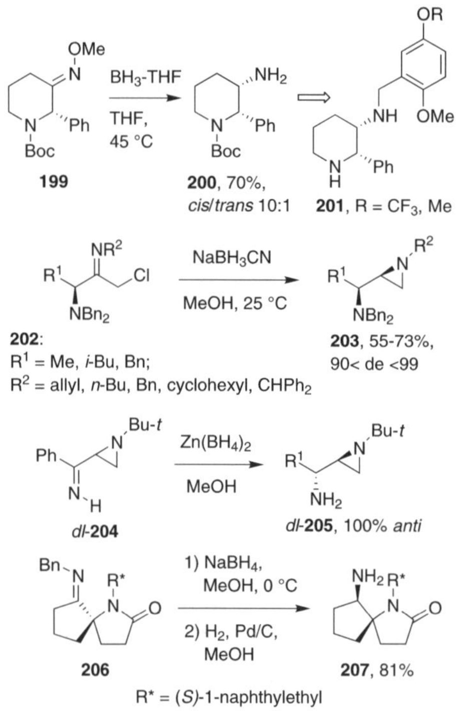

Scheme 30 Substrate-induced diastereoselective reductions of α-aminooximes and α-amino imines

asymmetric induction of a chiral auxiliary placed on the azomethine nitrogen, e.g., the cyclization of **195** to **196**. The possibility to use external chiral ligands to promote the enantioselective cyclization appears more remote. Initial attempts directed to the synthesis of 2-amino-3-hydroxy-1-cyclohexanes by intramolecular aldehyde/hydrazone coupling in the presence of chiral ligands similar to HMPTA gave very low levels of diastereoselectivity and enantioselectivity [91].

By an analogous route, the cyclization of chiral α-amino imines **197**, bearing the proper ω-functionality in the α-alkyl chain, will lead to 3-substituted 1,2-diaminocycloalkanes **198**. The intermolecular and intramolecular addition of allylic silicon and tin compounds to α-amino azomethine compounds should also be investigated.

2.4
Reduction of α-Amino Azomethine Compounds

2.4.1
Reduction of α-Amino Ketimines

The reduction of α-amino ketimines is a complementary procedure to the previously described addition of organometallic reagents to α-amino aldimines, but has been relatively less studied. Configurationally stable non-racemic α-aminoketones are required as starting materials. Moreover, the α-amino ketimines derived can exist as mixtures of *syn* and *anti* stereoisomers, and this makes more difficult the stereocontrol in the addition step. Nevertheless, the reduction of the cyclic oxime ether **199**, prepared by a multistep route from enantiopure *N*-Boc phenylglycinaldehyde, with borane in THF occurred with high diastereoselectivity, mainly affording the *cis* disubstituted piperidine **200**, which is the precursor of the neurokinin-1 receptor antagonists **201** [95] (Scheme 30).

N,*N*-Dibenzyl α'-amino α-chloroketimines **202** can be prepared from the corresponding ketones, which in turn are available by the addition of chloromethyllithium to esters of natural α-amino acids. Reduction of **202** with sodium cyanoborohydride directly afforded α-aminoalkyl-substituted aziridines **203** with high *syn* diastereoselectivity, which was only moderately affected by the size of the substituent R^1 [96] (Scheme 30). A complemen-

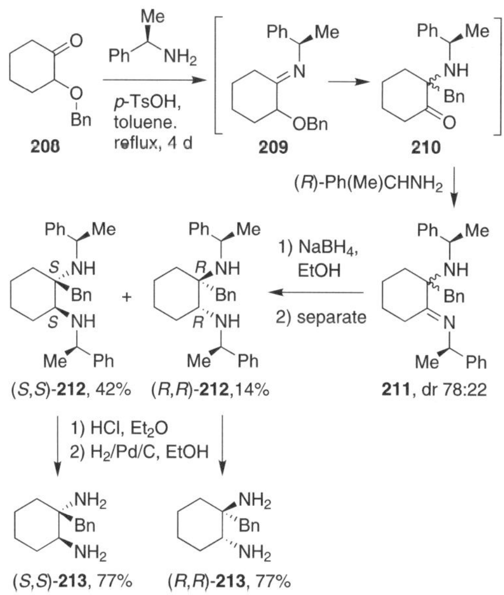

Scheme 31 Auxiliary-induced diastereoselective reduction of α-amino ketimines

tary protocol, based on the chelation-controlled reduction of α-aziridine ketimines, would give the *anti* diastereomers, but has been applied on a single racemic compound; see the reduction of (*d,l*)-**204** to (*d,l*)-**205** [97].

The spiro compound **206** was prepared in five steps from (*S*)-1-naphthylethylamine and was composed of a mixture of imine and enamine tautomers. Reduction of the imine function by sodium borohydride occurred on the less hindered *si* face, leading to the diamine with the *R* configuration of the newly formed stereocenter, then the *N*-benzyl substituent was removed by hydrogenolysis to give **207** with good overall yield [98] (Scheme 30).

By heating 2-benzyloxycyclohexanone **208** and (*R*)-1-phenylethylamine in refluxing toluene for 4 days in a Dean–Stark apparatus, the imine **209** was formed, then a rearrangement occurred to give first the α-aminocyclohexanone derivative **210** and finally the α,α-disubstituted imine **211** with moderate diastereoselectivity. Reduction of this imine with sodium borohydride gave a mixture of two *trans* diamines (*S,S*)-**212** and (*R,R*)-**212**, which were separated by chromatography. The enantiomers of 1-benzyl-1,2-diaminocyclohexanes **213** were then obtained by hydrogenolysis [99] (Scheme 31).

2.4.2
Reductive Amination of α-Aminoketones

Rather than preforming the α-amino ketimines to be reduced, it is often advantageous to form in situ the more reactive iminium ions from α-aminoketones and primary amines or ammonium salts in the presence of the reducing agent, e.g., sodium cyanoborohydride. Use of this procedure (reductive amination) with the enantiopure α-aminoketone **214** and benzylamine allowed the preparation of the *syn* diamines **215** with high yields and (almost) complete diastereoselectivities [100] (Scheme 32). Then, the primary diamines **216** were obtained by routine N-debenzylation. Similarly, the diamine **217** was prepared using ammonium acetate. In

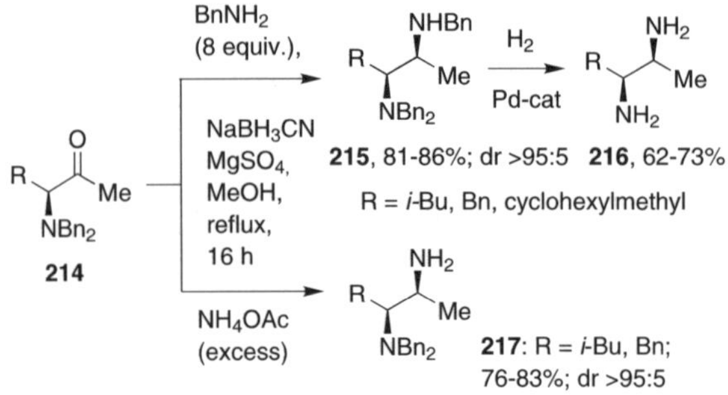

Scheme 32 Reductive amination of chiral α-aminoketones with sodium cyanoborohydride

this reaction, it is likely that the *syn* diastereoselectivity is controlled by the intramolecular hydrogen bonding between the iminium ion and the α-dibenzylamino function. In fact, *anti* diastereoselectivity was observed in the reaction of 2-dimethylaminocyclopentanone with dimethylamine and sodium borohydride in buffered THF–H_2O solution, affording *cis*-1,2-bis(dimethylamino)cyclopentane [101].

The reductive amination of ketones can be carried out under hydrogen pressure in the presence of palladium catalysts. However, if enantiopure α-aminoketones are used, partial racemization of the intermediate α-amino imine can occur, owing to the equilibration with the corresponding enamine [102]. Asymmetric hydrogenation of racemic 2-amidocyclohexanones **218** with Raney nickel in ethanol gave a mixture of *cis* and *trans* 1,2-diaminocyclohexane derivatives **219** in unequal amounts, presumably because the enamines are intermediates, but with excellent enantioselectivity. The two diastereomers were easily separated and converted to the mono-protected *cis*- and *trans*- 1,2-diaminocyclohexanes **220**. The receptor **221** has been also synthesized by this route [103] (Scheme 33).

The Leuckart–Wallach reaction is the oldest method of reductive amination of carbonyl compounds. It makes use of formamide, formic acid or ammonium formate at high temperature. The final product is a formamide derivative, which can be converted to an amine by reduction or hydrolysis. The method has been applied to the preparation of 1,2-diamines with a norbornane framework, which are interesting rigid analogues of 1,2-diaminocyclohexanes. As a matter of fact, starting from *N*-acetyl-2-oxo-1-norbornylamine **222**, the diamide **223** was obtained with excellent diastereoselectivity and then converted to the *N*-methyl-*N'*-ethyl derivative **224** by reduction with borane [104] (Scheme 34). On the other hand, when the reac-

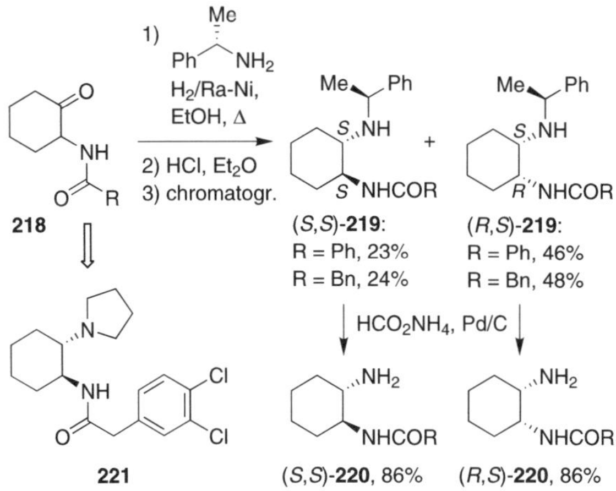

Scheme 33 Auxiliary-induced hydrogenative amination of 2-amidocyclohexanones

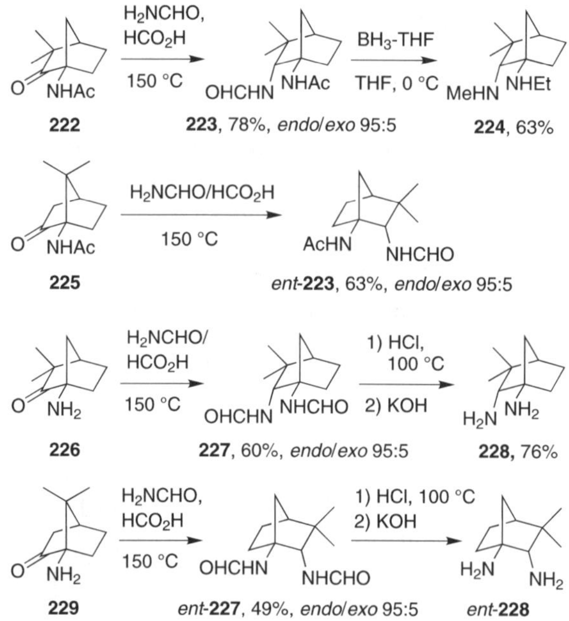

Scheme 34 Leuckart–Wallach reductive amination of aminobornanones

tion was performed on the aminoketone **225** the diamide *ent*-**223** was formed by a pathway that involved skeleton rearrangement (Wagner-Meerwein) and intramolecular transamidation. Unfortunately, the corresponding primary diamines could not be obtained by acid or basic hydrolysis. So, the same procedure was applied to the free aminoketone **226**, which is available in four steps from naturally occurring 1(*R*)-camphor. In this way, the bisformamide **227** was obtained as the sole product with moderate yield and good stereoselectivity, then converted to the primary diamine **228** by treatment with boiling concentrated hydrochloric acid. Similarly, starting from the aminoketone **229** derived from 1(*R*)-fenchone, the amide *ent*-**227** and the diamine *ent*-**228** were prepared.

2.4.3 Organometallic Addition/Reduction Sequence on Chiral α-Aminonitriles

Optically pure α-amino acids can be converted to 1,2-diamines by a route that involves the preliminary formation of N-protected α-aminonitriles through the intermediate amides. The addition of organometallic reagents to these α-aminonitriles gives α-amino ketimines, which are then reduced in situ to 1,2-diamines. However, this route has been scarcely applied to acyclic α-aminonitriles. As a matter of fact, the sequential addition of methylmagnesium bromide and lithium aluminum hydride to the *N*-Boc-α-aminonitrile derived from L-alanine gave the corresponding 1,2-diamine with low yield and almost no diastereoselectivity [105]. Moreover, the cyano group of α-aminonitriles can undergo displacement by Grignard reagents (Bruylants reaction). However, organolithium and cuprate reagents successfully added to the cyano group of α-aminonitriles with a bicyclic framework, e.g., **229**, which was prepared from (*R*)-phenylglycinol, glutardialdehyde and sodium cyanide (Scheme 35). The intermediate bicyclic imine **230** was reduced in situ with high diastereoselectivity by the sequential addition of sodium cyanoborohydride and lithium aluminum hydride; this was followed by hydrogenolysis of the chiral auxiliary to give ultimately 2-(1-aminoalkyl)piperidines **231** [106, 107]. On the other hand, reduction of the imines **230** directly with lithium aluminum hydride gave the 3-amino azepines **232**, especially with aryl substituents R^1, since the cleavage of the intermediate bicyclic aminal occurred with opposite regioselectivity. Moreover, an α-alkyl substituent, eventually bearing ω-halogen or masked carbonyl

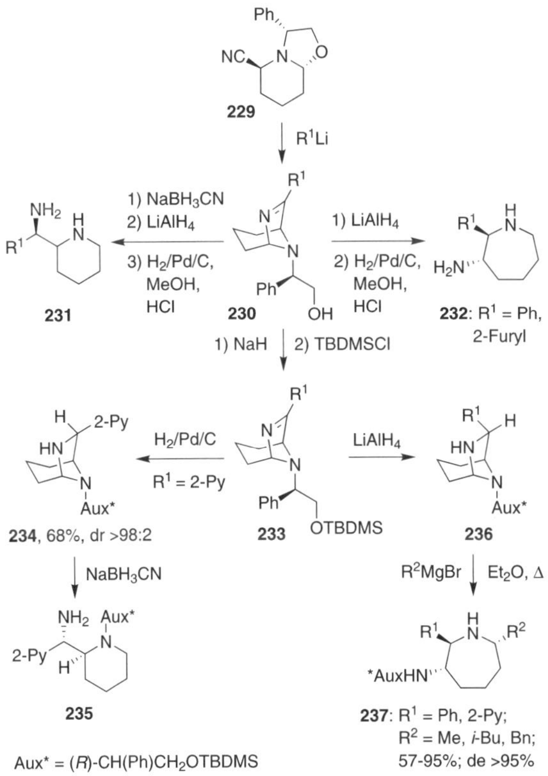

Scheme 35 Asymmetric syntheses of 2-(1-aminoalkyl)piperidines and 3-aminoazepins by the CN(*R*,*S*) method

functions, was introduced by exploiting the acidity of the α-cyano hydrogen atom in **229**; this allowed the preparation of spiropiperidines by a further cyclization step [108, 109]. After protection of the hydroxyl group in **230** to give the silyl ether **233**, the imine function can be hydrogenated with opposite diastereoselectivity; see the preparation of the bicyclic aminal **234** and successively the substituted piperidine **235** [110, 111]. On the other hand, reduction of the imines **233** with lithium aluminum hydride gave the bicyclic aminals **236**, from which the 2,7-disubstituted 3-aminoazepans **237** were obtained with excellent diastereoselectivity by addition of Grignard reagents [111].

Enantiopure 3-phenyl-2-cyanoazetidines (*S*)-**238** and (*R*)-**238**, which are epimeric at C2, are prepared in high yields from (*R*)-phenylglycinol. A one-pot sequence, including addition of organolithium or allylmagnesium bromide to the cyano group and in situ reduction of the resulting imine with sodium borohydride, allowed for the preparation of 2-(1-aminoalkyl)azetidines, which were then protected as *N*-Boc derivatives (*R*,*S*)-**239** and (*S*,*R*)-**239** [112] (Scheme 36). Complete *anti* diastereoselectivity (dr more than 95 : 5 by ^{1}H NMR) was observed in both cases. The same sequence

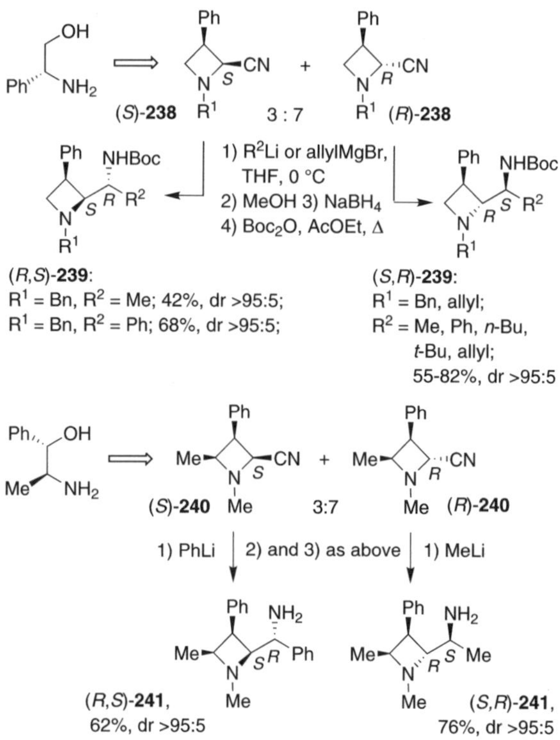

Scheme 36 Asymmetric synthesis of 2-(1-aminoalkyl)azetidines from enantiopure 2-cyanoazetidines

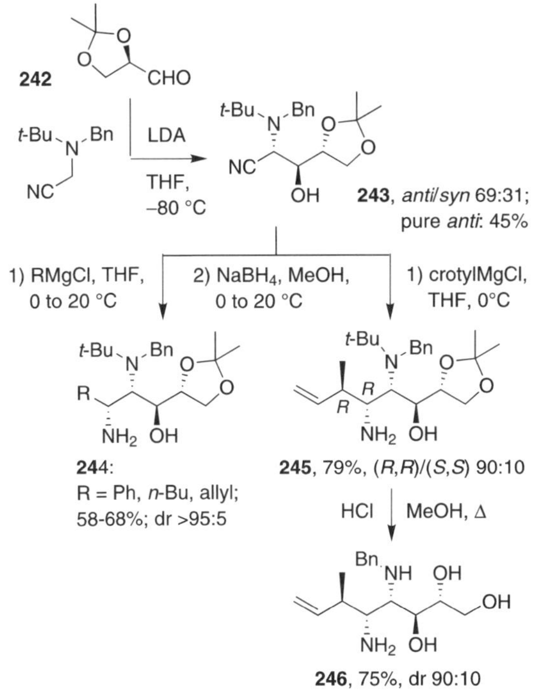

Scheme 37 Asymmetric synthesis of 1,2-diamines from chiral α-aminonitriles derived from diastereoselective aldolization

was also performed on ephedrine-derived cyano-azetidines. For example, (*S*)-**240** and (*R*)-**240**, coming from (1*S*,2*S*)-*pseudo*-ephedrine, were transformed to the *anti*-diamines (*R*,*S*)-**241** and (*S*,*R*)-**241**, respectively.

The aldol-type reaction of D-isopropylideneglyceraldehyde **242** with *N*-*tert*-butyl-*N*-benzyl aminoacetonitrile gave the α-aminonitrile **243** as a mixture of *syn* and *anti* diastereomers in 69 : 31 ratio, but the prevalent *anti* diastereomer was isolated pure by recrystallization with 40% yield [113] (Scheme 37). Subsequently, the addition of Grignard reagents to **243** in THF, followed by reduction of the intermediate ketimines with sodium borohydride in methanol, gave the polyhydroxylated diamines **244** with high yields and excellent stereocontrol. The successful addition to the nitrile group was attributed to the steric effect of the *N*-*tert*-butyl substituent, which prevents the formation of the immonium ion intermediate in the CN-substitution pathway. Moreover, the diamine **245** was prepared by the addition of crotylmagnesium chloride to **243** with complete *anti* diastereoselectivity and high substrate-induced diastereoselectivity. Cleavage of both the *N*-*tert*-butyl bond and the dioxolane ring was accomplished by heating **245** with hydrochloric acid in methanol.

3 Synthesis of 1,2-Diamines from 1,2-Diimines and 1,2-Dihydrazones

3.1 Addition of Organometallic Reagents

3.1.1 Glyoxal Diimines

The double addition of organometallic reagents to a glyoxal diimine provides a straightforward access to 1,2-diamines. However, this method has some limitations. First of all, common organometallic reagents, such as Grignard reagents and organolithium compounds, have limited applicability, owing to competitive reaction mechanism (single electron transfer) and pathway (N-alkylation) leading to different products and/or by-products. The reactions of diorganozinc and triorganoaluminum reagents to achiral glyoxal diimines give a picture of this complex reactivity [114–116]. The method is more usefully applied to the preparation of symmetrically 1,2-disubstituted 1,2-diamines by the sequential double addition of the same organometallic reagent. On the other hand, in order to prepare differently substituted 1,2-diamines, the reactivity of the two azomethine groups must be discriminated, but this can be achieved only in particular cases.

With the aim to control the absolute configuration of the newly formed carbon stereocenters, the asymmetric induction of a chiral N substituent on the imine must be exploited. Moreover, one is also faced with the problem of controlling the *syn*/*anti* diastereoselectivity. This is generally more efficiently achieved, since the stereochemistry in the second addition step, i.e., to the intermediate α-amido imine, is controlled by the chirality of the newly formed stereocenter. Until now, glyoxal diimines derived from optically pure 1-arylalkylamines have mainly been used. The first reported examples involved the addition of allylmagnesium chloride in THF [117], and phenylmagnesium and methylmagnesium [118] halides in Et_2O to the diimine **247** derived from (*S*)-1-phenylethylamine (Scheme 38). These reactions, when performed under strictly controlled experimental conditions (– 78 °C, syringe-controlled addition of RMg*X* over 6 h) provided the 1,2-diamines **248a–248c** with moderate to good yields and diastereoselectivities. The same sense of asymmetric induction was followed in all these reactions, i.e., the (*R*,*R*) configuration of the newly formed stereocenters was mainly obtained using the (*S*)-auxiliary (1,3-asymmetric induction and chelation-controlled *syn* diastereoselectivity). On the other hand, the double addition of *tert*-butylmagnesium chloride occurred only in pentane at 50 °C to give the diamine **248d** with complete diastereoselectivity [119]. Additions of other alkyl Grignard reagents were not described; in fact, the reaction of *n*-butylmagnesium chloride mainly oc-

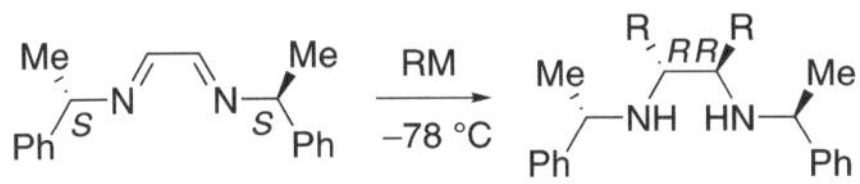

247 **248**

RM	Solvent, T (°C)	(*R,R*)/(*R,S*)/(*S,S*)	(*R,R*)-**248** (%)
allylMgCl	Et_2O, −78	86:0:14	**a**, –
PhMgCl	Et_2O, −78	90:0:10	**b**, 47
MeMgCl	Et_2O, −78	70:0:30	**c**, 35
t-BuMgCl	pentane, 50	>99:1	**d**, 95
2,5-$(MeO)_2C_6H_3$	THF, 0-25	50:23:27	**e**, 33
PhLi	THF, −78	69:2:29	**b**, 42
n-BuLi	THF, −78	71:1:28	**f**, 38
t-BuLi	THF, −78	67:2:31	**g**, 48
vinylLi	THF, −78	55:2:43	**h**, –
allylZnBr	THF, −78	93:3:4	**a**, 76
methallylZnBr	THF, 0	79:11:10	**i**, 55

248d,e

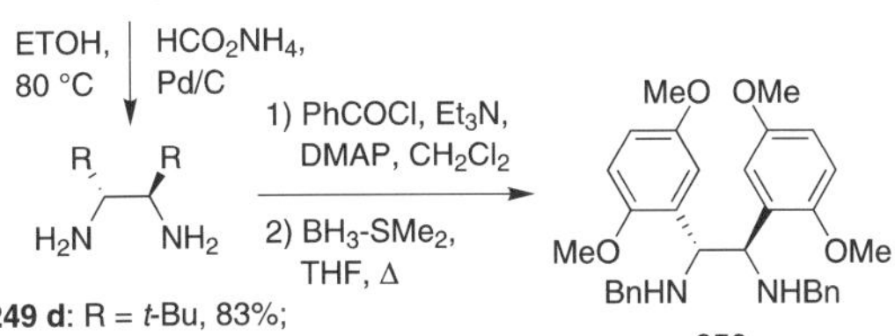

249 d: R = *t*-Bu, 83%;
e: R = 2,5-$(MeO)_2C_6H_3$, 77%

250

Scheme 38 Addition of organometallic reagents to a chiral glyoxal diimine

curred by the N-alkylation pathway, and this was followed by an elimination step [120]. The addition of the more reactive organolithium reagents to the diimine in THF at − 78 °C required quenching of the reaction mixtures at low temperature with deaerated water. In fact, since the dilithium 1,2-diamides are in equilibrium with the α-amido C radical above 0 °C (Scheme 3), the reaction of these intermediates with oxygen dissolved in water during quenching led ultimately to monoimines [121]. Generally, the reactions with the lithium reagents showed low stereocontrol; however, the two main *syn* diastereomers of the 1,2-diamines **248b**, **248f** and **248g**, but not **248h** (R is vinyl), could be separated by column chromatography and isolated in a pure state with moderate yields.

Primary 1,2-diamines can be obtained by hydrogenolysis of the benzylic N-auxiliary bond over Pd/C, e.g., the preparation of **249d**. The cleavage is not selective in the presence of other benzylic bonds in the molecule, e.g., aryl substituents at C1 – C2. However, in the presence of methoxyphenyl or dimethoxyphenyl C substituents, the N substituents can be selectively removed, as exemplified by the preparation of the primary 1,2-diamine

249e [122], which was then converted to the *N*,*N*′-dibenzyl derivative **250** by routine N-benzoylation and borane reduction steps.

The complete stereocontrol obtained in the double addition of *t*-BuMgCl to the diimine **247** in pentane at 50 °C is in marked contrast with the result of the same reaction performed in THF at low temperature: in the latter case, the α-amino imine was formed with lower diastereoselectivity after proton quenching of the monoadduct **251**. Hence, it is assumed that the double-addition product **248d** is formed at high temperature through a mechanism that involves the kinetic resolution of (*S*)-**251** and (*R*)-**251**, which are in equilibrium through the enamine **252**, or the corresponding magnesium endiamide formed by intramolecular or intermolecular deprotonation (Scheme 39). Since a single addition step occurs at low temperature, a hydride or a different Grignard reagent was used for the second addition step, so providing useful access to unsymmetrically substituted 1,2-diamines **253** and **254**, respectively [123]. By an analogous approach, the monoaddition of prenylzinc bromide to the diimine **247** occurred even at 0 °C, then the addition of the more reactive allylzinc bromide afforded the 1,2-diamine **255** with

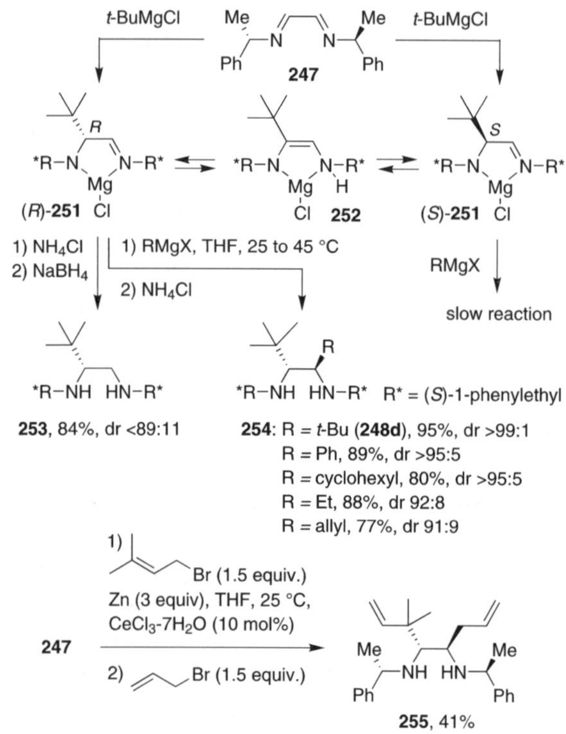

Scheme 39 Synthesis of C_1-symmetric 1,2-diamines from a chiral glyoxal diimine

moderate yield after column chromatography (SiO_2) (G. Martelli, S. Morri, D. Savoia, unpublished result).

The double addition of allylic organometallic reagents is of paramount importance because of the potential of the C_2-symmetric products, 4,5-diamino-1,7-octadienes, as intermediates for the construction of more complex or functionalized molecules exploiting the transformation of the alkene functions. Following the first addition of allylzinc bromide, which provided a better diastereoselectivity than allylmagnesium chloride [124], several substituted allylzinc halides, including pentadienyl and cinnamyl reagents, have been added to the diimine **247**. These reactions take place with allylic inversion, affording the branched diamines, namely, 3,6-disubstituted-4,5-diamino-1,7-octadienes **256**, with high stereocontrol [125] (Scheme 40). On the other hand, the corresponding lithium reagents gave the linear diamines **257** with moderate diastereoselectivities. It is noteworthy that the addition of prenylmagnesium chloride affords the corresponding disubstituted linear

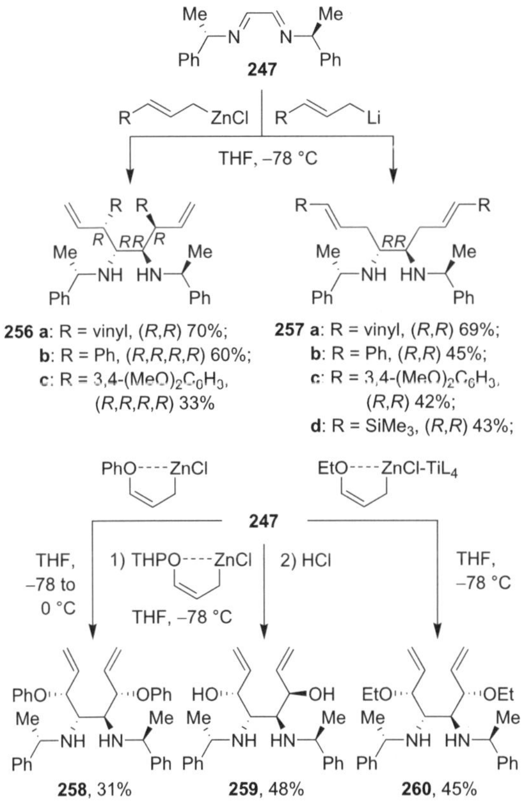

Scheme 40 Diastereoselective addition of γ-substituted allyllithium and allylzinc reagents to a chiral diimine

diamine, through the rearrangements of the initially introduced branched prenyl substituents. Oxygen-substituted allylic zinc reagents were prepared from O-substituted or O-protected allylic alcohols through the corresponding lithium reagents, then added to the diimine **247** to give the products **258** and, through previous deprotection in an acidic medium, **259**. On the other hand, allylic titanium reagents prepared from acrolein acetals were converted to the corresponding zinc reagents, by which the diamine **260** was obtained [126]. It appears that the stereocontrol in these reactions is affected by the O substituent and the presence of the acidic titanium salt.

The 1,2-diamines **256a** and **256b** with branched pentadienyl and cinnamyl substituents, especially the latter, can be used as "masked chiral glyoxal diimines" for the preparation of C_1- and C_2-symmetric 1,2-disubstituted 1,2-diamines by treatment with the proper amounts of organolithium reagents under strictly controlled experimental conditions [127, 128] (Scheme 41). For example, **256b** underwent isomerization to **257b** by treatment with 2 or even 4 equiv of methyllithium, by a sequence of steps that involves, for each homoallylic amine moiety, metallation of the N – H bond and subsequent stereospecific rearrangement. When other organolithium reagents were used, a *retro*-allylation step was observed after metallation, and this was followed

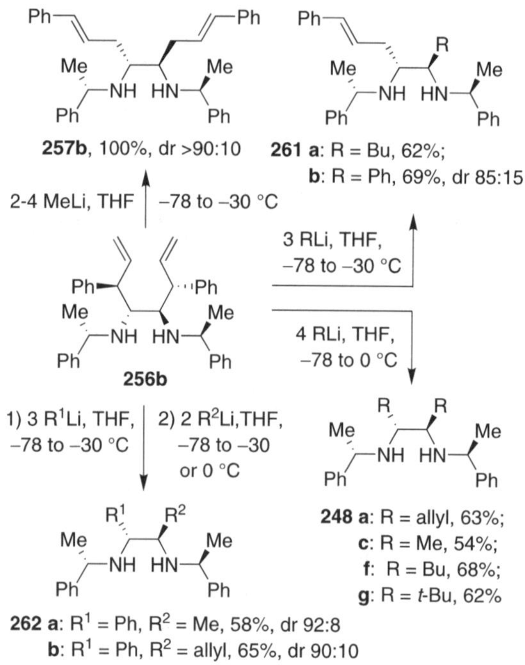

Scheme 41 Stereoselective synthesis of C_1- and C_2-symmetric 1,2-diamines from *N*,*N*′-di[(*S*)-1-phenylethyl] (3*R*,4*R*,5*R*,6*R*)-3,6-diphenyl-4,5-diamino-1,7-octadiene

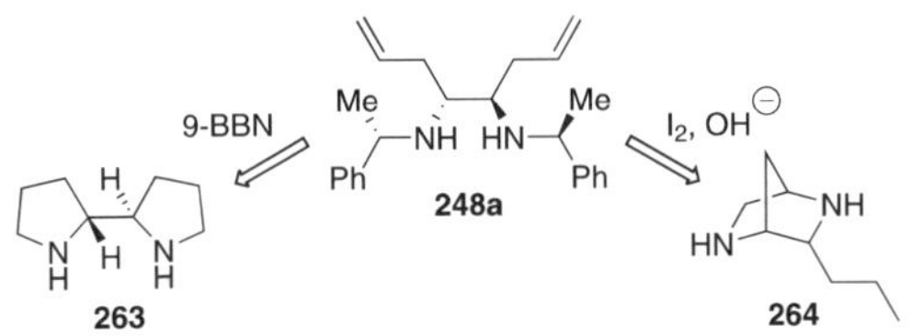

Scheme 42 Cyclizations of *N*,*N'*-di[(*S*)-1phenylethyl]-4*R*,5*R*-diamino-1,7-octadiene

by the addition of RLi to the reformed C = N bond. By controlling the amount of the RLi reagent, the substitution of only one or both of the cinnamyl substituent(s) could be obtained with satisfactory to good selectivity and stereoselectivity; see the preparation of the C_1- and C_2-symmetric diamines **261a**, **261b** and **248a**, **248c** **248f**, **248g**, respectively. Finally, two different organolithium reagents were used for the stepwise replacements of the two cinnamyl substituents, e.g., the preparation of the C_1-symmetric 1,2-diamines **262a** and **262b**.

Most importantly, the 4,5-diamino-1,7-octadienes **248a**, **256** and **257**, available by the double addition of allylic zinc reagents to the diimine **247**, are

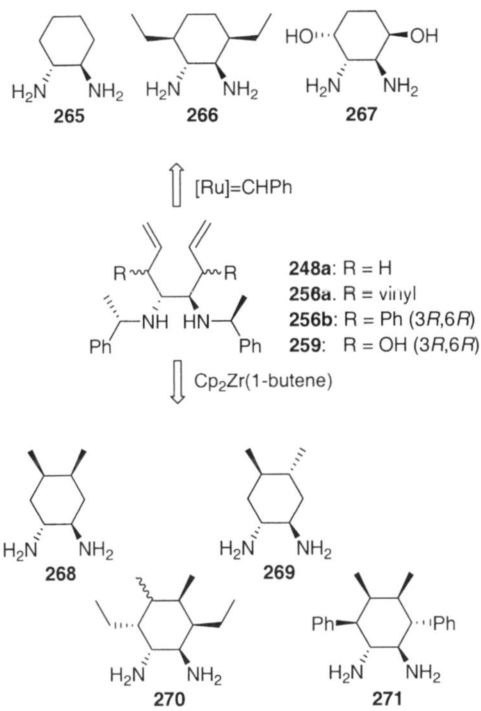

Scheme 43 Synthesis of enantiopure ring-substituted 1,2-diaminocyclohexanes by transition metal catalyzed and promoted cyclization of 1,7-octadienes

valuable intermediates for the preparation of 1,2-diamines with a cyclic or a bicyclic framework. Two different approaches have been developed so far. The electrophile-promoted cyclizations of the 4- or 5-aminoalkene moieties led to azaheterocycles (Scheme 42). For example, 2,2′-dipyrrolidine **263** was synthesized by two formal 4-aminoalkene cyclizations, through hydroboration of the two alkenes in the diamine **248a** [129]. On the other hand, the 5-aminoalkene moieties are involved in the iodine-promoted cyclization of **248a**, which provided access to the 2,5-diazabicyclo[2.2.1]heptane derivative **264** [130].

On the other hand, coupling reactions of the 1,7-diene moiety, catalyzed or promoted by transition-metal complexes, were carried out on the diaminodienes **248a, 256a, 256b** and **259** to prepare ring-substituted 1,2-diaminocyclohexanes. Grubbs' ruthenium carbene complexes catalyzed the ring-closing metathesis reactions of 4,5-diamino-1,7-octadienes. The nature of the substituents at C3 and C6 and the configuration of those stereocenters affected the reactivity. Up to now, the 1,2-diaminocyclohexanes **265–267** have been prepared from **248a, 256a** and **259** through the corresponding cyclohexenes [131, 132] (Scheme 43). On the other hand, methyl substituents at C4 and C5 of the cyclohexane ring are introduced from 4,5-diamino-1,7-dienes by the reaction with in situ formed [η^2-(1-butene)]zirconocene. The relative stereochemistry of the methyl substituents was dependent on the procedure (catalytic or stoichiometric in zirconium) [133, 134]. In this way, substituted 1,2-diaminocyclohexanes **268–271** have been prepared from **248a, 256a** and **256b**.

3.1.2
Glyoxal Dihydrazones

Enantiopure C_2-symmetric 1,2-diamines have been prepared from chiral glyoxal hydrazones, e.g., **272**. Organolithium–cerium trichloride reagents are the reagent of choice, as they react at − 100 °C in THF to afford the corresponding hydrazines **273** with quantitative yields and high diastereoselectivity, the *syn* diastereomers being prevalent. A drawback of this reaction is that the organometallic reagents must be used in large excess (8 equiv) [135, 136] (Scheme 44). Since purification of the hydrazines led to significant loss of material, and the cleavage of the N – N bond with Raney nickel was often accompanied by epimerization, the optimized protocol requires the cleavage of the N – N bond by borane (at reflux for several days!), subsequent protection of the crude products with a large excess of di-*tert*-butyl dicarbonate or benzyloxycarbonyl chloride and final purification by chromatography. Although the diastereomeric hydrazines were not perfectly separated and the yield of **274b** was low, the de was high for **274b** and **274c**, and the ee was excellent for all the *syn* diastereomers.

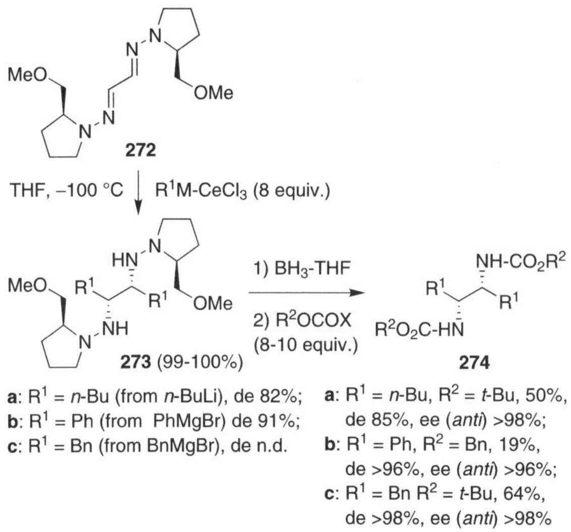

Scheme 44 Diastereoselective synthesis of protected 1,2-diamines from glyoxal Ramp hydrazone

3.1.3
Perspectives

As a final comment on the applicability of the methodology described in this section, two points should be remarked upon. First of all, the use of chiral auxiliaries might be implemented, e.g., investigating the applicability and the level of asymmetric induction provided by 1-arylalkylamines, analogous to 1-phenylethylamine. In fact, it was reported that in the monoaddition of *t*-butylmagnesium chloride (THF at 20 °C) to glyoxal diimines prepared from different primary amines (auxiliaries) the diastereoselectivity decreased in the order 1-phenylpropylamine (de 80%), 1-(4-chlorophenyl)ethylamine (de 76%), 1-phenylethylamine (de 66%) [114]. However, the use of valine esters and O-silylated valinol (G. Martinelli, D. Savoia, unpublished results) as well as oxime ethers [137] gave less satisfactory or disappointing results. Nevertheless, the applicability of enantiopure sulfinamides should be worth studying. Finally, it is noteworthy that no examples of enantioselective addition of organolithium reagents to achiral glyoxal diimines in the presence of optically pure bidentate or polydentate ligands have been reported. In this case, the product itself (metal diamide) might act as a ligand of the organometallic reagent and affect the stereocontrol in competition with the added ligand.

3.2 Reduction of 1,2-Diketimines

3.2.1 Diastereoselective Reduction of Chiral 1,2-Diketimines

1,2-Diketimines can be prepared by condensation of 1,2-diketones with 2 equiv of an amine, or 1 equiv of a 1,2-diamine, by azeotropic removal of water. Either a chiral diketone or a chiral amine/diamine can be used in order to obtain a chiral diimine. In both cases, the use of 1,2-diamines is expected to provide better stereocontrol, because of the rigidity of the derived cyclic diimines. For example, the reaction of camphor 1,2-diketone **275** and racemic 1,2-diphenylethylenediamine (*d,l*)-**26** gave the diimine **276** as a mixture of two diastereomers (Scheme 45) [138]. Reduction of **276** with sodium borohydride followed by hydrogenolysis of the N substituents afforded the camphordiamine, which was isolated as the dihydrochloride

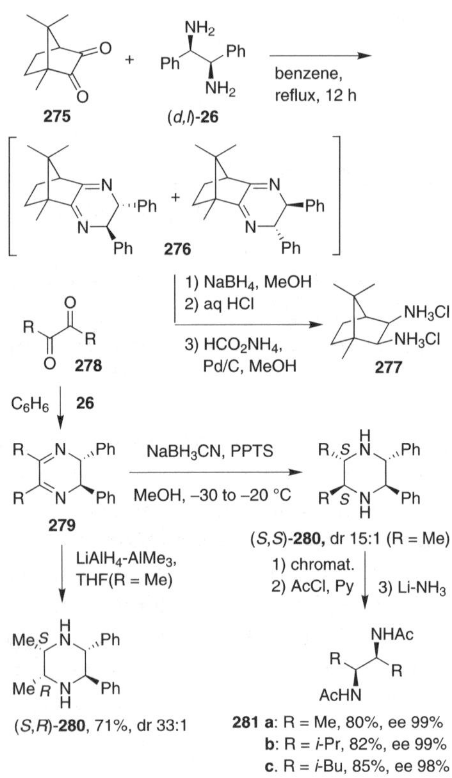

Scheme 45 Substrate- and auxiliary-induced diastereoselective reduction of chiral 1,2-diimines

277. A single stereoisomer of **277** was obtained, as the attack of the hydride reagent occurred to the *endo* face of both the azomethine groups (substrate-induced diastereoselectivity), independent of the configuration of the N substituents. On the other hand, condensation of the same diketone with *meso*-1,2-diphenylethylenediamine gave a mixture of two unstable diastereomeric diimines, which decomposed at different rates by an electrocyclic ring opening.

By the alternative approach, one can exploit the asymmetric induction of a chiral auxiliary. For example, the aliphatic 1,2-diketones **278** were condensed with (*R*,*R*)-1,2-diphenylethylenediamine **26** to give the cyclic diimines **279** [139] (Scheme 45). Reduction of **279** with sodium cyanoborohydride in the presence of pyridinium *p*-toluenesulfonate mainly gave the C_2-symmetric 2,3,5,6-tetrasubstituted piperazines (*S*,*S*)-**280**, having the *S* configuration of the newly formed stereocenters. After chromatographic separation of diastereomers, the cleavage of the benzylic bonds (auxiliary) was best accomplished by previous acylation of the amine functions and subsequent treatment with lithium in ammonia, to give the protected 1,2-diamines **250** with good yields and excellent ees. On the other hand, reduction of **279** with lithium aluminum hydride in the presence of trimethylaluminum occurred with the opposite diastereoselectivity, affording (*S*,*R*)-**280**; hence, this procedure is only suitable for the preparation of C_1-symmetric substituted piperazines, but not linear aliphatic 1,2-diamines.

3.2.2
Enantioselective Reduction of 1,2-Diketimines

Whereas the enantioselective reduction/hydrogenation of ketimines has been extensively explored, the corresponding reactions of 1,2-diketimines have been seldom investigated. In one case, reduction of the aromatic 1,2-diimine **282** with the oxaazaborolidine-borane reagent formed in situ from borane and a catalytic amount of the enantiopure β-amino alcohol **283** gave the 1,2-diamine **23** with excellent yield as a mixture of *R*,*R* and *meso* diastereomers, where the former was highly prevalent and almost enantiomerically pure (ee 99%) [140] (Scheme 46). In order to obtain the pure primary diamine, the 1,3-imidazolidin-2-one **25** was first prepared and at this stage the diastereomers could be separated by column chromatography or by crystallization from ethyl acetate. Then, the *N*-(4-methoxyphenyl) substituents were oxidatively removed and the imidazolidinone ring was cleaved by heating in HBr/AcOH mixture, so giving the diamine **26** with good overall yield. Despite this promising result, no further examples of enantioselective reduction of 1,2-diketimines have been reported.

By an alternative approach, the enantioselective reduction of 1,2-diketimines was achieved by using hydride reagents in the presence of stoichiometric amounts or an excess of enantiopure carboxylic acids. An

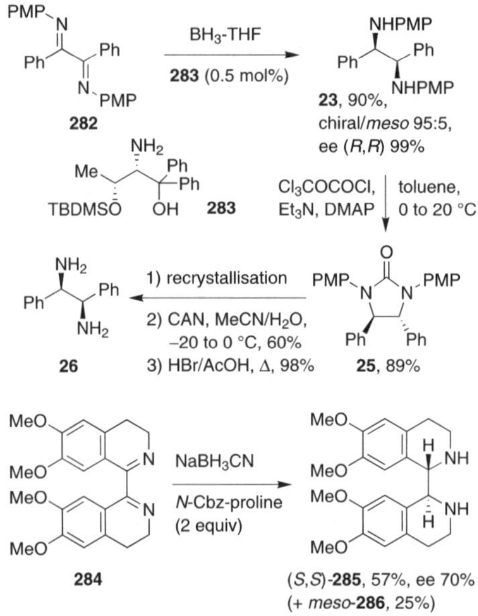

Scheme 46 Enantioselective reduction of 1,2-diketimines

example of this procedure is the reduction of the ring-substituted 1,1′-bis(dihydroisoquinoline) **284** with sodium cyanoborohydride in the presence of (*S*)-*N*-Cbz-proline (2 equiv): in this case, the diamine (*S*,*S*)-**285** was mainly obtained with moderate enantioselectivity and diastereoselectivity [141] (Scheme 46). Interestingly, the reduction performed with sodium borohydride gave only *meso*-**285**. Even for this procedure, other examples were not provided. So, it appears that there is room for further investigations. Especially, one should take into account the results already obtained in the enantioselective reduction of monoimines which are capable of acting as bidentate ligands towards metal ions, e.g., 2-pyridyneimine.

References

1. Lucet D, Le Gall T, Mioskowski C (1998) Angew Chem Int Ed Engl 37:2581
2. Bennani YL, Hanessian S (1997) Chem Rev 97:3161
3. Alexakis A, Aujard I, Kanger T, Mangeney P (1999) Org Synth 76:23
4. Denmark SE, Su X, Nishigaiki Y, Coe DM, Wong KT, Winter SBD, Choi JY (1999) J Org Chem 64:1958
5. Mattson A, Orrenius C, Öhrner N, Unelius R, Hult K, Norin T (1996) Acta Chem Scand 50:918
6. Alfonso I, Rebolledo F, Gotor V (1999) Tetrahedron Asymmetry 10:367
7. Luna A, Alfonso I, Gotor V (2002) Org Lett 4:3627

8. Robertson GM (1991) In: Trost BM, Fleming I (eds) Comprehensive organic synthesis, vol 3. Pergamon, Oxford, p 563
9. Enholm EJ, Forbes DC, Holub DP (1990) Synth Commun 20:981
10. Tanaka H, Dhimane H, Fujita H, Ikemoto Y, Torii S (1988) Tetrahedron Lett 29:3811
11. Smith JG, Ho I (1972) J Org Chem 37:653
12. Alexakis A, Aujard I, Mangeney P (1998) Synthesis 875
13. Fry AJ (1989) Synthetic organic electrochemistry. Wiley, New York
14. Horner L, Skaletz DH (1975) Liebigs Ann Chem 1210
15. Kise N, Moriyama H, Kitagishi Y, Ueda N (2003) J Am Chem Soc 125:11591
16. Kise N, Oike H, Okazaki E, Yoshimoto M, Shono T (1995) J Org Chem 60:3980
17. Shimizu M, Iida T, Fujisawa T (1995) Chem Lett 609
18. Annunziata R, Benaglia M, Caporale M, Raimondi L (2002) Tetrahedron Asymmetry 13:2727
19. Yanada R, Negoro N, Okaniwa M, Miwa Y, Taga T, Yanada K, Fujita T (1999) Synlett 537
20. Yanada R, Okaniwa M, Kaieda A, Ibuka T, Takemoto Y (2001) J Org Chem 66:1283
21. Taniguchi N, Uemira M (1998) Tetrahedron 54:12775
22. Taniguchi N, Hata T, Uemura M (1999) Angew Chem Int Ed Engl 38:1232
23. Larsen AO, Taylor RA, White PS, Gagné MR (1999) Organometallics 18:5157
24. Shono T, Kise N, Shirakawa E, Matsumoto H, Okazaki E (1991) J Org Chem 56:3063
25. Mercer G, Sigman MS (2003) Org Lett 5:1591
26. Portner K, Nieger M, Vögtle F (2004) Synlett 1167
27. Chin J, Mancin F, Thavarajah N, Lee D, Lough A, Chung DS (2003) J Am Chem Soc 125:15276
28. Siu T, Li W, Yudin AK (2001) J Combinat Chem 3:554
29. Dahlén A, Hilmersson G (2003) Chem Eur J 9:112
30. Aurrecoechea JM, Fernández-Acebes A (1992) Tetrahedron Lett 33:763
31. Makoto S, Suzuki I, Makino H (2003) Synlett 1635
32. Cameron TM, Ortiz CG, Abboud KA, Boncella JM, Baker RT, Scott BL (2000) Chem Commun 573
33. Makioka Y, Taniguchi Y, Fujiwara Y, Takaki K, Hou Z, Wakatsuki Y (1996) Organometallics 15:5476
34. Takaki K, Komeyama K, Takehira K (2003) Tetrahedron 59:10381
35. Hill JE, Fanwick PE, Rothwell IP (1992) Organometallics 11:1175
36. Takai K, Ishiyama T, Yasue H, Nobunaka T, Itoh M, Oshiki T, Mashima K, Tani K (1998) Organometallics 17:5128
37. Gao Y, Yoshida Y, Sato F (1997) Synlett 1353
38. Roskamp EJ, Pedersen SF (1987) J Am Chem Soc 109:6551
39. Ito H, Taguchi T, Hanzawa Y (1992) Tetrahedron Lett 33:4469
40. Buchwald SL, Watson BT, Wannamaker MW, Dewan JC (1989) J Am Chem Soc 111:4486
41. Zhong Y-W, Xu M-H, Lin G-Q (2004) Org Lett 6:3953
42. Park YS, Boys ML, Beak P (1996) J Am Chem Soc 118:3757
43. v Keyserlingk NG, Martens J (2002) Eur J Org Chem 301
44. Baricordi N, Benetti S, Biondini G, De Risi C, Pollini GP (2004) Tetrahedron Lett 45:1373
45. Foresti E, Calmieri G, Petrini M, Profeta R (2003) Org Biomol Chem 1:4275
46. Shibasaki KY, Harwood SJ, Gröger H, Shibasaki M (1999) Angew Chem Int Ed Engl 38:3504
47. Yamada K, Moll G, Shibasaki M (2001) Synlett 980

48. Knudsen KR, Risgaard T, Nishiwaki N, Gothelf KV, Jørgensen KA (2001) J Am Chem Soc 123:5843
49. Nishiwaki N, Knudsen KR, Gothelf KV, Jørgensen KA (2001) Angew Chem Int Ed Engl 40:2992
50. Okino T, Nakamura S, Furukawa T, Takemoto Y (2004) Org Lett 6:625
51. Bernardi L, Bovini BF, Capito E, Dessole G, Comes-Franchini M, Fochi M, Ricci A (2004) J Org Chem 69:8168
52. Nugent BM, Yoder RA, Johnston JN (2004) J Am Chem Soc 126:3418
53. Davis FA, Deng J (2004) Org Lett 6:2789
54. Viso A, Fernández de la Pradilla R, García A, Guerriero-Strachan C, Alonso M, Tortosa M, Flores A, Martínez-Ripoll M, Fonseca I, André I, Rodríguez A (2003) Chem Eur J 9:2867
55. Viso A, Fernández de la Pradilla R, López-Rodríguez ML, García A, Flores A, Alonso M (2004) J Org Chem 69:1542
56. Viso A, Fernández de la Pradilla R, López-Rodríguez ML, García A, Tortosa M (2002) Synlett 755
57. Alker D, Harwood LM, Williams CE (1998) Tetrahedron Lett 39:475
58. Soloshonok VA, Avilov DV, Kukhar VP, Van Meervelt L, Mischenko N (1997) Tetrahedron Lett 38:4671
59. Zhou X-T, Lin Y-R, Dai L-X (1999) Tetrahedron Asymmetry 10:855
60. Bernardi L, Gothelf AS, Hazell, Jørgensen KA (2003) J Org Chem 68:2583
61. DeGoey DA, Chen H-J, Flosi WJ, Grampovnik DJ, Yeung CM, Klein LL, Kempf DJ (2002) J Org Chem 67:5445
62. Zanardi F, Battistini L, Rassu G, Pinna L, Mor M, Culeddu N, Casiraghi G (1998) J Org Chem 63:1368
63. Dudot B, Chiaroni A, Royer J (2000) Tetrahedron Lett 41:6355
64. Reetz MT, Jaeger R, Drewlies R, Hübel M (1991) Angew Chem Int Ed Engl 30:103
65. Reetz MT, Lee WK (2001) Org Lett 3:3119
66. Suh M-J, Kim SW, Beak SI, Ha H-J, Lee WK (2004) Synlett 489
67. Tanner D, Harden A, Johansson F, Wyatt P, Andersson PG (1996) Acta Chem Scand 50:361
68. Hwang G-I, Chung J-H, Lee WK (1996) J Org Chem 61:6183
69. Giri N, Petrini M, Profeta R (2004) J Org Chem 69:7303
70. Prakash GKS, Mandai M (2002) J Am Chem Soc 124:6538
71. Davidsen SK, Chu-Moyer MY (1989) J Org Chem 54:5558
72. Cao B, Xsiao D, Joullié MM (1999) Org Lett 1:1799
73. Gensini M, Kozhushkov SI, Yufit DS, Howard JAK, Es-Sayed M, de Meijere A (2002) Eur J Org Chem 2499
74. Gensini M, de Mejiere A (2004) Chem Eur J 10:785
75. Enders D, Chelain E, Raabe G (1997) Bull Soc Chim Fr 134:299
76. Hirabayashi R, Ogawa C, Sugiura M, Kobayashi S (2001) J Am Chem Soc 123:9493
77. Merino P, Lanaspa A, Merchan FL, Tejero T (1987) Tetrahedron Asymmetry 8:2381
78. Dondoni A, Merchan FL, Merino P, Tejero T, Bertolasi V (1994) J Chem Soc Chem Commun 1731
79. Merino P, Franco S, Merchan FL, Tejero T (1998) J Org Chem 63:5627
80. Merino P, Lanaspa A, Merchan FL, Tejero T (1998) Tetrahedron Asymmetry 9:629
81. Fujisawa T, Nagai M, Koike Y, Shimizu M (1994) J Org Chem 59:5865
82. Macdonald SJF, Clarke GDE, Dowle MD, Harrison LA, Hodgson ST, Inglis GGA, Johnson MR, Shah P, Upton RJ, Walls SB (1999) J Org Chem 64:5166

83. Andrews DM, Carey SJ, Chaignot H, Coomber BA, Gray NM, Hind SL, Jones PS, Mills G, Robinson JE, Slater MJ (2002) Org Lett 4:4475
84. Nagasaka T, Koseki Y (1998) J Org Chem 63:6797
85. Reetz MT, Hübel M, Jaeger R, Schwickardi R, Goddard R (1994) Synthesis 733
86. Seki M, Hatsuda M, Mori Y, Yoshida S, Yamada S, Shimizu T (2004) Chem Eur J 10:6102
87. Li B-F, Yuan K, Zhang M-J, Wu H, Dai L-X, Wang QR, Hou XL (2003) J Org Chem 68:6264
88. Fondekar KPP, Volk F-J, Khaliq-uz-Zaman SM, Bisel P, Frahm AW (2002) Tetrahedron Asymmetry 13:2241
89. Tanaka K, Sawanishi H (1995) Tetrahedron Asymmetry 6:1641
90. Miyabe H, Torieda M, Inoue K, Tajiri K, Kiguchi T, Naito T (1998) J Org Chem 63:4397
91. Riber D, Hazell R, Skrydstrup T (2000) J Org Chem 65:5382
92. Miyabe H, Ueda M, Naito T (2004) Synlett 1140
93. Miyabe H, Naito T(2004) Org Biomol Chem 1267
94. Makabe M, Sato Y, Mori M (2004) Synthesis 1369
95. Atobe M, Yamazaki N, Kibayashi C (2004) J Org Chem 69:5595
96. Concellón JM, Bernad PL, Riego E, García-Granda S, Forcén-Acebal Á (2001) J Org Chem 66:2764
97. Bartnik R, Laurent A, Lesniak S (1982) J Chem Res S 287
98. Planas L, Pérard-Viret J, Royer J (2004) Tetrahedron Asymmetry 15:2399
99. Bisel P, Schlauch M, Weckert E, Sin K-S, Frahm AF (2001) Chirality 13:89
100. Reetz M, Schmitz A (1999) Tetrahedron Lett 40:2741
101. Fraenkel G, Pramanik P (1984) J Org Chem 49:1316
102. Dominguez MJ, García-López MT, Herranz R, Martín-Martínez M, González-Muñiz R (1995) J Chem Soc Perkin Trans 1:2839
103. Schlichter WH, Frahm AW (1993) Arch Pharm 326:429
104. García-Martínez A, Teso Vilar E, García-Fraile A, Martínez-Ruiz P (2001) Tetrahedron Asymmetry 12:2153
105. Costello GF, James R, Shaw JS, Slater AM, Stutchbury NCJ (1991) J Med Chem 34:181
106. Zhu J, Quirion J-C, Husson H-P (1989) Tetrahedron Lett 30:5137
107. Froelich O, Desos P, Bonin M, Quirion J-C, Husson H-P (1996) J Org Chem 61:6700
108. Zhu J, Quirion J-C, Husson H-P (1989) Tetrahedron Lett 30:6323
109. Zhu J, Quirion J-C, Husson H-P (1993) J Org Chem 58:6451
110. Cutri S, Bonin M, Micouin L, Froelich O, Quirion J-C, Husson H-P (2000) Tetrahedron Lett 41:1179
111. Cutri S, Bonin M, Micoin L, Husson H-P, Chiaroni A (2003) J Org Chem 68:2645
112. Couty F, Evano G, Prim D, Marrot J (2004) Eur J Org Chem 3893
113. Leclerc E, Vrancken E, Mangeney P (2002) J Org Chem 67:8928
114. van Koten G, Vrieze K (1982) Adv Organometal Chem 21:151
115. Rjnberg E, Boersma J, Jasztrebski JTBH, Lakin MT, Spek AL, van Koten G (1997) Organometallics 16:3158
116. Gibson VC, Redshaw C, White AJP, Williams DJ (1998) J Organometal Chem 550:453
117. Neumann WL, Rogic MM, Dunn TJ (1991) Tetrahedron Lett 32:5865
118. Bambridge K, Begley MJ, Simpkins NS (1994) Tetrahedron Lett 35:3391
119. Roland S, Mangeney P, Alexakis A (1999) Synthesis 228
120. Martelli G, Savoia D (2003) Curr Org Chem 7:1049
121. Martelli G, Morri S, Savoia D (2002) Tetrahedron 56:8367

122. Bette V, Mortreux A, Ferioli F, Martelli G, Savoia D, Carpentier J-F (2004) Eur J Org Chem 3040
123. Roland S, Mangeney P (2000) Eur J Org Chem 611
124. Alvaro G, Grepioni F, Savoia D (1997) J Org Chem 62:4180
125. Alvaro G, Grepioni F, Grilli S, Maini L, Martelli G, Savoia D (2000) Synthesis 58
126. Fiorelli C, Maini L, Martelli G, Savoia D, Zazzetta C (2002) Tetrahedron 58:8679
127. Grilli S, Martelli G, Savoia D (2001) Eur J Org Chem 2917
128. Martelli G, Savoia D (2003) Collect Czech Chem Commun 68:1531
129. Alexakis A, Tomassini A, Chouillet C, Roland S, Mangeney P, Bernardinelli G (2000) Angew Chem Int Ed Engl 39:4093
130. Fiorelli C, Marchioro C, Martelli G, Monari M, Savoia D (2005) Eur J Org Chem (in press)
131. Alvaro G, Grilli S, Martelli G, Savoia D (1999) Eur J Org Chem 1523
132. Grilli S, Martelli G, Savoia D, Zazzetta C (2002) Adv Synth Catal 344:1068
133. Grepioni F, Grilli S, Martelli G, Savoia D (1999) J Org Chem 64:3679
134. Grilli S, Martelli G, Savoia D, Zazzetta C (2003) Synthesis 1083
135. Enders D, Meiers M (1996) Angew Chem Int Ed Engl 35:2261
136. Enders D, Meiers M (2002) Synthesis 2542
137. Moody CJ (2004) Chem Commun 1341
138. Busacca CA, Campbell S, Dong Y, Grossbach D, Ridges M, Smith L, Spinelli E (2000) J Org Chem 65:4753
139. Nantz MH, Lee DA, Bender DM, Roohi AH (1992) J Org Chem 57:6653
140. Shimizu M, Kamei M, Fujisawa T (1995) Tetrahedron Lett 36:8607
141. Busato S, Craig DC, Judeh ZMA, Read RW (2003) Tetrahedron 59:461

Top Organomet Chem (2005) 15: 59–92
DOI 10.1007/b136348
© Springer-Verlag Berlin Heidelberg 2005
Published online: 14 September 2005

Sparteine as a Chiral Ligand for Asymmetric Catalysis

Olivier Chuzel · Olivier Riant (✉)

Laboratoire de Chimie Organique et Médicinale, Department de Chimie,
Université Catholique de Louvain, place Louis Pasteur 1, 1348 Louvain la Neuve, Belgium
riant@chim.ucl.ac.be

Abstract While the use of stoichiometric amounts of sparteine and related ligands in various asymmetric reactions often lead to highly enantioselective transformations, there have been far fewer applications of sparteine to asymmetric catalysis. The aim of this review is to highlight recent advances in the field of asymmetric transformations that use sparteine as chiral auxiliary, emphasizing the use of substoichiometric or catalytic amounts of this ligand.

1
Introduction

Sparteine **1** is a well known alkaloid from the lupine family with a semi-rigid bisquinolizidine structure. This cage-like ligand and its isomers **2** and **3** (Fig. 1) have been extremely successfully used as chiral ligands for lithium in a wide range of enantioselective transformations [1]. However, in most of the reported applications, the reaction requires the use of a stoichiometric amount of chiral ligands. Far fewer applications of sparteine as a chiral ligand to catalysis have been published. The aim of this review is to highlight recent advances in the field of asymmetric transformations that uses sparteine as chiral auxiliary, emphasizing the use of substoichiometric or catalytic amounts of this ligand.

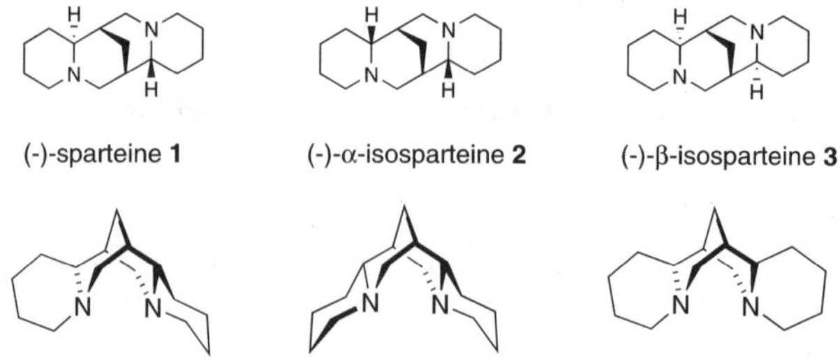

Fig. 1

2 Sparteine as a Chiral Ligand for Organolithium Reagents

The utilization of sparteine as a chiral ligand for organolithium reagents has already been thoroughly reviewed for reactions with a stoichiometric sparteine ligand on lithium generating configurally stable chiral carbanions [1]. We will focus here on recent applications in which a catalytic amount of a chiral ligand can be used to control the reactivity and stereoselectivity

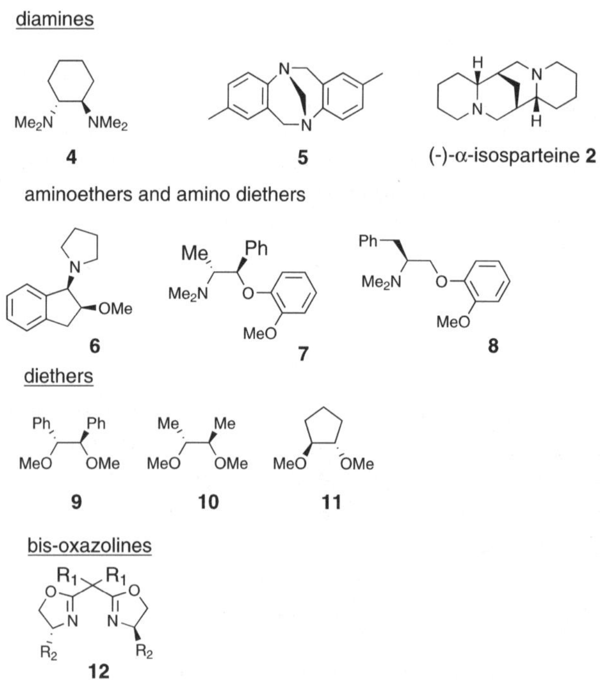

Fig. 2

of organolithium complexes. The use of sparteine as a chiral ligand under stoichiometric and catalytic conditions will also be compared to other chiral chelating ligands that are potentially useful as replacements for sparteine in asymmetric organolithium chemistry. For this purpose, most of the examples presented here will focus on research performed over the past ten years. Figure 2 shows some of the most useful structures used in various reactions to activate organolithium compounds.

2.1 Organolithium Reagents/Sparteine Combinations as Chiral Nucleophiles for Enantioselective Additions

When organolithium carbanions are generated in non-polar solvents, their low reactivity is usually ascribed to their oligomeric structures (such as dimers, tetramers or higher aggregates). Decreasing the level of oligomerization by complexation of the lithium with a coordinating ligand present as an additive or used as a solvent is the usual way to increase the reactivity of the carbanion. Enantioselective control can thus occur when a chiral chelating ligand such as sparteine is used. When a configurationally stable chiral center is created after activation by complexation with the chiral ligand, it is possible, in some cases, to decrease the amount of ligand used for the activation.

The enantioselective carbolithiation of cinnamyl derivatives described by the group of I. Marek and J.F. Normand [2] is one of the few reports on the use of sparteine as a catalytic activator of organolithium carbanions

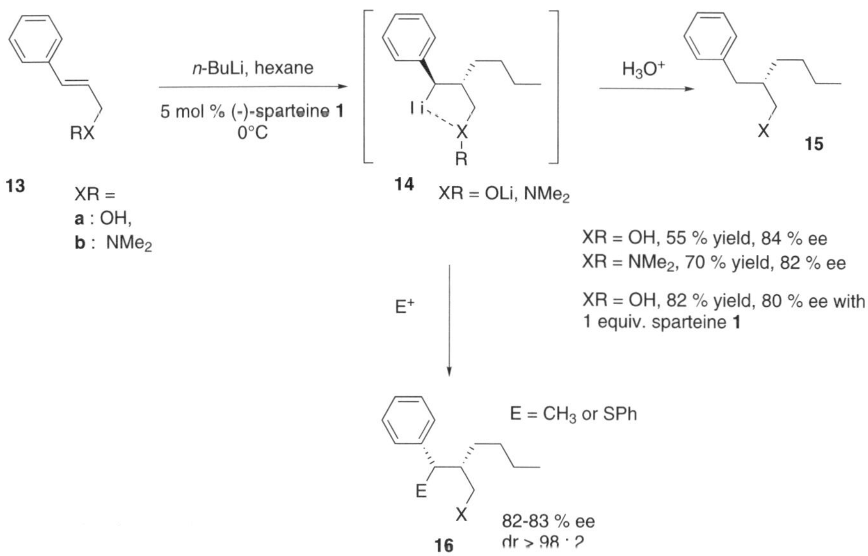

Scheme 1

(Scheme 1). Initial experiments used one equivalent of sparteine **1** to aid the asymmetric addition of *n*-BuLi to cinnamyl alcohol **13a** at 0 °C in cumene. The chiral thermodynamically stable benzylic anion **14a** was then generated and trapped with various electrophiles to yield the corresponding adducts **15a** and **16a** with complete diastereoselectivity and high enantioselectivity (80–83%) Since the organolithium reagent *n*-BuLi is unreactive toward the cinnamyl alcohol **13a**, activation by a chelating ligand is necessary to achieve the addition and it is then theoretically possible to decrease the amount of chiral ligand required. Indeed, the use of a 5 mol % amount of sparteine in similar reaction conditions gives, after acidic quenching, the alcohol **15a** in reduced yield, but without any decrease in enantioselectivity. The same degree of enantioselection was observed (82%) when a cinnamyl amine **13b** was used instead of the alcohol **13a**.

This methodology was later extended to the preparation of chiral disubstituted cyclopropanes when chiral cinnamyl acetals such as **17** and **21** were used for carbolithiation with various organolithiums [3]. When **17** was used as the substrate, the addition was carried out in a non-polar solvent such as hexane or cumene at – 50 °C in the presence of a catalytic or stoichiometric amount of sparteine to generate the chiral benzylic carbanion **18** (Scheme 2). The use of the acetal as a directing group gave increased enantioselectivity (92–95% ee) and the corresponding adduct **19** was isolated in good yields after hydrolysis at this temperature. No decrease in the enantioselectivity was observed when the amount of sparteine used was lowered from 1 to 0.1 equivalent, and only a slight decrease was obtained when only 1 mol % of the chiral ligand was used. After generation of the chiral benzylic organolithium **18**, intramolecular elimination of the acetal can occur when the temperature of the reaction mixture is allowed to return to room temperature. In that case, thermodynamic equilibration causes the benzylic carbanion to epimerize and to promote the formation of the more thermodynamically stable

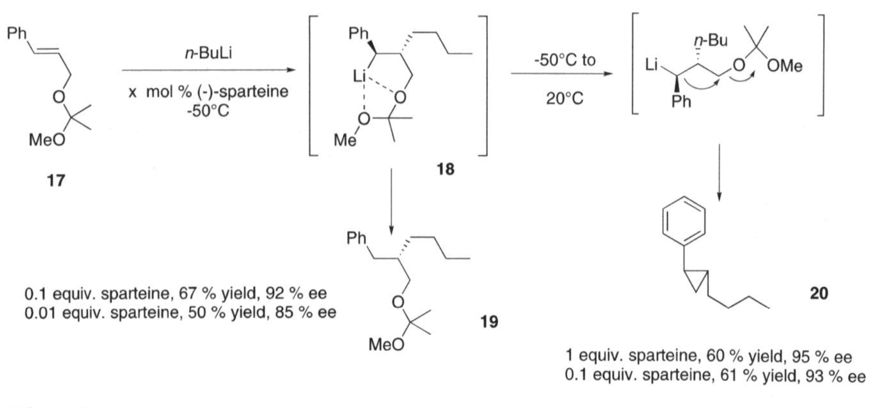

Scheme 2

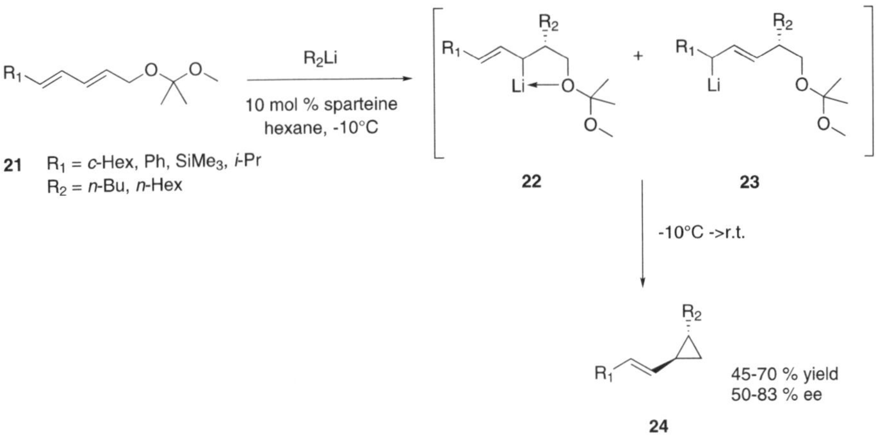

Scheme 3

trans-cyclopropane **20**. When this methodology was extended to the dienol acetals **21** (Scheme 3), the enantioenriched vinyl cyclopropanes **24** bearing various substituents were prepared in good yields and fair to good enantioselectivities (55–83%).

An early example of the use of a subcatalytic amount of sparteine for the activation of an organolithium nucleophile was reported by Lautens et al. in the carbometallation of a *meso*-unsaturated oxabicycle **25**, with ring opening leading to the substituted cycloheptene derivative **26** (Scheme 4) [4]. Both yield and enantiomeric excess remained virtually unchanged when the ratio *n*-BuLi : sparteine was lowered to 1 : 0.15. However, when a 3 mol % amount of the ligand **1** was used, a 20% decrease in enantioselectivity was observed.

Carbolithiation of an arene ring can also occur when the ring is activated by complexation with a transition metal such as chromium. Kundig et al. [5] reported the enantioselective addition of organolithium to a prochiral tricarbonyl arene chromium complex **27** in the presence of various chiral ligands (Scheme 5). The anionic intermediate **28** was trapped by an electrophile such as propargyl bromide to yield, after decomplexation of chromium, the *trans*-cyclohexadiene **29**. A non-coordinating solvent such as toluene was necessary to allow proper complexation of the chiral ligand with the lithium nucleophile and to achieve fair to high enantioselectivities. All of the chiral C_2 symmetric

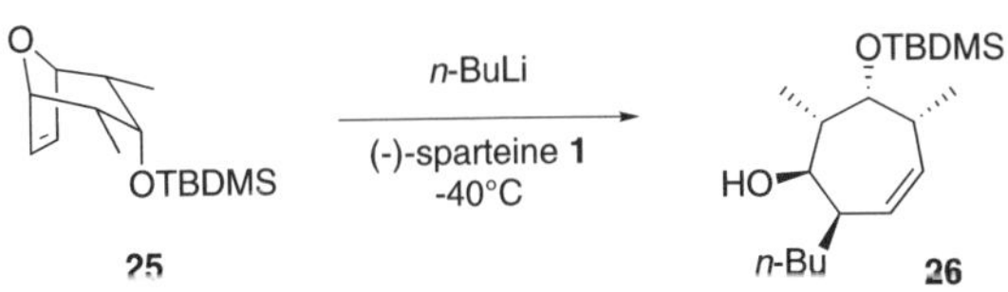

sparteine	: *n*-BuLi	:substrate	
25	5	1	48 % ee
2.5	5	1	44 % ee
0.15	5	1	52 % ee
0.03	5	1	30 % ee

Scheme 4

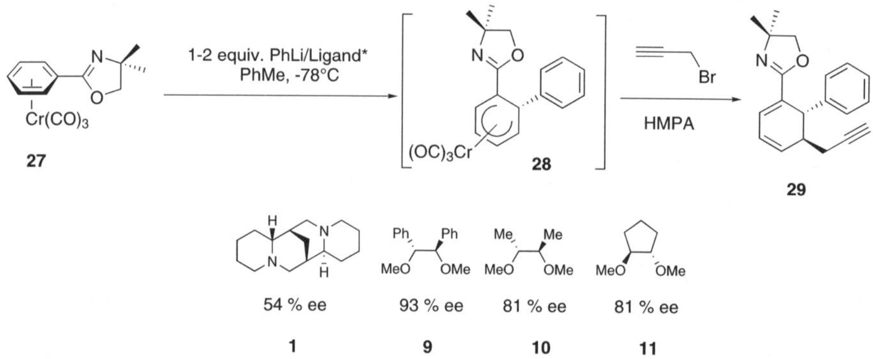

Scheme 5

diethers **9–11** gave better enantioselectivities than sparteine **1** and a maximum of 93% was achieved when the diether **9** was used with phenyl lithium as the nucleophile.

More recently, a very efficient asymmetric carbolithiation of *N*,*N*-dimethylaminofulvene **30**, leading to a chiral cyclopentadienide anion, was reported by Hayashi et al. [6] for the synthesis of chiral metallocenes (Scheme 6). By adding an aryl lithium such as **31** complexed with a chiral ligand on fulvene **30**, a cyclopentadienide ion **32** bearing a stereogenic center at the α position was generated. This anion was reacted with $[RhCl(nbd)]_2$ to yield

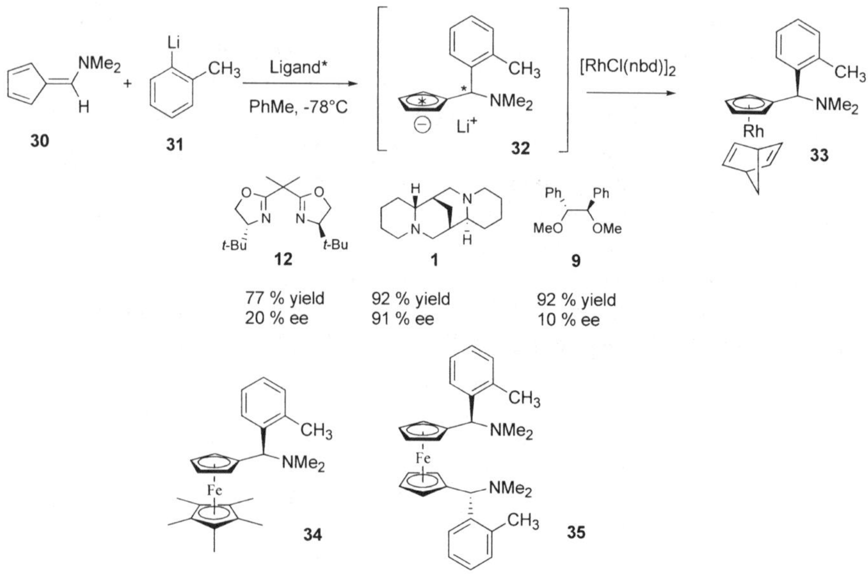

Scheme 6

the chiral cyclopentadienyl rhodium (I) complex **33**. When compared with bisoxazolines **12** and the bisether ligand **9**, sparteine **1** gave much higher enantioselectivities (up to 91%), and this ligand was retained for the application of this method to the synthesis of new chiral metallocenes. Thus, treatment of the anion **32** with FeCp*(acac) gave a 95% yield of the ferrocene complex **34** with a 93% ee, while reaction of the same anion with $FeCl_2$ gave the C_2-symmetric 1,1′-disubstituted ferrocene **35**, with ee > 99%.

Unsaturated esters have also been used as electrophiles for enantioselective carbon-carbon bond formation through the addition of organolithium nucleophiles in the presence of chiral ligands (Scheme 7). Tomioka et al. [7] reported the conjugate addition of various organolithiums to BHA alkenoates in the presence of sparteine **1** or the chiral diether **9**. The chiral ligand was usually used in a very large excess (1.4–4.2 equiv.) in order to achieve high enantioselectivity. The chiral diether **9** gave superior results, with enantioselectivities in the range of 91–99% for different alkenoates and organolithium nucleophiles. In a study by the group of Xu [8], *ortho*-substituted aryl alkenoates such as **38** were used as the acceptors (Scheme 8). Several ligand structures were compared to sparteine, such as diether **9**, amino ether **6**, and

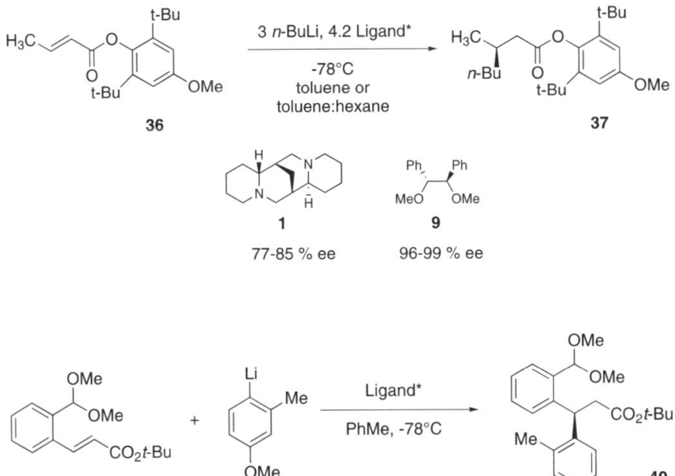

Scheme 7

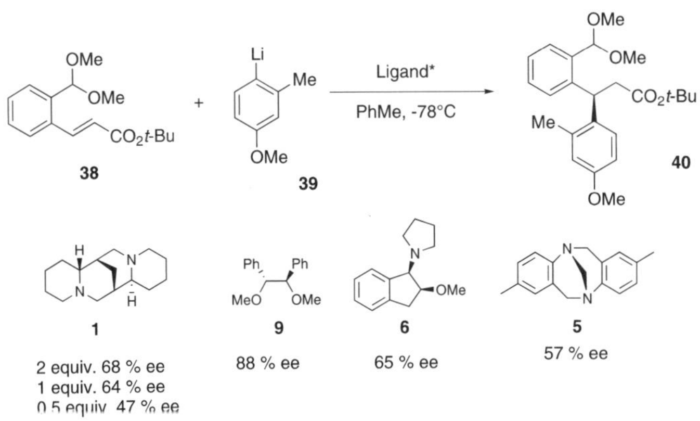

Scheme 8

a chiral Tröger's base **5**. However, the most versatile ligand for this reaction was sparteine **1**, giving fair to high enantioselectivities (68–92%) with various alkenoate esters and aryl lithium nucleophiles. It was also shown that a stoichiometric amount of sparteine **1** was necessary to achieve a good selectivity, as the use of a substoichiometric amount of the ligand produced a decrease in the selectivity. ^{6}Li NMR analysis of a 1 : 2 aryllithium/sparteine mixture showed a major signal attributed to a LiBr/sparteine complex and at least seven signals, indicating a complex mixture of species.

One of the first significant examples of asymmetric addition of organolithium reagents to imines was reported by the group of Denmark in 1994 (Scheme 9) [9]. This study also described the first comparison of the efficiency of bisoxazoline ligands compared to that of sparteine as chiral ligands for lithium. As already seen in earlier examples, a clear solvent effect was observed for the reaction of organolithium reagents in the presence of chiral chelating ligands. Low polar ethers such as Et_2O and *i*-Pr_2O gave the best results with most of the ligands studied. For the model reaction involving the imine **41** and *n*-butyl lithium, sparteine **1** gave the highest enantioselectivity, even when a substoichiometric amount of the ligand was used. However, for all ligands, a slight decrease of selectivity was observed when the amount of the ligand was lowered. When considering stoichiometric conditions for the ligands, the optimization for each substrate depended on the choice of the ligand and both sparteine and bisoxazolines **12** could be used to reach enantioselectivities > 90%.

A similar method was later used by the group of Tomioka [10] for the asymmetric addition of thiazolyllithium **44** to prochiral aldimines (Scheme 10) for the preparation of (*S*)-Boc-Doe **46**, a component of antileukemic marine natural product dolastatin 10. In this case, sparteine **1** was

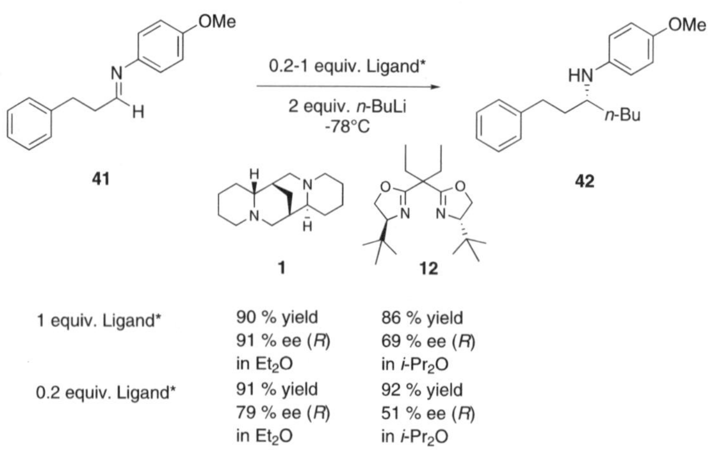

Scheme 9

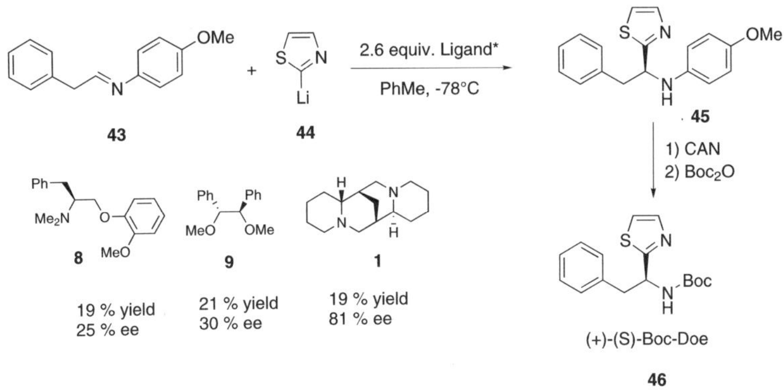

Scheme 10

compared to the bisether **9** and the amino diether **8**. Sparteine gave the best enantioselectivity (81%) for the addition to the imine **43**, albeit with a very modest yield (19%). After deprotection of the PMP group and formation of the Boc derivative, the enantioenriched (+)-(*S*)-Boc-Doe **46** was isolated with a 25% overall yield and with an 82% ee.

The scope of the addition of organolithium reagents to prochiral imines was also studied by the groups of North [11] and Tomioka [12]. North et al. (Scheme 11) used unsaturated imines such as **47** for the asymmetric addition of organolithium reagents in the presence of various chiral ligands. Sparteine **1** was the ligand of choice for those substrates, and after optimisation enantioselectivities as high as 88% could be reached. The unsaturated amine **48** could then be oxidized to give the *N*-protected amino acid **49**. In the study by Tomioka et al., *o*-vinyl substituted benzaldimines such as **50** were used as the substrates for the asymmetric addition of methyllithium and phenyllithium in the presence of a stoichiometric amount of a chiral ligand (Scheme 12). Similar results were reached for most of the ligands studied, and the enan-

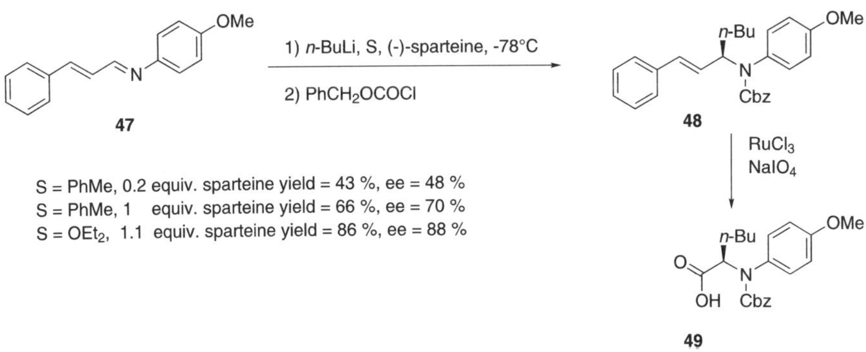

Scheme 11

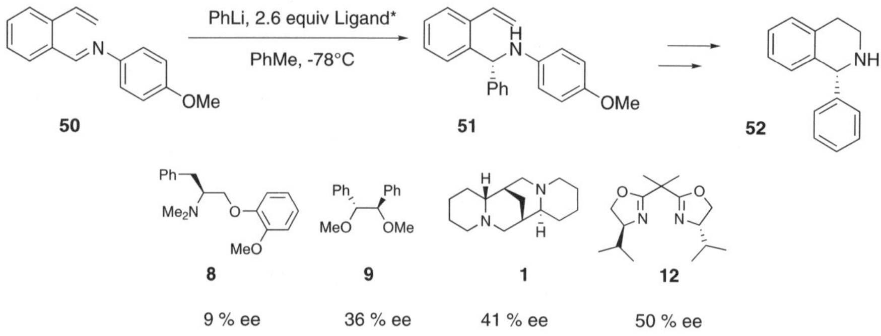

Scheme 12

tioselectivities remained modest in most cases. The presence of the vinyl group was also designed to transform the amine adducts into optically active 1-substituted tetrahydroisoquinolines such as **52**.

Isoquinoline **53** was also used as a substrate for the addition of organolithium reagents by Alexakis and Amiot (Scheme 13) [13]. While remaining quite modest, the best enantioselectivities (48%) were still reached with sparteine after quenching with methyl chloroformate. However, a mixture of mono- and bisacylated products **54** and **55** were obtained in all cases and the use of a catalytic amount of sparteine lowered the selectivity of the adducts.

There are few reports in the literature of chiral ligand-directed asymmetric addition of organomagnesium nucleophiles to prochiral electrophiles. An elegant study was recently reported by Sintani and Fu [14] for the enantioselective desymmetrization of *meso*-anhydrides by organomagnesium reagents in the presence of chiral ligands (Scheme 14). In the model reaction, the addition of phenylmagnesium chloride to the anhydride **56** was studied with various chiral N-O, and N-N ligands such as diamines, bisoxazolines and amino alcohols. The highest enantioselectivity (88%) was reached when a stoichiometric amount of sparteine **1** was used with toluene as solvent. The optimized conditions were also applied to various anhydrides and aryl magnesium halides, yielding the corresponding ketoacids in good yields (51–91%) and high enantioselectivities (78–92%).

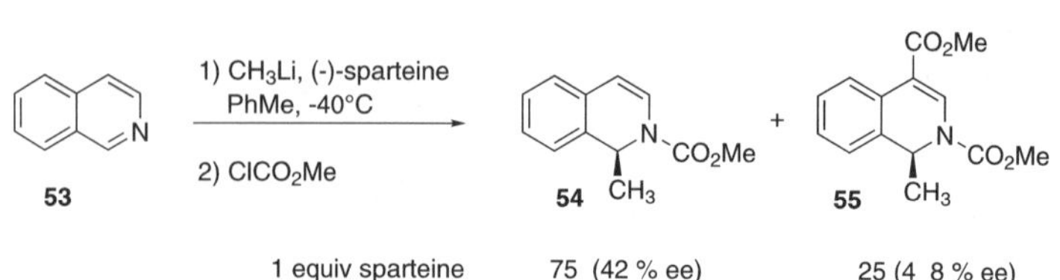

Scheme 13

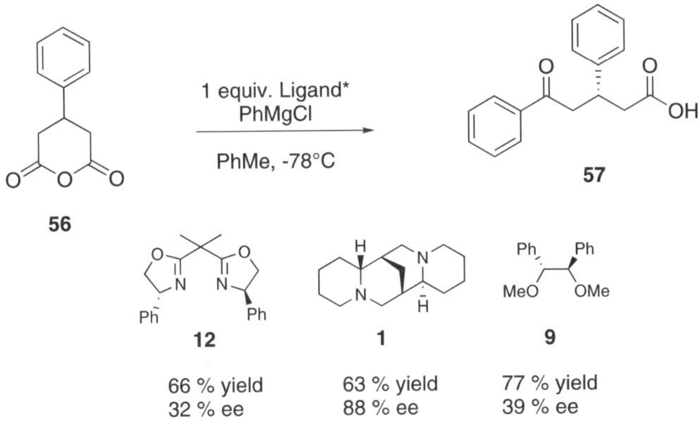

Scheme 14

The last example of the use of sparteine as a chiral ligand for the nucleophilic addition of organolithium nucleophiles concerns the use of prochiral acetals as the electrophilic counterparts (Scheme 15) [15]. Thus, nucleophilic substitution with a ring opening of the dioxolane **58** was carried out using an aryllithium reagent such as **59** in the presence of $BF_3 : OEt_2$ as a Lewis acid activator and a chiral ligand. Sparteine **1** proved superior in terms of selectivity compared to other chelating ligands such as the diether **9**, and the monosubstitution product **60** was isolated with an 81% ee. Attempts to broaden the scope of the acetals and aryllithium nucleophile showed that the enantioselectivity was strongly influenced by steric effects on both partners of the reaction. For instance, the synthesis of (*S*)-neobenodine **65** was carried out starting from acetal **62** and *p*-tolyllithium **63** (Scheme 16). The monoadduct **64** was isolated in a low ee (12%) when sparteine was used as a chiral ligand and a 49% ee with the chiral diether **9**. A two-step procedure was then used to transform the adduct into enantioenriched (*S*)-neobenodine **65**.

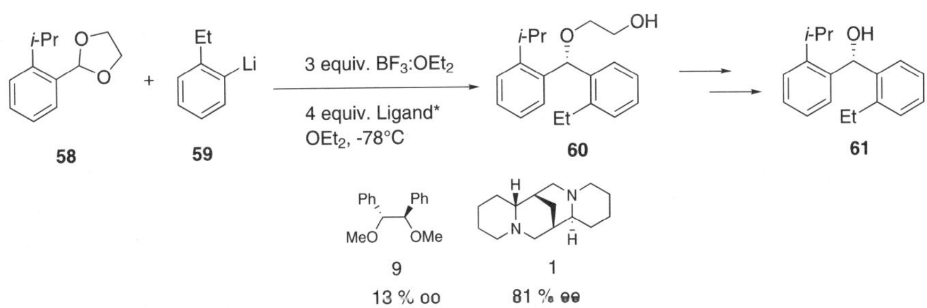

Scheme 15

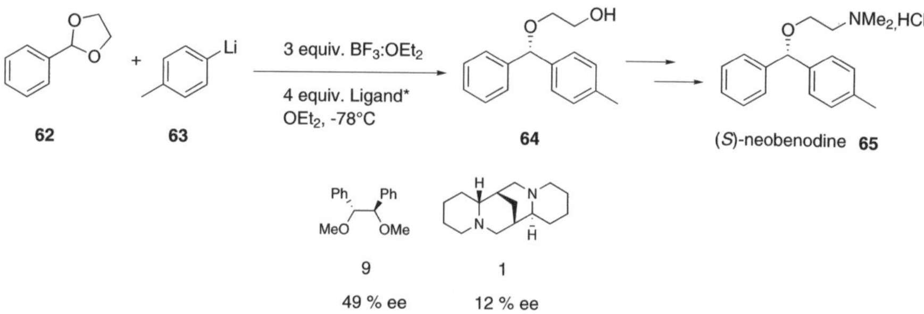

Scheme 16

2.2 Organolithium Reagents: Sparteine as Chiral Bases

As stated earlier, the combination of an organolithium reagent with sparteine has been widely used to generate chiral carbanions. We will focus here on recent examples where the efficiency of sparteine was compared with other chiral chelating ligands.

Chiral bases such as RLi:sparteine has been previously used for the enantioselective ortholithiation of ferrocenyl complexes bearing a non-chiral *ortho*-directing group [16]. This concept was recently used by the group of Iwao [17] to introduce planar chirality onto the azaferrocenes **66** and **68** via an ortholithiation-electrophilic capture sequence (Scheme 17). The first system studied involved a direct lithiation on the azacyclopentadienyl ring with a combination of *sec*-butyl lithium and a chiral ligand in diethyl ether. The corresponding substituted adducts were isolated in good yields after capture of the lithio intermediate with various electrophiles. Sparteine **1** proved to be superior compared to bisoxazoline **12** and bisamine **4** and enantioselectivities of up to 79% could be reached with this ligand. However, when substrate **68** was used, deprotonation occurred at the benzylic position of one of the enantiotopic methyl groups next to the nitrogen, and the lithiation gave high enantioselectivity (99%) when a bisoxazoline **12** was used. In that case, a disappointing result was obtained with sparteine (57% ee).

Sparteine has been widely used for the generation of configurationally stable benzylic carbanions on various substrates bearing carbamates as stabilizing groups. Simple substrates such as benzyl ether **70** [18] and benzyl sulfide **72** [19] were recently studied for the generation of a chiral center at the benzylic position by deprotonation with an alkyl lithium: chiral ligand combination (Schemes 18–19). In both cases, sparteine gave very low induction compared to bisoxazolines such as **12**. In particular, high enantioselectivities were reached with the benzyl sulfide **72** (up to 98%).

An elegant enantioselective [2,3] sigmatropic rearrangement of bisalkynyl ethers such as **75** was reported by Manabe in 1997 [20]. The deprotonation

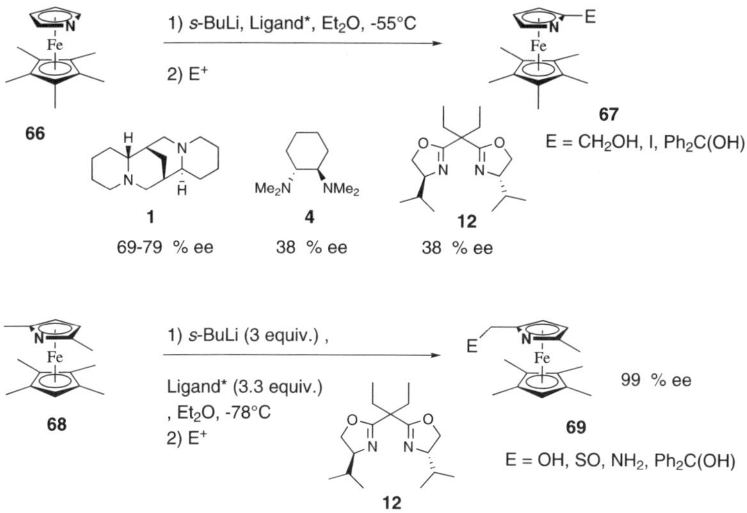

Scheme 17

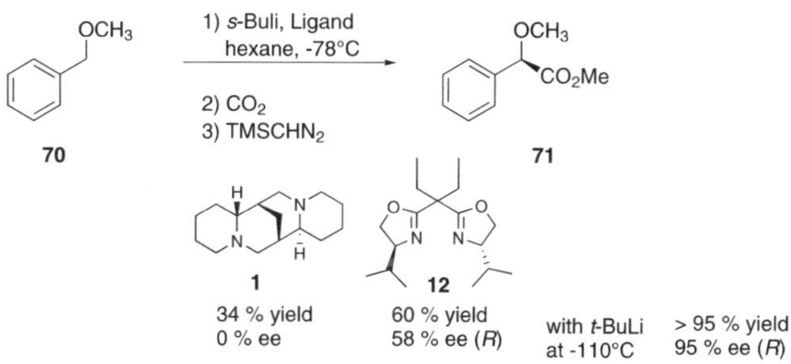

Scheme 18

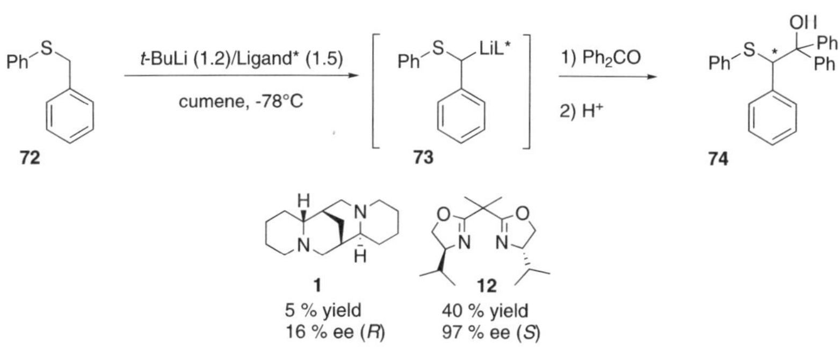

Scheme 19

of this bispropargylic ether **75** by a BuLi/ligand combination gives a chiral carbanion which undergoes rearrangement to yield the allenic alcohol **76** in moderate yield (50%) (Scheme 20). While a disappointing level of enantioselectivity (9% ee) was achieved when sparteine was used as the ligand, the chiral tridentate ligand **7** gave a more promising level of enantioselectivity of 61%. A similar strategy was also used by Nakai et al [21] for the [2, 3] Wittig rearrangement of crotyl propargyl ethers **77** (Scheme 21). Disappointing ee's were obtained for the propagylic alcohols **79** when sparteine **1** and isosparteine **2** were used as ligands. However, the bisoxazoline ligand **12** gave ee's of up to 89% when the reaction was carried out at – 95 °C in pentane.

The enantioselective α-deprotonation of prochiral epoxide was extensively studied by the group of Hodgson. When using medium-sized epoxides and organolithium as bases, deprotonation occurs at the α position to give transient carbenoid species that undergoes a fast rearrangement [22]. Thus, starting from the eight-membered ring epoxide **80**, the bicyclic alcohol **81** was

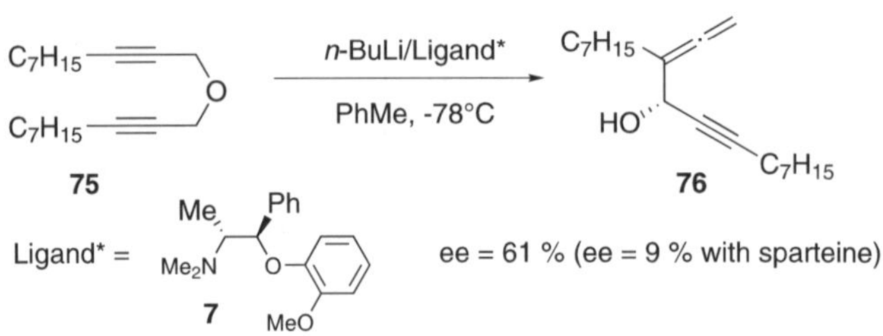

Scheme 20

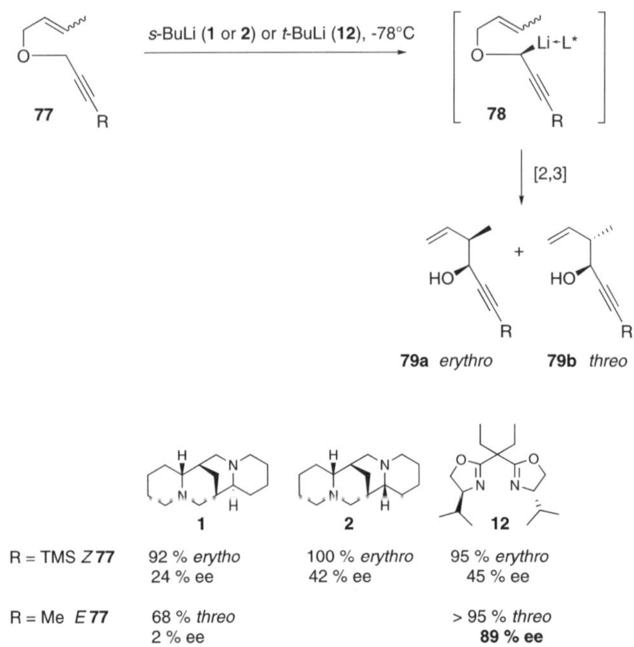

Scheme 21

obtained in good yields and high ee when isopropyl lithium was used in combination with a chiral ligand (Scheme 22). Sparteine **1** and isosparteine **2** gave better enantioselectivities than bisoxazoline **12** and it was shown that a catalytic amount of the ligand could be used with little modification of the enantioselectivity. The introduction of alkoxy substituents to the epoxide precursor was also studied in order to give access to polyfunctionalized chiral synthons (Scheme 23) [23, 24]. Optimized conditions for the desymmetrization of the epoxide **82** used isopropyl lithium-isosparteine **2** as a base and yielded the alcohol **83** in a 55% yield and a 80% ee. This alcohol was used in the total synthesis of (–)-xialenon A, a naturally occurring secondary metabolite from the culture broth of the *Streptomyces* genus. A different pathway for the reactivity of the α-metallated epoxide was observed in the case of epoxydihydrofuran **85** (Scheme 24) [25]. Insertion of a second equivalent of the organolithium into the carbenoid proceeds via a 1,2-metallate shift followed by elimination. The enantioselectivities remained modest for the resulting allylic alcohol **87** when using either sparteine **1** (40% ee) or bisoxazoline **12** (38% ee). Different bicyclic substrates were also tested in this reaction and an optimum 84% ee was reached with the epoxide **88** (Scheme 25).

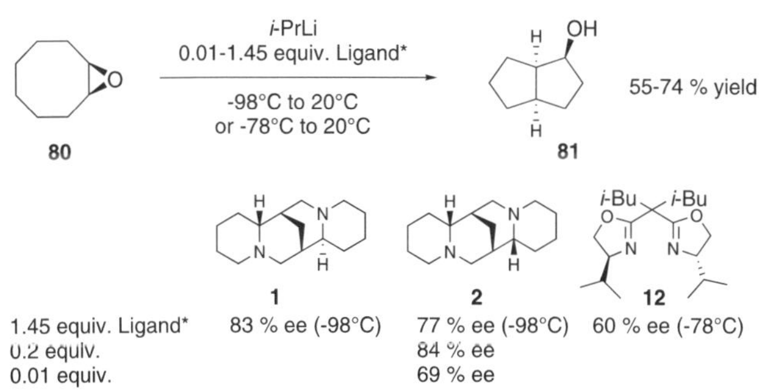

Scheme 22

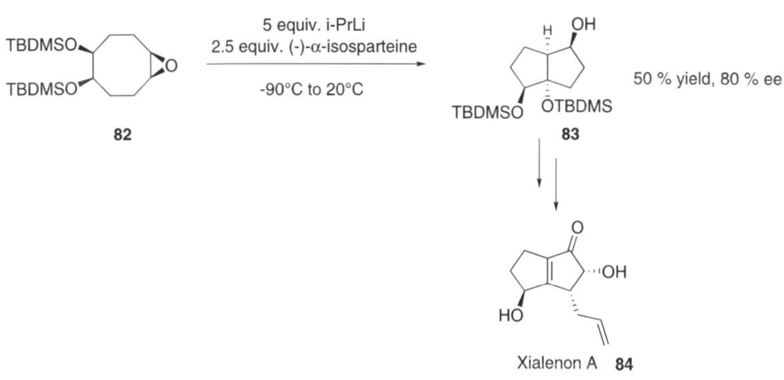

Scheme 23

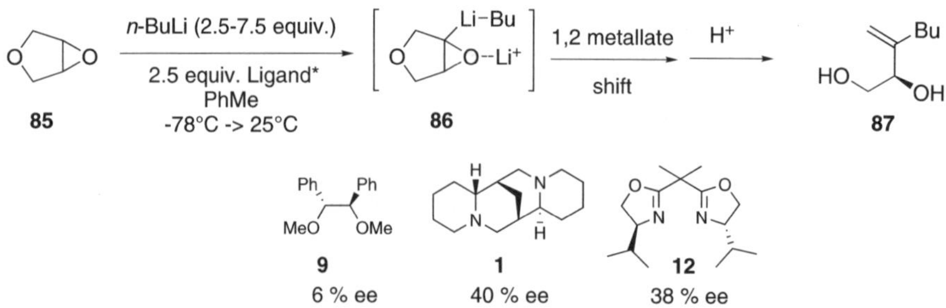

Scheme 24

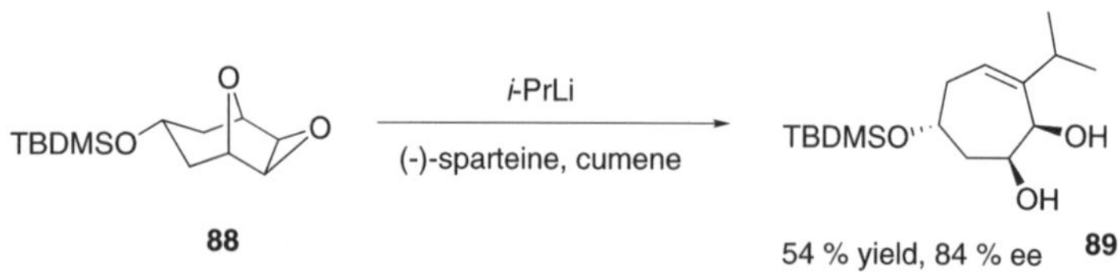

Scheme 25

3 Sparteine as a Chiral Ligand for Copper Catalysts

3.1 Stoichiometric Processes with Copper Complexes

There is a growing interest in chiral π-conjugated molecules, directed at the design of organic materials. The chirality may provide not only additional electronic properties but may also allow for the control of packing and $\pi - \pi$ interactions in the solid as well. To this end, Rajca et al. [26] have described an efficient asymmetric synthesis of chiral tetra-*o*-phenylenes, based upon a (–)-sparteine-$CuBr_2$-mediated coupling of 2,2′-dilithiobiaryls. Two aryl-aryl C – C bonds are formed and the configuration of four chiral axes is set. In the key step, the 5,5′-diarylated compounds (**90a–c**) are treated with *t*-BuLi (4 equiv.) and (–)-sparteine **1** (2 equiv.) in ether/pentane at – 78 °C. After addition of $CuBr_2$ (6 equiv.), the vigorously stirred reaction mixture is allowed to attain ambient temperature overnight (Scheme 26). Ee's of around 50% are reproducibly obtained. The tetra-*o*-phenylene adducts **91** are isolated in aproximately 80% yields, with only minor amounts of the biphenylene side products **92**. The reaction conditions required for the optimization of both reactivity and enantioselectivity were also studied. A 2–3 fold increase in the number of equivalents of *t*-BuLi and/or (–)-sparteine and/or $CuBr_2$ had no effect on ee's or yields.

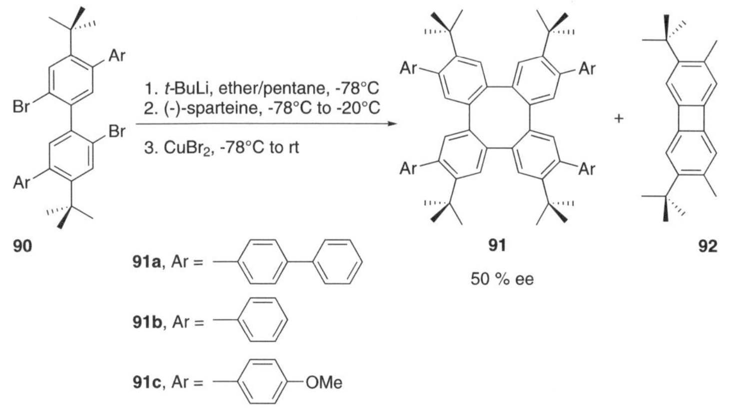

Scheme 26

Recently, Wulff and coworkers demonstrated that a copper-mediated deracemization of the vaulted biaryl ligands VANOL and VAPOL can be readily achieved in the presence of (–)-sparteine **1** (Scheme 27) [27].

The first studies leading to the deracemization of VANOL **93** and VAPOL **94** were carried out with 1.4 equiv. of copper chloride and 2.8 equiv. of (–)-sparteine **1** in a mixture of methanol and methylene chloride (1 : 3.4) deoxygenated by an argon purge, which gave a 64% yield of (*S*)-VAPOL with 99% ee, and a 77% yield of (*S*)-VANOL with 99% ee. However, in the last case, the co-solvent was deoxygenated via a freeze-thaw method.

The optimal procedure involves the in situ generation of copper(II)-sparteine species. When the sonification of CuCl and (–)-sparteine is conducted in the presence of air, there is a rapid dissolution of CuCl along with concomitant formation of a green solution, presumably a copper(II) specie which was obtained twenty times faster than with commercial copper(II) chloride. The deracemization of VANOL **93** occurs in 1.75 h to give a 79% yield of (*S*)-VANOL and 99% ee. Under these conditions, the reaction is also reproducible on a larger scale (5 g), giving a 69% yield of (*S*)-VANOL with 99% ee in 2 h.

Smrčina and Kočovský [28] have described a straightforward method for racemic homocoupling of the 2-naphthol **95** and the 2-naphthylamine **96** under controlled conditions, using a $CuCl_2$-benzylamine complex in methanol (Fig. 3). The authors also focused on enantioselective oxidative coupling using chiral amines. (–)-Sparteine **1** and (*R*)-(+)-α-methylbenzylamine were then selected as inexpensive commercially available candidates. The coupling of 2-naphthol, using a complex of $CuCl_2$ (1 equiv.) and (–)-sparteine (2 equiv.) generated in situ, resulted in the formation of a precipitate and a mother liquor. Working up the precipitate led to the isolation of enantiomerically pure (–)-binaphthol **99** with a 14% yield. The mother liquor afforded a 42% yield in (–)-binaphthol **99** in low enantiomeric excess (20%)

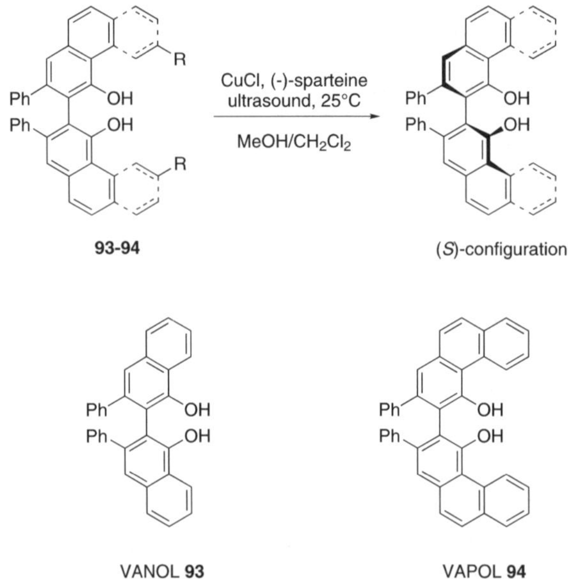

Scheme 27

after work-up. The isolation of (–)-binaphthol was due to a second-order asymmetric transformation of the racemate formed **99** rather than being due to asymmetric coupling. In order to confirm their theory, racemic binaphthol **99** was treated with a $CuCl_2$-sparteine (1 : 2 equiv.) complex under the same conditions as those used for the coupling experiment. The formation of a precipitate was observed again, and both the precipitate and the mother liquor were worked up separately. The precipitate furnished a 36% yield of enantiomerically pure (–)-binaphthol, while the mother liquor gave the same enantiomer of 59% ee with a 60% yield. Work-up of the whole mixture (without the separation) gave a crude product of 80% ee with 94% yield. This experiment confirmed the operation of a second-order asymmetric transformation and demonstrated that the deracemization of binaphthol is superior to the attempted asymmetric coupling. In contrast to binaphthol, the bisnaphthylamine **100** was obtained in much higher enantiomeric purity from the coupling rather than from the deracemization experiment. Work-up of the whole mixture after the deracemization experiment afforded the racemic diamine **100** in 95% yield, while the coupling experiment gave an 84% ee and 19% yield of the (–)-diamine **100** from the precipitate, and a 31% ee and 49% yield of the (+)-diamine from the solution. After a kinetic crystallization they obtained the enantiomerically pure diamine with a 13% yield for the (–)-diamine and a 22% yield for the (+)-diamine. The authors have rationalized these observations by assuming diastereoselective crystallization

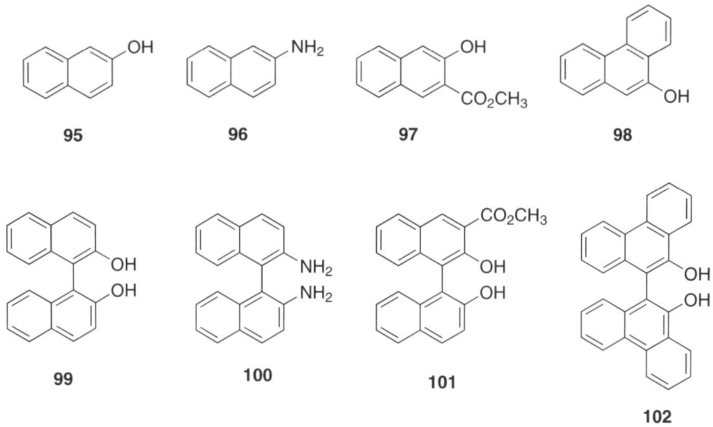

Fig. 3

of the Cu(II)-amine-product complex as being the dominant mechanism of stereodifferentiation.

Smrčina and Kočovský [29] have generalized their $CuCl_2$-mediated oxidative homo- or heterocoupling of naphthols to substituted binols such as **101** and **102** (Fig. 3). Dramatic differences have been observed in the stereodifferentiation processes, which can be interpreted as stemming from three different mechanisms. Experimental evidence has increasingly suggested that a second-order asymmetric transformation is the dominant process controlling the production of **99** and **102** (80% yield, 76% ee). In contrast, the stereodifferentiation observed in the case of **100** ((−)-**100** 19% yield, 84% ee; (+)-**100** 49% yield, 31% ee) can be ascribed to a diastereoselective crystallization, while formation of the enantiomerically enriched **101** (Cu(II)/sparteine, 1 : 2.4 equiv.) (45% yield, 71% ee) can be best explained by a direct enantioselective coupling reaction.

A catalytic version of the coupling was also developed, by using 10 mol % of $CuCl_2$ and 20 mol % of sparteine **1** (silver chloride was used as a stoichiometric oxidant to regenerate the copper (II) oxidant). This catalytic system was applied to the asymmetric cross-coupling leading to **101** in a 41% yield and 32% ee.

3.2 Catalytic Processes with Copper Complexes

In recent years, axially chiral binaphthalene derivatives have emerged as important ligands and chirality inducers in organic synthesis. Oxidative coupling of 2-naphthols represents a well established method for the preparation of binaphthols. The couplings are usually carried out by treating naphthols with more than an equimolar amount of a metal such as Fe(III), Mn(III), or Cu(II). Very recently, catalytic processes have been developed using the $CuCl_2$-amine/AgCl system or $FeCl_3$ in the solid state.

Nakajima and coworkers [30] observed that, in the presence of CuCl(OH).-TMEDA with oxygen or air as the oxidant, 2-naphthol **103a** is transformed into 1,1′-bi-2,2′-naphthol **104a.** A wide variety of substrates undergo oxidative coupling in excellent yields (Scheme 28). It is worth noting that the reaction requires as little as 1 mol % of the catalyst.

After they established an efficient process for the catalytic oxidative coupling of naphthols, Nakajima and coworkers [31] turned their attention to the enantioselective oxidative coupling of naphthols with (−)-sparteine **1** as a chiral ligand (Scheme 29). According to the above procedure, 2-naphthol **103a** was added to a solution of the chiral catalyst prepared in situ from CuCl (10 mol %) and sparteine **1** (11 mol %) in dichloromethane, and the reaction mixture was heated at reflux under oxygen. Chromatography afforded unreacted starting material (58%) and binaphthol **104a** (32%) in low enantiomeric excess (18%). Among several substrates surveyed, the reaction of methyl 3-hydroxy-2-naphthoate **103e** afforded the corresponding binaphthols in 47% ee.

Copper complexes of sparteine have also been used for the catalysis of asymmetric carbon-carbon bond formation. The copper-catalyzed reaction

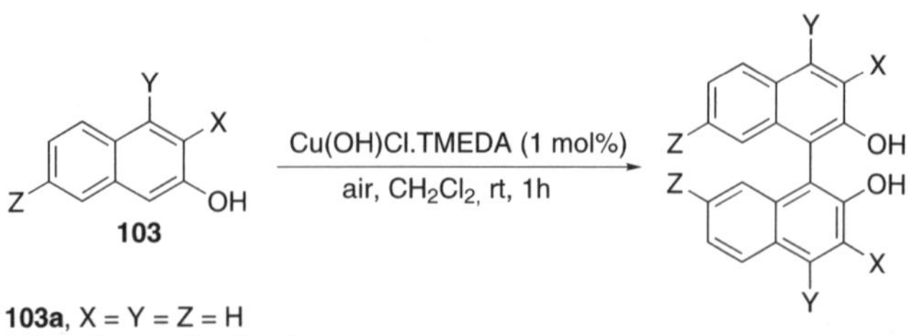

103a, X = Y = Z = H
103b, X = Me, Y = Z = H
103c, X = OMe, Y = Z = H
103d, X = Y = H, Z = OMe
103e, X = COOMe, Y = Z = H
103f, X-Y = -(CHCH)$_2$-, Z = H

Scheme 28

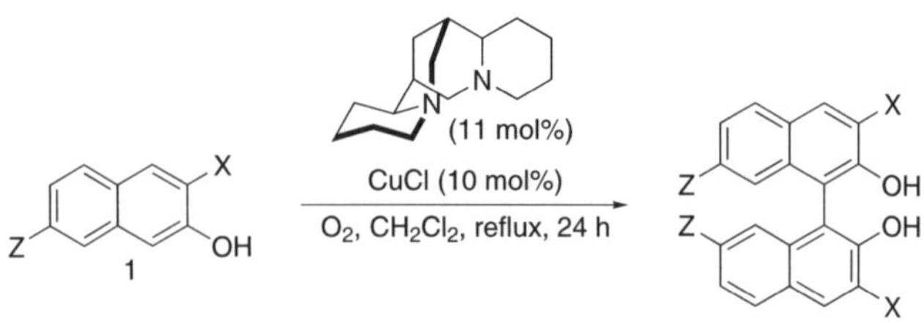

103a, X = Z = H
103e, X = COOMe, Z = H

104a ee = 18 %
104e ee = 47 %

Scheme 29

of terminal alkynes with nitrones is an elegant method for the synthesis of 1-aza-1-buten-3-yne and 2-azetidinone derivatives [32].

In the racemic version, the reaction of various terminal alkynes **105** with nitrones **106** was carried out using 10 mol % of CuI in pyridine-DMF at room temperature. As expected, the corresponding azetidinones **107** were formed as a mixture of *trans* and *cis*-isomers, along with the imines **108** (Scheme 30).

The authors also postulated that asymmetric induction in the formation of **107** could be achieved if the reaction was carried out in the presence of certain chiral ligands. While an excess of pyridine was usually used in the normal reaction, the reaction also proceeds with a catalytic amount of 1,10-phenanthroline. Therefore, chiral nitrogen-containing bidentate ligands could be suitable for this purpose. The reaction using CuI-(–)-sparteine (0.1 equiv./1 equiv.) gave the (–)-*trans*-**107** adduct ($R_1 = R_2 = R_3 = Ph$) in a yield of 47% with a 23% ee. In this reaction, bisoxazoline ligands gave superior results with ee's of up to 68%. The different results suggest that in the catalytic reaction, complete chelation of the ligands to the copper acetylide intermediate may be hindered by the alkyne substrate, which generally has a high affinity for copper(I) species, since it exists in high concentrations in the early stages. This may imply that the complexation affinity with the ligands must be higher that with the acetylide intermediate.

The enantioselectivity of the copper-catalyzed intramolecular cyclopropanation of allyl diazomalonates **110a–c** was investigated with a series of chiral ligands [33]. In this study, the approach of Müller et al. consisted of optimizing the enantioselectivity of the diazo decomposition by screening a large number of Cu-catalysts, and varying the reaction conditions. Cu-catalyzed cyclopropanation of the allyl diazomalonates has been previously investigated by Koskinen et al., with the objective to synthesize optically pure aminocyclopropane-1-carboxylic acids. The cyclopropanation of **110b** with CuI-$P(OEt)_3$ required raising the temperature to > 150 °C overnight, and it provided **111b** in 76% yield.

The starting diazo esters **110** were prepared by diazo transfer from the corresponding malonate esters **109**. A selection of chiral ligands in conjunction with 2 mol % (with respect to the diazo compound) of $[Cu(OTf)_2]$ in $(CH_2Cl)_2$ was then examined at 65 °C (Scheme 31). All of the ligands tested were sufficiently reactive to produce diazo decomposition at 65 °C, although the yields of cyclopropanation products were quite variable. Even tertiary

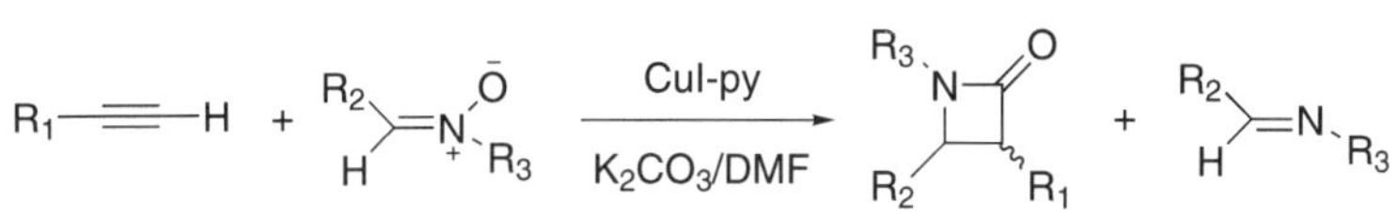

105 106 107 108

Scheme 30

amines such as sparteine, which is not generally applied in diazo decompositions, were found to be effective. Along with the (–)-sparteine **1**, **111b** was obtained in a yield of 55% with 22% ee, but **111c** was obtained in a yield of 35% with a really poor enantiomeric excess (4%). Similarly, the most efficient ligand **112** for *tert*-butyl ester **110b** (74% ee) gave a low enantioselectivity with the corresponding benzyl ester **110c** (4% ee). With most of the other ligands, the enantioselectivity in the decomposition of *ter*-butyl malonate **110b** was higher than that of the benzyl malonate **110c**. The enantioselectivities observed with the catalysts varied enormously in an unpredictable manner.

Müller et al. have also examined the enantioselectivity and the stereochemical course of copper-catalyzed intramolecular CH insertions of phenyliodonium ylides [34]. The decomposition of diazo compounds in the presence of transition metals leads to typical reactions for metal-carbenoid intermediates, such as cyclopropanations, insertions into X – H bonds, and formation of ylides with heteroatoms that have available lone pairs. Since diazo compounds are potentially explosive, toxic, and carcinogenic, the number of industrial applications is limited. Phenyliodonium ylides are potential substitutes for diazo compounds in metal-carbenoid reactions. Their photochemical, thermal, and transition-metal-catalyzed decompositions exhibit some similarities to those of diazo compounds.

The authors have examined the enantioselectivities of intramolecular insertions upon the Cu(I)-catalyzed decomposition of phenyliodonium ylides in comparison to those of the corresponding diazo compounds (Scheme 32). The phenyliodonium ylide **113b** was exposed to [Cu(OTf)$_2$] in CH_2Cl_2 at 0 °C in the presence of chiral ligands (2 mol %). The intramolecular CH insertion afforded the cyclopentanone carboxylate **114**, which was subjected to hydrolysis and decarboxylation (HBr/EtOH) to furnish the ketone **115**, for which the enantiomeric excess and absolute configuration were determined. Owing to the low reactivity of the Cu(I) catalysts towards diazo compounds, the diazo decomposition had to be carried out in 1,2-dichloroethane at 65 °C. Unfortunately, the ylide **113b** decomposed at this temperature, and it was impossible to carry out the reactions of **113b** and **113c** under identical conditions. In general, the Cu-catalyzed insertions proceed with acceptable yields from the

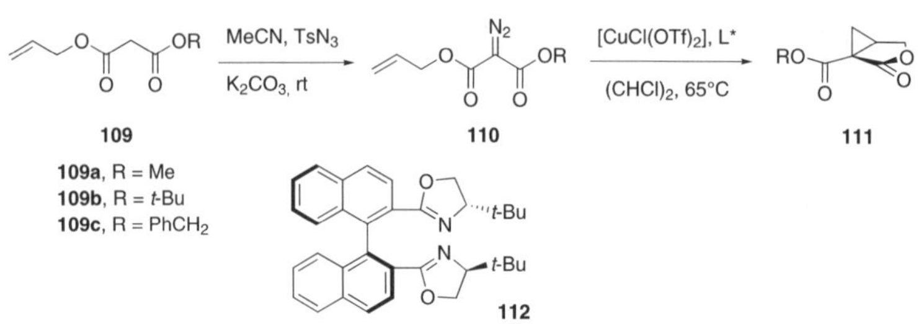

Scheme 31

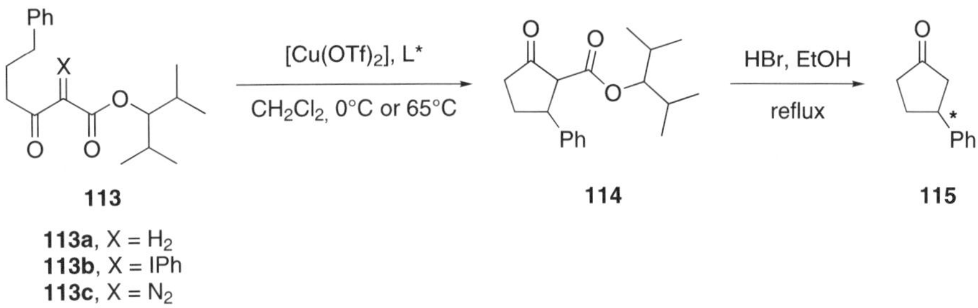

Scheme 32

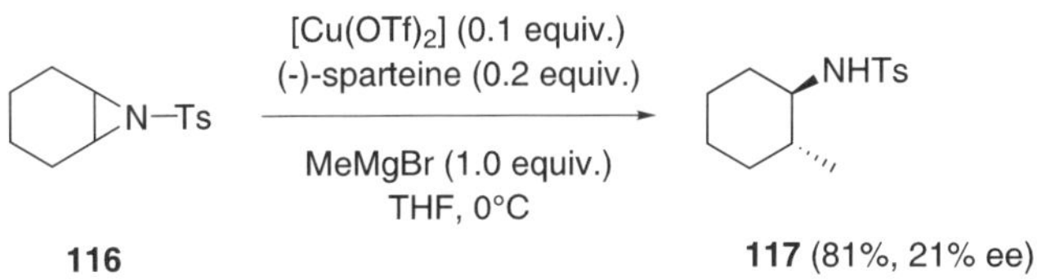

Scheme 33

ylide. In the case of (–)-sparteine **1**, a low yield of 11% was obtained from **113b**, albeit with a good enantiomeric excess of (57%). Starting from **113c**, (–)-sparteine **1** gave a yield of 32% with an 18% ee.

Müller et al. also found some conditions under which desymmetrization of *N*-sulfonated aziridines such as **116** occurred under catalytic conditions with organometallic reagents in the presence of chiral Cu-complexes [35]. In the case of (–)-sparteine, aziridine opening was effected by preparing the catalysts in situ by adding 0.2 equiv. of the ligand to 0.1 equiv. of $[Cu(OTf)_2]$. The desymmetrizations were carried out by adding the organometallic reagent (1.0 equiv.) in THF at 0 °C to a THF solution containing the aziridine **116** and the catalyst. Using (–)-sparteine **1** as the ligand, the reaction led to the formation of **117** in a good yield (81%) but with a moderate ee (21%) (Scheme 33).

4 Sparteine as a Chiral Ligand for Palladium Catalysts

The use of chiral transition-metal complexes as catalysts for stereoselective C-C bond forming reactions has developed into a topic of fundamental importance. The allylic alkylation is one of the best known of this type of reaction. It allows the Pd-catalyzed substitution of a suitable leaving group in the allylic position by a soft nucleophile.

Trost et al. [36] showed, in an early report on the stoichiometric allylic alkylation of $[Pd(\pi^3\text{-MeCHCHCHMe})Cl]_2$ **118** with $Na[CH(COOMe)_2]$ in the presence of various chiral ligands, that sparteine would compete (amongst others) with (+)-DIOP for asymmetric induction. The π-allylpalladium chloride dimer **118** was mixed with the ligand (ratio 1 : 2) in THF at the stated temperature, after which a solution of diethyl sodiomalonate in THF is added (Scheme 34). **119** was obtained with an optical purity of up to 20%.

One of the first efficient catalytic palladium systems for asymmetric allylic alkylation that does not involve phosphines has been reported by Pfaltz [37], who used bisoxazolines.

Togni's [38] approach was therefore to test the ability of sparteine to act as an ancillary ligand in Pd(II)-allyl complexes—susceptible to nucleophilic attack by stabilized anions such as Na[CH(COOMe)$_2$]—which could be employed as catalyst precursors. In addition he speculated that the rather rigid and bulky sparteine would be able to induce significant differentiation between the two diastereotopic sites of 1,3-disubstituted allyl ligand, thus leading to enantioselection upon nucleophilic attack.

Starting from either racemic or achiral allylic acetates **120**, the general procedure involves (Scheme 35) the addition of diethyl sodiomalonate (2 equiv.) to a mixture containing **122** (0.1 equiv.), **120** (1 equiv.) and 0.2 equiv. of (–)-sparteine **1** in THF. The optically active alkylated products **121** obtained showed enantiomeric excesses of up to 85% for **121c$_2$** and **124** in moderate yields (13–61%) (Scheme 36). These are comparable to those obtained with most efficient catalysts containing chiral chelating phosphine ligands. However, the applicability of this system seems to be restricted to either cyclic substrates or those bearing aryl substituents. The substrate **120c** illustrates the general problem with the regioselectivity in the allylic alkylation of unsymmetrically substituted allylic substrates. As found for most Pd-catalyzed

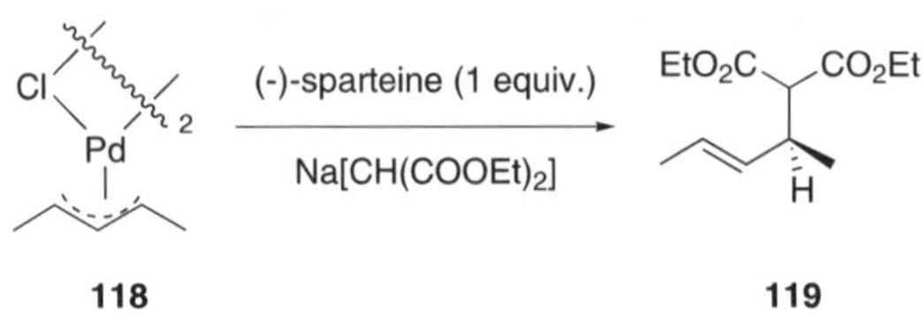

Scheme 34

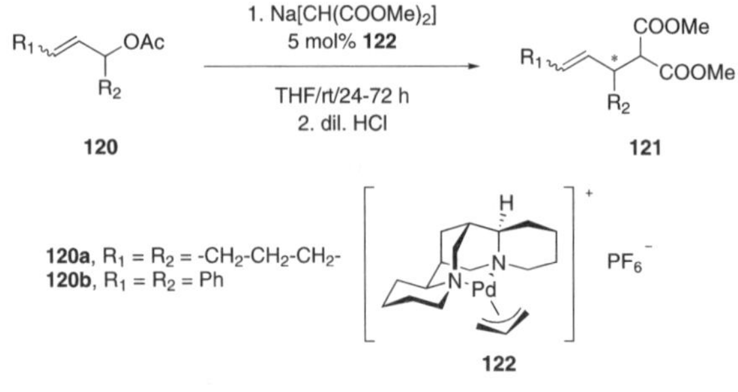

Scheme 35

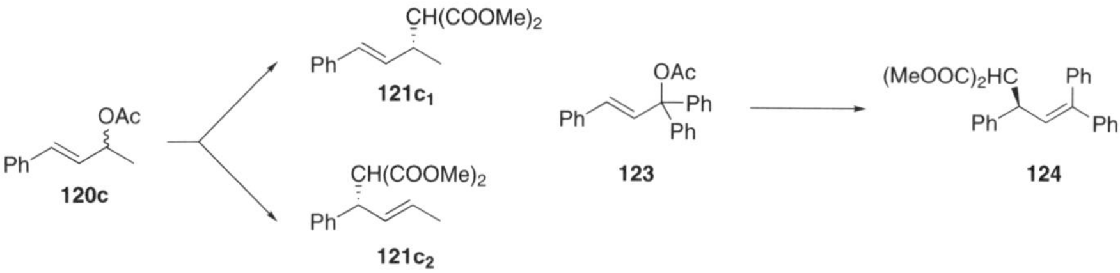

Scheme 36

substitution reactions, the Pd-sparteine catalyst preferentially directs the nucleophilic attack to the less hindered end of the allyl fragment. A further limitation of this catalyst is the rather low reactivity, as reflected by the turnover frequency of ca. 0.5 h^{-1}.

The results obtained with sparteine reported by Togni et al. were not representative. Consequently, Kang et al. [39] compared the results obtained using the Pd-sparteine complex **125** with those obtained using the Pd-α-isosparteine complex **126** (Fig. 4) in the alkylation of various allylic acetates by $Na[CH(CO_2Me)_2]$ in several solvents in parallel reactions. Contrary to the previous report [38], the reaction was best carried out in a solvent at reflux, which ensured rapid reaction to completion and also good enantiomeric purity. THF and DMF were the solvents of choice for universal use. The results show that both sparteine **1** and isosparteine **2** can indeed behave as a chiral bidentate ligand, although isosparteine **2** is preferred overall, presumably because its pocket depth is deeper than that of sparteine. This fact is reflected in the stability of the π-allyl-palladium complex and the shorter reaction time.

(–)-Sparteine **1** was also used in a palladium complex-catalyzed enantioselective benzoylation of alcohols using monoxide and the organobismuth(V) compound (Scheme 37). The carbonylative acylation of alcohols using carbon monoxide (CO) is known to be an alternative tool for the prepar-

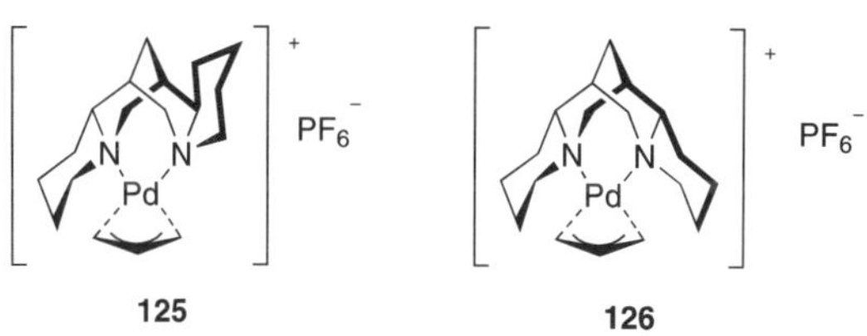

Fig. 4

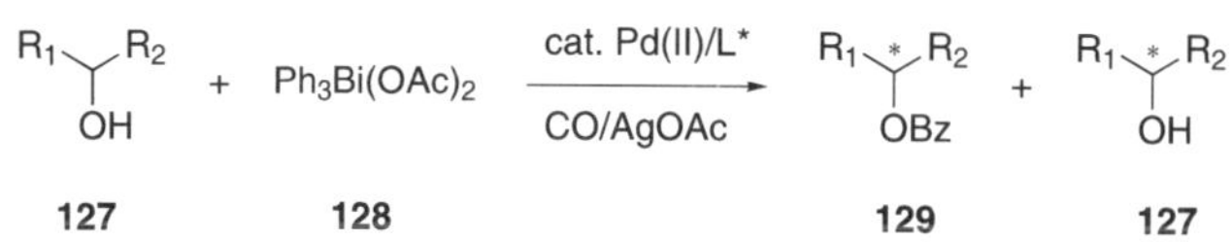

Scheme 37

ation of esters, and various transition metal-catalyzed reaction systems have been reported. Uemura et al. described the enantioselective benzoylation of racemic alcohols **127** catalyzed by a Pd(II) complex using CO, an organobismuth(V) compound **128** and AgOAc as the oxidant in the presence of a chiral ligand [40]. Bidentate ligands like diphosphine or dinitrogen or hybrid phosphine-nitrogen ligands were tested in the reaction. In the case of (–)-sparteine **1**, a poor yield (9%) and a poor ee (3%) were obtained. The most efficient ligands in terms of both reactivity and enantioselectivity were phosphinooxazoline ligands, although they gave only modest enantioselectivities (ee up to 19%).

The complex Pd-(–)-sparteine was also used as catalyst in an important reaction. Two groups have simultaneously and independently reported a closely related aerobic oxidative kinetic resolution of secondary alcohols. The oxidation of secondary alcohols is one of the most common and well-studied reactions in chemistry. Although excellent catalytic enantioselective methods exist for a variety of oxidation processes, such as epoxidation, dihydroxylation, and aziridination, there are relatively few catalytic enantioselective examples of alcohol oxidation. The two research teams were interested in the metal-catalyzed aerobic oxidation of alcohols to aldehydes and ketones and became involved in extending the scopes of these oxidations to asymmetric catalysis.

Stoltz et al. [41] and Sigman et al. [42] initiated their investigation into an oxidative kinetic resolution catalyst by screening various chiral amine ligands in addition to common ligands for Pd-mediated asymmetric reactions. The authors came to agree that sparteine may be the best ligand for this type of reaction (Scheme 38). The selectivity factor *s* measured by Stoltz was 8.8 in the unoptimized reaction with the 1-phenylethanol **128** and only 1 with the other ligands tested (such as binap, cinchonidine or bisoxazoline ligands). In order to improve both the reaction rate and the selectivity factor *s*, the reaction parameters of the (–)-sparteine-Pd(II) catalyst system were optimized. The nature of the palladium source was found to be critical, as it can induce a marked increase in the selectivity factor and in the conversion. The two teams found two different systems. Stoltz proposed a system using 5 mol % $Pd(nbd)Cl_2$, 20 mol % of (–)-sparteine **1** and molecular sieves (3 Å) under 1 atm of O_2 in toluene at 80 °C. Two sets of conditions were evaluated by Sigman, but the best conditions were given by the system using 5 mol % $Pd(MeCN)_2Cl_2$, 20 mol % of (–)-sparteine **1**, molecular sieves (3 Å) under

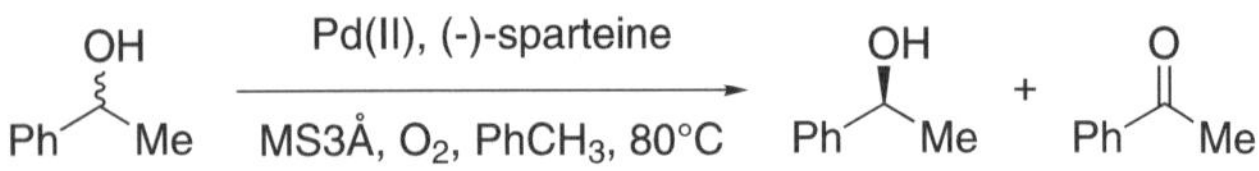

128 129

Scheme 38

1 atm of O_2 in 1,2-dichloroethane at 60 °C. Stoltz's system gave the best results with the same substrates, with a selectivity factor *s* that can reach up to 47 in the case of *α*-methyl-2-naphthalenemethanol (44% yield, 99% ee in alcohol). Sigman applied his system to the desymmetrization of a 1,3-*meso*-diol **130**. Under the quoted conditions, the enantioselective oxidation provided **131** in 82% ee and 69% yield (Scheme 39).

Both research teams then focused on optimizing their own asymmetric catalytic systems. We will describe those studies separately, although the progress of one of the groups aid the hypotheses of the other.

In an effort to develop more active catalyst systems for the oxidative kinetic resolution of non-activated alcohols, Stoltz et al. discovered a modified set of conditions that accomplishes similar resolutions in a fraction of the time [43].

During the course of their investigation, a four-fold excess of sparteine relative to the palladium source was used to form the complex in situ. Interestingly, the reactivity of the system could be restored upon introduction of an additional 3 equiv. of sparteine relative to the complex. Intrigued by this finding, the authors hypothesized that the excess sparteine was serving as a general base for the ultimate neutralization of HCl liberated from the system, and thus investigated the effects of other bases on the reaction. Indeed, Sigman's kinetic work (*vide supra*) has confirmed this hypothesis. Specifically, inclusion of 1.0 equivalent of Cs_2CO_3 resolved benzylic alcohol in just 13 h, with 68% conversion and 99% ee (previously reported conditions: 96 h, 67% yield, 98% ee). It was found that alcohols that are not readily oxidized are also beneficial to the reactivity. In particular, the addition of *t*-BuOH increased the selectivity of the reaction at slightly decreased temperatures (50 °C). A selectivity factor *s* of 20 was observed over 11.5 h, with resolution of the alcohol **132** to 57% conversion and 94% ee (Scheme 40). The combined effect of the additives Cs_2CO_3/*t*-BuOH encourage high conversion, and enantiopurity can now achieved in less than 16 h for all of the benzyl alcohols tested. The opti-

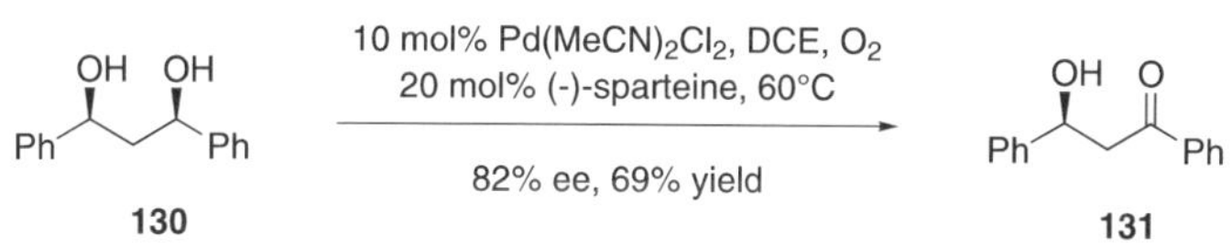

Scheme 39

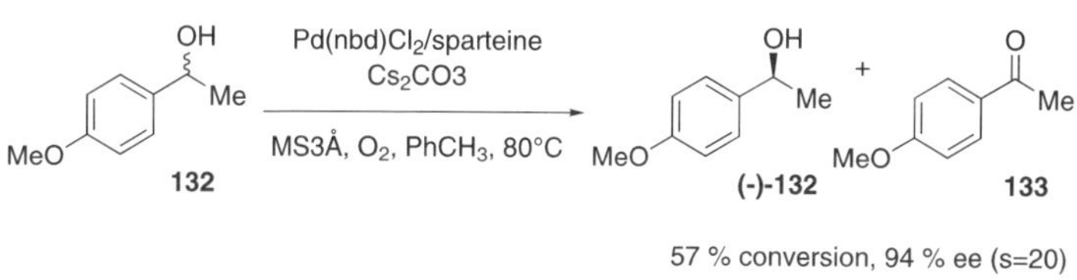

Scheme 40

mized conditions are 5 mol % Pd(nbd)Cl_2, 20 mol % (–)-sparteine, 0.5 equiv. of Cs_2CO_3, 1.5 equiv. of *t*-BuOH, 1 atm. of O_2, 0.25 M substrate concentration in $PhCH_3$.

The effect of *t*-BuOH in the model of the mechanism proposed by the authors for the resolution is rather subtle [44]. They hypothesized that the exogenous alcohol was affecting reaction rates by forming hydrogen bonds when aiding the solvation of halide anions (Fig. 5). With these hypotheses in mind, a solvent screen was used to explore the effect of hydrogen bond donation. Having identified chloroform as the most effective solvent for kinetic resolution, supporting rapid reaction rates even at room temperature, it was determined that the additional *t*-BuOH no longer imparted any benefit to the reaction. But the authors reconciled this observation by reasoning that the hydrogen bond-donating solvents may have supplanted the donating effect of the *t*-BuOH additive. The carbonate base, however, still enhanced the reaction rate. After the ratio of the different additives in chloroform was optimized, the best system was found to be: 5 mol % Pd(nbd)Cl_2, 12 mol % (–)-sparteine, molecular sieves (3 Å), ambient air (1 atm.), Cs_2CO_3 (0.4 equiv.), $CHCl_3$, 23 °C. Stoltz et al. have discovered the most mild and selective conditions

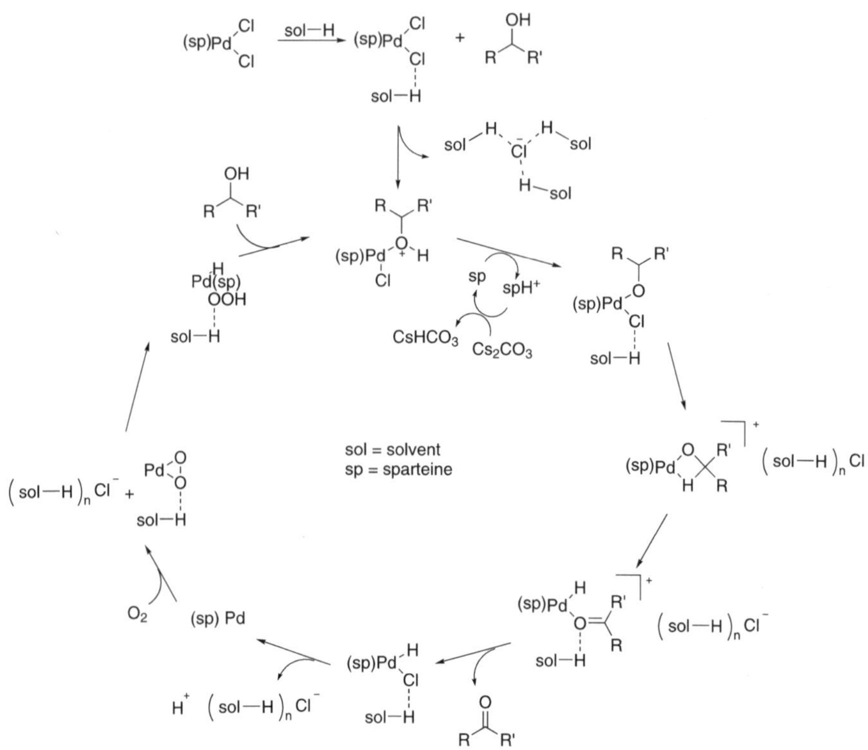

Fig. 5

for the oxidative kinetic resolution of secondary alcohols by catalytic palladium to date, and the first palladium-catalyzed oxidative process employing ambient air as the stoichiometric oxidant (instead of oxygen).

Sigman et al. have optimized their system too [45]. A study of different solvents showed that the best solvent was *t*-BuOH instead of 1,2-dichloroethane, which increased the conversion and the ee. To ensure that the best conditions were selected, several other reaction variables were evaluated. Reducing the catalyst loading to 2.5 mol % led to a slower conversion, and varying temperature from 50 °C to 70 °C had little effect on the selectivity factor *s*. Overall, the optimal conditions for this oxidative kinetic resolution were 5 mol % of Pd[(–)-sparteine]Cl_2, 20 mol % of (–)-sparteine, 0.25 M alcohol in *t*-BuOH, molecular sieves (3 Å) at 65 °C under a balloon pressure of O_2.

Sigman continued his investigation into the nature of ligand/base interactions aimed at improving the oxidative kinetic resolution catalyst system. A significant limitation of the system is the requirement of (–)-sparteine, which is only available as a single antipode and is a difficult template to optimize through systematic structural variations. Since interactions between the base and the ligand influence the Pd-catalyzed oxidative kinetic resolution, Sigman chose to examine two approaches that exploit this interplay. In the first approach, a Pd complex with an achiral ligand is used in combination with exogenous (–)-sparteine, and the second approach uses a Pd complex with a chiral ligand with exogenous (–)-sparteine to see if this scenario provides the ability to enhance the "matched" kinetic resolution (Scheme 41) [46]. The ligand should comply with two criteria: the ligand must form a Pd(II) complex that is competent at oxidizing alcohols, and the ligand must be not displaced by (–)-sparteine over the course of the reaction. *N*-heterocyclic carbenes were selected as a possible class of ligands due to the inertness of the derived metal complexes toward ligand substitution. Experiment showed that these complexes were competent for oxidative kinetic resolution when the chiral base (–)-sparteine was added, and this resulted in 64.7% conversion with a ee of 96% and *s* of 11.6 for 1-phenylethanol

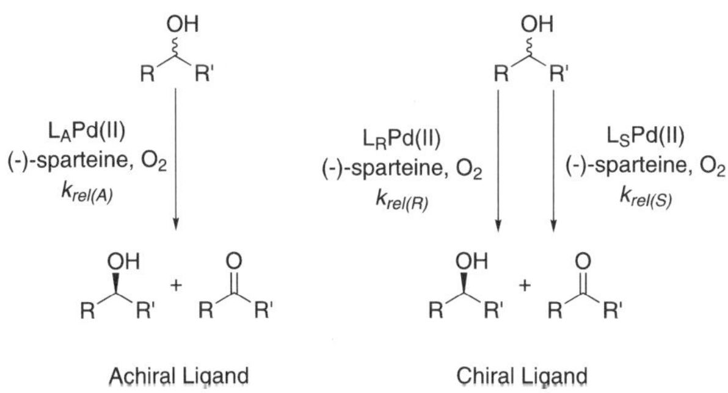

Scheme 41

and **134a** as the achiral ligand. Both enantiomers of **134b** were evaluated in the oxidative kinetic resolution of an alcohol using (–)-sparteine as the base (Scheme 42). A significantly higher k_{rel} value of 11.8 was observed for catalyst (*S*,*S*)-**136b** versus (*R*,*R*)-**136b** ($k_{rel} = 4.5$). This observation of a matched interaction showcases the approach in which the chiral ligand and chiral base can act in concert to enhance the kinetic resolution.

The details of the metal-catalyzed alcohol oxidation sequence and the precise role of additives are poorly understood. Mechanistic work from Sigman et al. [47, 48] suggested that the exogenous sparteine acts as a Brönsted base, deprotonating Pd-bond alcohol (Scheme 43). High concentrations of (–)-sparteine provided faster rates and higher k_{rel} values. Under these conditions, kinetic experiments are consistent with a rate-limiting β-hydride elimination. Asymmetric induction is proposed to arise from a combination of two factors: a thermodynamic difference in the diastereomeric alkoxides formed, and a kinetic difference in the reactions of these alkoxides. Exogenous (–)-sparteine only plays the role of a Brönsted base and does not directly influence the asymmetric induction of the process. Therefore, an achiral base that allows for equilibration of the alkoxides and rate-limiting β-hydride

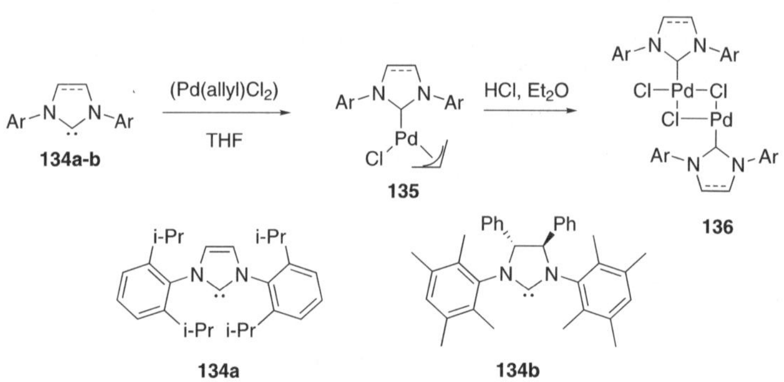

Scheme 42

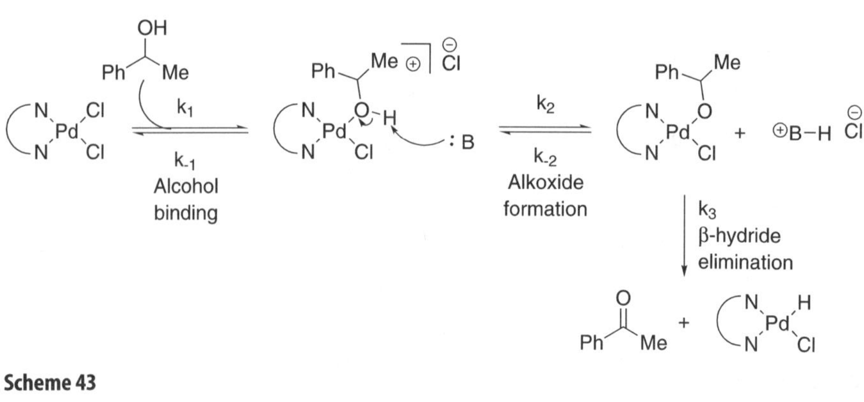

Scheme 43

elimination should give similar k_{rel} values to those above, avoiding the need for exogenous (–)-sparteine.

Following these hypotheses, the authors replaced the exogenous (–)-sparteine with an achiral base. They therefore chose to evaluate weakly coordinating bases in a basicity range similar to that of acetate and tertiary amines (pk_a of conjugate acids from 3 to 11 in H_2O). The carbonate bases provided effective oxidative resolution but higher loadings were necessary (50 mol %). The scope of the aerobic kinetic resolution of secondary alcohols was explored using 50 mol % of Na_2CO_3 as exogenous base with 5% of Pd[(–)-sparteine]Cl_2 in *t*-BuOH at 65 °C, and the new system performs as well as the previously reported system with exogenous (–)-sparteine. These conditions offer a more practical method for oxidative kinetic resolution, replacing a valuable chiral agent with an inexpensive achiral base.

In conclusion, both systems presented by both teams are rather competitive, even if the selectivity factors given by the system of Stoltz is better for a given set of substrates. In the case of 1-phenylethanol, Stoltz's system led to a 99% ee and *s* of 31, while in Sigman's procedure the resolution led to a 98.5% ee and *s* of 19. Both teams synthesized a range of model substrates that gave good ee's and good selectivity factors [44, 45, 49].

An application of this was described by Sudalai et al. [50] for a five-step synthesis of (*R*)-tomoxetin hydrochloride **141** and (*S*)-fluoxetine hy-

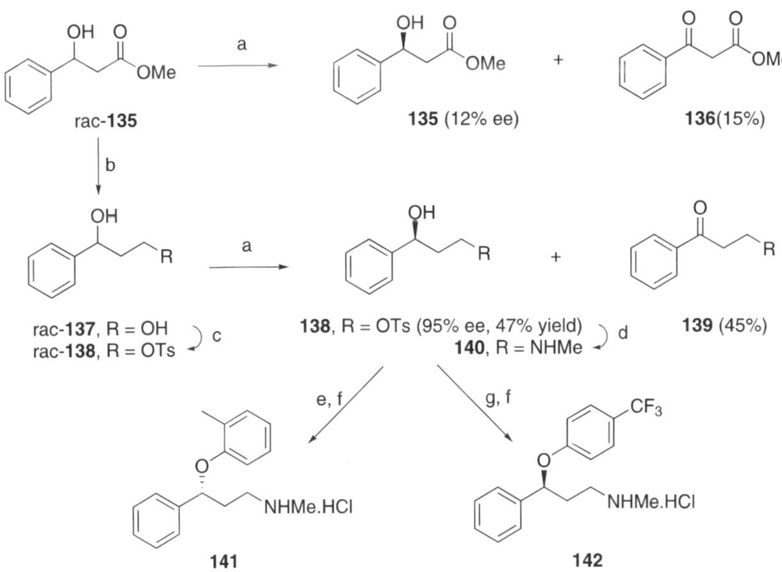

(a) 5 mol% $Pd(OAc)_2$, 20 mol% (-)-sparteine/O_2 (1 atm), MS3Å, $PhCH_3$, 80°C, 36 h; (b) $LiALH_4$, THF; (c) TsCl,Et_3N, DCM, -10 to 0°C; (d) 40% aq. $MeNH_2$, THF, 65°C; (e) o-cresol, PPH_3, DEAD, ether, -10 to 0°C; (f) HCl (gas), ether; (g) NaH, DMAC, 90°C, *p*-chlorobenzotrifluoride, 100-105°C.

Scheme 44

Scheme 45

drochloride **142** by using a kinetic resolution of benzylic alcohol, with the Sigman/Stoltz Pd(II)-(−)-sparteine/O_2 catalytic system as the key step for introducing the stereogenic center into the molecule (Scheme 44). Two strategies gave the enantioenriched benzylalcohols **135** and **138**, but the second path, leading to **138**, offered a very good ee (95%) for a yield of 47%.

Stoltz et al. [51] demonstrated the use of a simple system (Pd catalyst, ligand, $PhCH_3$, O_2) for constructing a range of heterocycles by catalytic oxidative cyclization, that could be incorporated into a Wacker cyclization. After having optimized their racemic procedure, the system provided a route to enantioselective catalysis using a readily available Pd-sparteine system similar to that used in the previously reported enantioselective alcohol dehydrogenation. The best conditions were as follows: 10 mol % of Pd[(−)-sparteine]TFA_2, $Ca(OH)_2$ (2 equiv.), molecular sieves (3 Å), O_2 (1 atm.), toluene at 80 °C (Scheme 45). Although the asymmetric results were somewhat limited in scope, the authors established for the first time that aerobic cyclizations of this kind can give high levels of enantioselectivity (57% yield, 90% ee).

5 Conclusion

While sparteine has efficiently demonstrated its utility as a chiral ligand for a wide range of asymmetric transformations, some drawbacks remain to be resolved. Although (+)-sparteine **1** is a natural product, it is not readily available. Furthermore, it is not possible to make some structural modifications to sparteine in order to optimize some of the reactions in which it is used as a chiral ligand. Some research groups have been working on those problems and recently some elegant solutions have been proposed. Kozlowski et al have designed a new family of chiral diaza-ligands **145** based on a 1,5-diaza-*cis*-decalin skeleton. The diamine **145** ($R_1 = R_2 = H$) is prepared in a three-step procedure from commercially available 3-aminopyridine and can be further functionalized by mono- or bisalkylation of the two nitrogen atoms [52]. Those ligands were successfully applied to copper-catalyzed enantioselective biaryl couplings of naphthols. The synthesis of cage-like diamines of from naturally occurring alkaloids is another pathway for the design of sparteine surrogates.

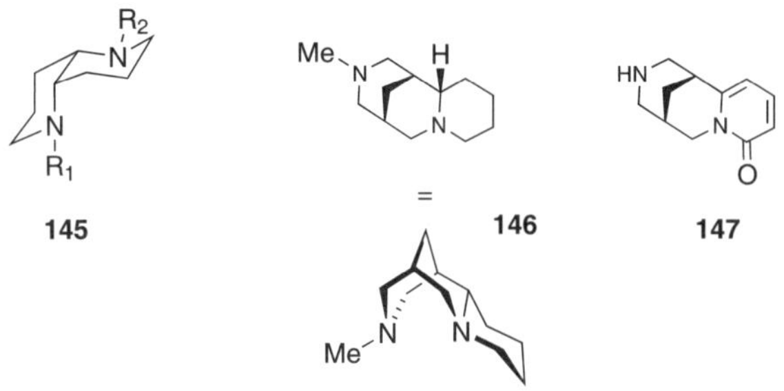

Fig. 6

O'Brien et al. recently reported the straightforward preparation of the chiral diamine **146** from (–)-cystine **147**. This alkaloid can be obtained by extraction from the seeds of *Laburnum anagyroides*, and this gives access to multigram quantities of the (+)-sparteine surrogate **146** [53]. Those results show that potential applications of sparteine-like ligands will open up new pathways in the field of asymmetric catalysis. However, studies on correlations between enantioselection and structural parameters for sparteine and related ligands are still required in order to enable the best choice of chiral ligand to be made for a given reaction. Very recently, the groups of M.C. Kozlowski [54] and P. O'Brien [55] have independently reported experimental and computational studies on the asymmetric lithiation of N-Boc-pyrrolidine using sparteine-like ligands in order to gain a better understanding of the structural requirements of the sparteine ligands used for enantioselectivity optimisation.

References

1. Hoppe D, Hense T (1997) Angew Chem Int Edit 36:2282
2. Klein S, Marek I, Poisson JF, Normant JF (1995) J Am Chem Soc 117:8853
3. (a) Norsikian S, Marek I, Poisson JF, Normant JF (1997) J Org Chem 62:4898, (b) Norsikian S, Marek I, Klein S, Poisson JF, Normant JF (1999) Chem Eur J 5:2055, (c) Majumdar S, De Meijere A, Marek I (2002) Synlett 423
4. Lautens M, Gajda C, Chiu P (1993) J Chem Soc Chem Commun 15:1193
5. Amurrio D, Khan K, Kundig EP (1996) J Org Chem 61:2258
6. Suzuka T, Ogasawara M, Hayashi T (2002) J Org Chem 67:3355
7. Asano Y, Iida A, Tomioka K (1997) Tetrahedron Lett 38:8973
8. Xu F, Tillyer RD, Tschaen DM, Grabowski EJJ, Reider PJ (1998) Tetrahedron Asymmetry 9:1651
9. Denmark SE, Nakajima N, Nicaise OJC (1994) J Am Chem Soc 116:8797
10. Tomioka K, Satoh M, Taniyama D, Kanai M, Iida A (1998) Heterocycles 47:77
11. Jones CAG, North M (1997) Tetrahedron Asymmetry 8:3789
12. Taniyama D, Hasegawa M, Tomioka K (1999) Tetrahedron Asymmetry 10:221
13. Alexakis A, Amiot F (2002) Tetrahedron Asymmetry 13:2117
14. Shintani R, Fu GC (2002) Angew Chem Int Edit 41:1057

15. (a) Müller P, Nury P (2000) Org Lett 2:2845, (b) Müller P, Nury P, Bernardinelli G (2001) Eur J Org Chem 21:4137
16. Tsukazaki M, Tinkl M, Roglans A, Chapell BJ, Taylor NJ, Snieckus V (1996) J Am Chem Soc 118:685
17. Fukuda T, Imazato K, Iwao M (2003) Tetrahedron Lett 44:7503
18. Komine N, Wang LF, Tomooka K, Nakai T (1999) Tetrahedron Lett 40:6809
19. Nakamura S, Nakagawa R, Watanabe Y, Toru T (2000) J Am Chem Soc 122:11340
20. Manabe S (1997) Chem Commun 737
21. Tomooka K, Komine N, Nakai T (1998) Tetrahedron Lett 39:5513
22. Hodgson DM, Lee GP (1997) Tetrahedron Asymmetry 8:2303
23. Hodgson DM, Cameron ID (2001) Org Lett 3:441
24. Hodgson DM, Galano JM, Christlieb M (2003) Tetrahedron 59:9719
25. Hodgson DM, Stent MAH, Stefane B, Wilson FX (2003) Org Biomol Chem 1:1139
26. Rajca A, Wang H, Bolshov P, Rajca S (2001) Tetrahedron 57:3725
27. Zhang Y, Yeung SM, Wu H, Heller DP, Wu C, Wulff WD (2003) Org Lett 5:1813
28. Smrčina M, Lorenc M, Hanus V, Sedmera P, Kočovský P (1992) J Org Chem 57:1917
29. Smrčina M, Poláková J, Vyskocyl S, Kočovský P (1993) J Org Chem 58:4534
30. Noji M, Nakajima M, Koga K (1994) Tetrahedron Lett 35:7983
31. Nakajima M, Miyoshi I, Kanayama K, Hashimoto SI, Noji M, Koga K (1999) J Org Chem 64:2264
32. Miura M, Enna M, Okuro K, Nomura M (1995) J Org Chem 60:4999
33. Müller P, Boléa C (2001) Helv Chim Acta 84:1093
34. Müller P, Boléa C (2002) Helv Chim Acta 85:483
35. Müller P, Nury P (2001) Helv Chim Acta 84:662
36. Trost BM, Dietsch TJ (1973) J Am Chem Soc 95:8200
37. Muller D, Umbricht G, Weber B, Pfaltz A (1991) Helv Chim Acta 74:232
38. Togni A (1991) Tetrahedron Asymmetry 2:683
39. Kang J, Cho WO, Cho H (1994) Tetrahedron Asymmetry 5:1347
40. Iwata T, Miyake Y, Nishibayashi Y, Uemura S (2002) J Chem Soc Perkin T 1 13:1548
41. Ferreira EM, Stoltz BM (2001) J Am Chem Soc 123:7725
42. Jensen DR, Pugsley JS, Sigman MS (2001) J Am Chem Soc 123:7475
43. Bagdanoff JT, Ferreira EM, Stoltz BM (2003) Org Lett 5:835
44. Bagdanoff JT, Stoltz BM (2004) Angew Chem Int Edit 43:353
45. Mandal SK, Jensen DR, Pugsley JS, Sigman MS (2003) J Org Chem 68:4600
46. Jensen DR, Sigman MS (2003) Org Lett 5:63
47. Mueller JA, Jensen DR, Sigman MS (2002) J Am Chem Soc 124:8202
48. Mueller JA, Sigman MS (2003) J Am Chem Soc 125:7005
49. Ferreira EM, Stoltz BM (2002) US Patent 02 072 514
50. Sayyed Ali I, Sudalai A (2002) Tetrahedron Lett 43:5435
51. Trend RM, Ramtohul YK, Ferreira EM, Stoltz BM (2003) Angew Chem Int Edit 43:2892
52. (a) Li X, Schenkel LB, Kozlowski MC (2000) Org Lett 2:875, (b) Li X, Yang J, Kozlowski MC (2001) Org Lett 3:1137, (c) Li X, Hewgley JB, Mulrooney CA, Yang J, Kozlowski MC (2003) J Org Chem 68:5500, (d) Mulrooney CA, Li X, DiVirgilio ES, Kozlowski MC (2003) J Am Chem Soc 125:6856
53. (a) Dearden MJ, Firkin CR, Hermet JPR, O'Brien P (2002) J Am Chem Soc 124:11870, (b) Hermet JPR, Porter DW, Dearden MJ, Harrison JR, Koplin T, O'Brien P, Parmene J, Tyurin V, Whitwood AC, Gilday J, Smith NM (2003) Org Biomol Chem 1:3977
54. Phuan PW, Ianni JC, Kozlowski MC (2004) J Am Chem Soc 126:15473
55. O'Brien P, Wiberg KB, Bailey WF, Hermet JP, McGrath MJ (2004) J Am Chem Soc 126:15480

Top Organomet Chem (2005) 15: 93–148
DOI 10.1007/b136349
© Springer-Verlag Berlin Heidelberg 2005
Published online: 14 September 2005

Use of *N*,*N*-Coordinating Ligands in Catalytic Asymmetric C–C Bond Formations: Example of Cyclopropanation, Diels–Alder Reaction, Nucleophilic Allylic Substitution

Emmanuelle Schulz

Laboratoire de Catalyse Moléculaire, Institut de Chimie Moléculaire
et des Matériaux d'Orsay, Université Paris-Sud, Bât. 420, 91405 Orsay Cedex, France
emmaschulz@icmo.u-psud.fr

Abstract Chiral *N*,*N*-coordinating ligands have been particularly synthesized for their use as metal chelates catalyzing asymmetric C – C bond formations. Bis(oxazoline) structures have been the most studied, probably due to their straightforward synthesis from the chiral pool, their easy handling, and their high stability. Efficient methods have been reported for the intermolecular cyclopropanation of olefins (mainly styrene derivatives) with alkyldiazoacetates involving various bis(oxazolines)-Cu(I) catalysts. Bis(oxazolinyl)-pyridine ruthenium complexes proved also highly active and selective catalysts for this reaction.

Associated to copper(II) pre-catalysts, bis(oxazolines) also allowed the asymmetric Diels–Alder and hetero Diels–Alder transformations to be achieved in nearly quantitative yield and high diastereo- and enantioselectivities. Optically active sulfoximines, with their nitrogen-coordinating site located at close proximity to the stereogenic sulfur atom, have also proven their efficiency as copper ligands for these asymmetric cycloadditions. Other precursors for this Lewis acid-catalyzed transformation have been described (e.g., zinc salts, ruthenium derivatives, or rare earth complexes) which, when associated to bis(oxazolines), pyridine-oxazolines or pyridine-bis(oxazolines), led to efficient catalysts.

Chiral pyridine-based ligands were, among various *N*,*N*-coordinating ligands, more efficient associated to palladium for asymmetric nucleophilic allylic substitution. Asymmetric molybdenum-catalyzed alkylations, especially of non-symmetric allylic derivatives as substrates, have been very efficiently performed with bis(pyridylamide) ligands.

The heterogeneous versions of most of these efficient nitrogen-containing complexes have been described via immobilization on organic or inorganic supports by covalent or non-covalent interactions. Such catalysts have been used, leading to activities and selectivities comparable to their homogeneous counterparts, and some recycling experiments have been successfully performed. Nevertheless, these methodologies need to be further optimized in order to enhance the activity and stability of the solid catalysts.

Keywords *N*,*N*-Containing ligands · Asymmetric catalysis · Cyclopropanation · Diels–Alder reaction · Nucleophilic allylic substitution

1 Introduction

Asymmetric metal-catalyzed transformations [1] are important tools for the targeted preparation of optically active synthons in an economical way. Research into the development of such methodologies are ongoing in both the academic and industrial worlds. Hydrogenation reactions of unsaturated carbon–carbon or carbon–heteroatom bonds have been intensively studied and have found satisfactory solutions for preparing reduced substrates with very high yields and selectivities, using low catalyst loadings. These reactions are thus performed at an industrial scale with mainly asymmetric diphosphine ligands (BINAP, BIPHEP, DUPHOS etc.), tartaric acid- and cinchonine-derivatives as chiral auxiliaries [2]. The formation of C – C bonds has been less developed at a preparative scale probably because of the large substrate/catalyst ratios (inducing high costs for both precious metal and optically pure ligand) required to achieve valuable yields in the expected products. On the other hand, *N*-containing chiral ligands, as easily accessible and stable compounds, are being more and more developed for their use as chelates in asymmetric metal catalysis [3]. They have proved particularly efficient for specific C – C bond formations, where phosphorus- or oxygen-containing ligands remained less active. Some representative examples of such C – C bond formations (i.e., cyclopropanation, Diels–Alder reaction, and nucleophilic allylic substitution) will be described here using chiral *N,N*-chelates (bis-amines, imines, oxazolines, pyridines and related compounds). Special attention will be also paid to reports dealing with the heterogenization of efficient *N,N*-liganded complexes and their recycling in these catalytic transformations.

2
Asymmetric Cyclopropanation

Asymmetric catalytic cycloaddition of electrophilic metal carbenes to prochiral olefins (i.e., asymmetric cyclopropanation [4]) is one of the catalytic C – C bond-formation reactions that has been the most often studied, and successfully performed, with *N,N*-containing chiral ligands. For this transformation, these ligands proved to lead to better activities than all other types of ligands and afforded excellent control of both diastereo- and enantioselectivity. Indeed, few examples of efficient and selective systems have been described using phosphine ligands. This facile methodology for highly enantioselective cyclopropane synthesis consists in the metal-catalyzed decomposition of substituted diazo compounds in the presence of various alkenes, proceeding via the initial formation of a metal carbene complex. The most successful catalysts are complexes of ruthenium, rhodium, and especially copper derivatives. Very efficient Co(III)-salen complexes were prepared by Katsuki [5]. Salen derivatives were also used as chelates for ruthenium complexes, but as chiral *N,O*-ligands they will not be discussed here. The reaction of styrene with alkyldiazoacetate (Scheme 1) has been mostly studied with the new

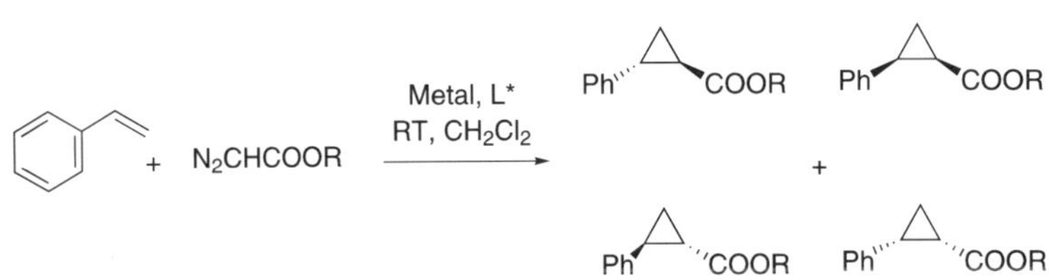

Scheme 1

ligands that have been developed, allowing an easy classification of their efficiency. This intermolecular transformation is quite difficult to achieve with high levels of both diastereo- and enantioselectivity. The most pronounced problem is to control the formation of the thermodynamically favored *trans*-cyclopropanes with high yield.

Several examples of this methodology, in which the reactions are classified according to the nature of the ligand, are given below.

2.1 Bis(oxazoline) Ligands

2.1.1 Copper Complexes

Very recently, ligands based on the bis(oxazoline) backbone have been very successfully introduced for this transformation, especially associated to ruthenium or copper salts. Asymmetric cyclopropanation of olefins is by far the most successful application of the Cu(I)-bis(oxazoline) catalytic system. More stable $Cu(OTf)_2$ has also often been used as the copper source, but the resulting Cu(II)-ligand complex has to be reduced (often by phenylhydrazine, DIBAL, or a slight excess of the diazo reagent itself) to provide an active species. Few articles deal with results obtained by using monodentate oxazoline ligands [6, 7], which are always less selective than their bidentate counterparts. Zhou et al. [8, 9] have prepared chiral quinolinyl- or pyridinyl-oxazoline ligands of type **1** and **2** (Scheme 2). Tested in asymmetric cyclo-

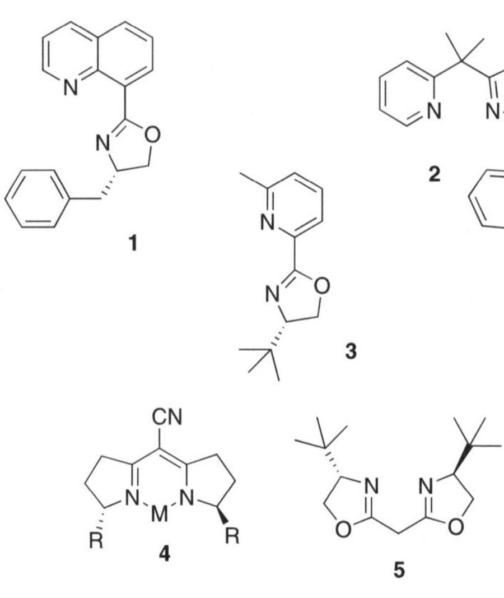

Scheme 2

propanation of styrene with dicyclohexylmethyl diazoacetates in the presence of copper(I)triflate, however, they led to moderate stereoselectivities (72% de using ligand **1** or 64% de with **2** in favor of the *trans* isomer) and enantioselectivities (45% ee and 14% ee, respectively, for the *trans* product). The authors argued that the presence of the six-membered chelate ring upon coordination was important for high chemical yield and that the conjugation between the quinolinyl and the oxazoline ring was responsible for the higher enantioselectivity in the case of ligand **1**. Chelucci et al. similarly developed C_1-symmetric oxazolinylpyridines, leading to the formation of a five-membered chelate ring upon coordination with copper. They prepared up to 28 different ligands bearing alkyl substituents on various positions of the pyridine ring. The best result was obtained with structure **3** (Scheme 2) in the cyclopropanation of styrene [10] (60% ee for the *trans* isomer isolated as major isomer with 24% diastereoselectivity).

Chiral C_2-symmetric semicorrins (structure **4**), developed by Pfaltz [11], were proven to be highly efficient ligands for the copper-catalyzed enantioselective cyclopropanation of olefins. Variations of the substituents at the stereogenic centers led to optimized structures and very high enantioselectivities [12].

Evans [13] has reported the synthesis of the catalyst generated in situ from Cu(I)triflate and bisoxazoline **5** (Scheme 2) for the cyclopropanation of styrene with ethyl diazoacetate with high asymmetric induction (*trans*/*cis* selectivity is 77/23 and 98 and 93% ee, respectively). Much research has been performed on improving the structure of the C_2-symmetric chiral bis(oxazoline) **5**, namely by varying the size of the oxazoline substituent and the length and/or nature of the bridge between the two heterocycles. Mosset and Saalfrank [14] prepared C_2-symmetric bis(oxazolines) from tartaric-derived diamines leading to vicinal bis(oxazolines). These ligands revealed, however, poorly efficient chelates for copper-catalyzed cyclopropanation, except when structure **6** (Scheme 3), bearing the bulky 1-adamantyl group, was used. Catalytic cyclopropanation of styrene with ethyl diazoacetate was thus performed with up to 79% ee in the *cis* isomer obtained; however, with a low diastereoselectivity (34% in favor of the *trans* derivative).

Andersson et al. [15, 16] (and simultaneously Knight et al. [17]) studied the effect of the length and structure of the bridge between the two oxazoline moieties. They prepared and tested a new class of bis(oxazolines) in which

6

Scheme 3

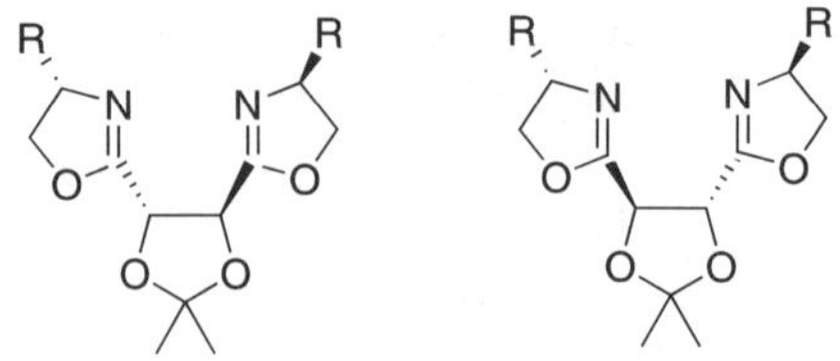

7 8

Scheme 4

both heterocycles were separated by a tartrate backbone and thus formed a seven-member chelate with the metal (Scheme 4).

Higher enantioselectivities, compared to ligand **7**, were obtained with **8** ($R = CHMe_2$; 50% ee and up to 84% ee for the *trans* product, Scheme 4). The authors argued that in these rigid tartrate- derived ligands, both oxazoline rings were slightly twisted, pushing the oxazoline substituent *R* either near the copper metal (as in **8**) or away from it (as in **7**). Accordingly, if the *R* group is more sterically demanding ($R = CMe_3$), there is not sufficient space for copper coordination to the bis(oxazoline) with ligand **8** and no more enantioselection is observed. In this case, however, the use of ligand **7** provided the *trans* adduct with 84% ee.

Hayashi et al. [18] have synthesized two diastereoisomers of 2,2′-bis[4-(alkyl)oxazol-2-yl]-1,1′-binaphthyl, bis(oxazoline) derivatives possessing both binaphthyl axial chirality and carbon centered chirality (structures **9** and **10**, Scheme 5).

These two compounds with *S* configuration on their oxazoline rings were tested as copper(I) catalysts for the cyclopropanation of styrene, the ligand **9** with *S* axial chirality being much more enantioselective than **10** with the *R* configuration. Thus, the catalytic system CuOTf–(*S*,*S*)-bis(oxazolyl)-binaphthyl (**9**, $R = {}^tBu$) led to excellent enantioselectivities, particularly for the cyclopropanation of styrene with *l*-menthyldiazoacetate 95% ee for the *trans*-cyclopropane and 97% ee for the *cis*, with *trans*/*cis* = 68/32.

Corey et al. [19] simultaneously reported similar studies using a 2,2′-bis(oxazolyl)-6,6′-dimethyl-1,1′-biphenyl as copper(I)triflate chelate. This ligand afforded a stable monomeric chiral Cu(I) complex providing a highly

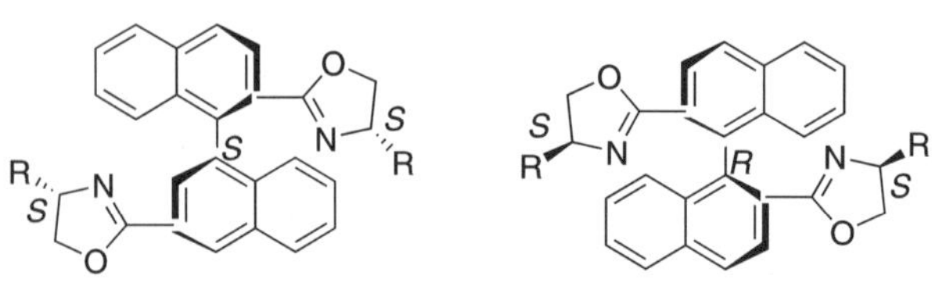

9 10

Scheme 5

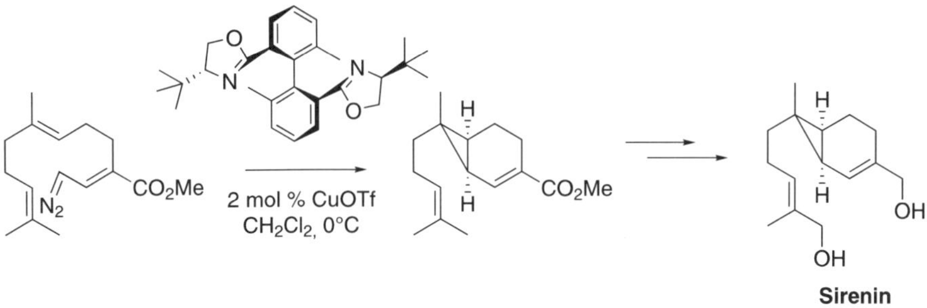

Scheme 6

enantioselective (90% ee) intramolecular cyclopropanation as a key step for the enantioselective synthesis of sirenin (Scheme 6).

Ahn et al. [20] reported the synthesis of homochiral bis(oxazolinyl)-biferrocene ligands (structure **11** in Scheme 7), which also have both planar and central chirality. With these complexes, 2-(phenyl)cyclopropane carboxylates were obtained in up to 99% ee and a *trans/cis* ratio of 88/12.

Several homochiral bis(oxazolines), with stereogenic centers on the oxazoline rings and on the side chains (containing hydroxy functionalities, see **12** in Scheme 8), were prepared by Aït-Haddou and Balavoine [21]. The most effective ligands were found to be **12a** and **12b**, which gave the product in good yields and up to 85% ee (for the *cis* adduct) in the cyclopropanation of styrene. These ligands did not lead to higher ee levels than those described in the literature, although they possess an additional chelating functionality.

Reiser et al. [22] modified bis(oxazoline)-copper catalysts for the preparation of "tailor-made" structures creating secondary binding sites with specific substrates. They thus modified the usual bis(oxazoline) ligands by introducing protected chiral α-amino acid side chains (see structure **13** in Scheme 8 as an example). This ligand led to disappointing results for the cyclopropanation of either styrene, 2,3-dihydrofuran, or furan. The authors indicated that the flexibility of the side chains of the ligands was not favorable for a specific substrate approach. They then studied the cyclopropanation of methyl furan-2-carboxylate with these catalysts and observed higher

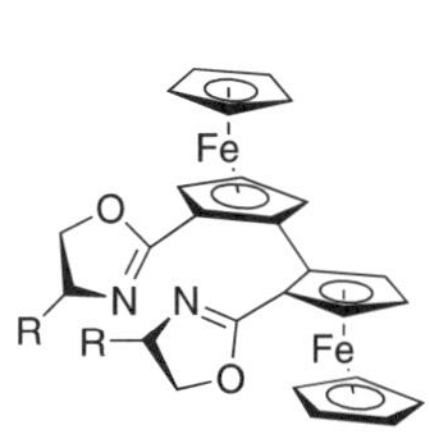

11

Scheme 7

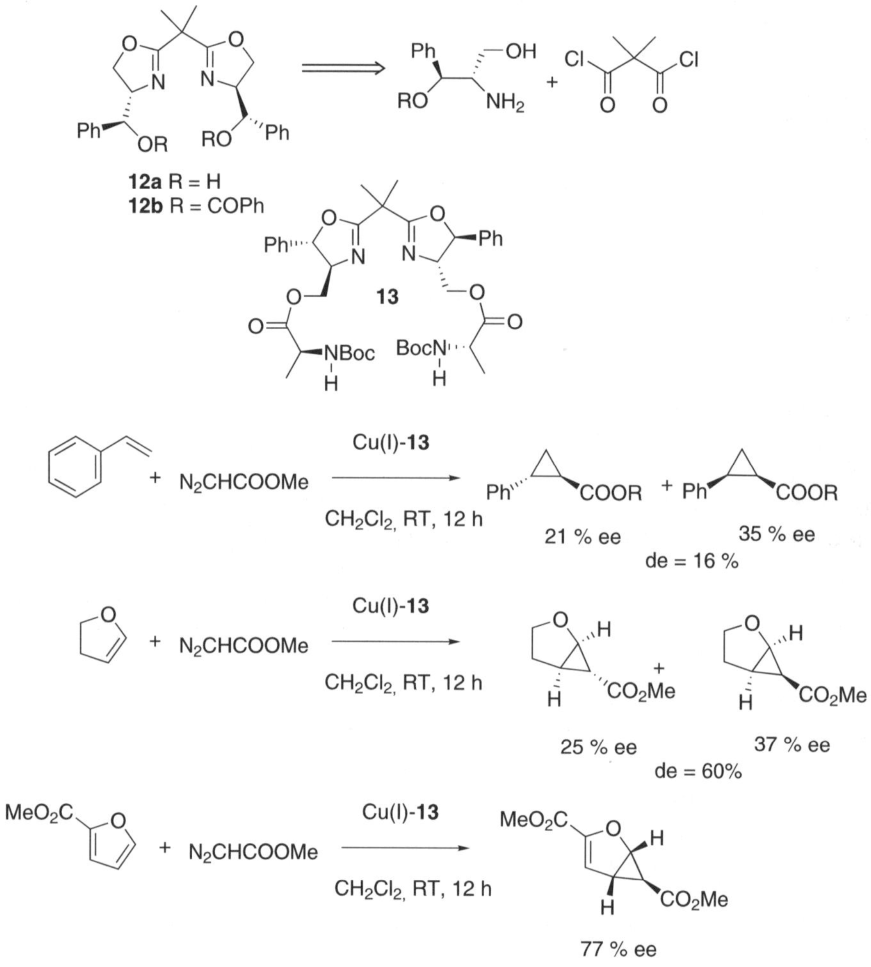

Scheme 8

enantio- and complete diastereoselectivities, with the *exo*-adduct formed exclusively. They attributed these results to secondary interactions from the ligand (i.e., hydrogen bonding), interacting with the carbomethoxy group in the substrate.

Glos and Reiser [23] introduced aza-bis(oxazolines) as new chiral ligands for copper and palladium catalysts. Because of the structural flexibility of these compounds they also prepared an immobilized ligand by covalent grafting to methoxypoly(ethyleneglycol) (structures **14** and **15** in Scheme 9).

The aza-bis(oxazoline) **14**, bearing sterically hindering groups, led to very good results in terms of activity and selectivity, comparable to those obtained from corresponding aza-semicorins or bis(oxazolines). For the enantioselective cyclopropanation of styrene, the *trans* isomer was obtained in 92% ee

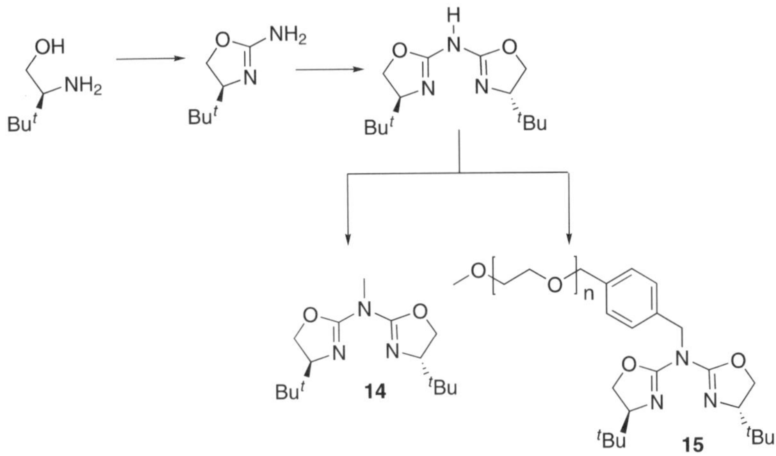

Scheme 9

and 46% de. Furthermore, the polymer-bound oxazoline (structure **15** in Scheme 9) was successfully used and recycled ten times with no loss of enantiomeric selectivity or activity. Methodologies allowing the recycling of the catalytic species will be specifically detailed at the end of this section.

Charrette and Lebel [24] developed a catalytic enantioselective cyclopropanation of *trans*-cinnamate esters with diazomethane. Their procedure involves an argon-flow-mediated diazomethane addition, leading to high yields (up to 80%) in products with up to 80% ee, by using the bis(oxazoline) arising from phenylglycinol (Scheme 10).

Some other groups have studied the opportunity to enhance the diastereoselectivity of the transformation using the usual copper-bis(oxazoline) catalysts but modifying the carbene source. France et al. [25] observed that the use of (trimethylsilyl)diazomethane associated with a bis(oxazoline) and $[Cu(CH_3CN)_4]PF_6$ as catalyst precursor allowed the formation of the *trans* isomer with high yield and selectivity, probably due to the steric bulk of the trimethylsilyl group.

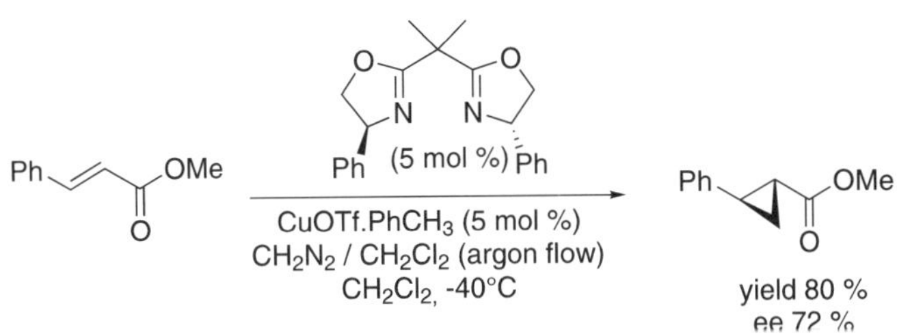

Scheme 10

In conclusion, the system copper-bis(oxazoline) has been demonstrated as highly efficient in performing intermolecular cyclopropanation of olefins, with high levels of enantioselectivity and favored *trans*-selectivity. Furthermore, many groups have modified the bis(oxazoline) ligands by introducing substituents, thus preparing tailor-made ligands for the cyclopropanation of various substrates. This catalytic transformation is by far the most important application for bis(oxazoline) ligands.

2.1.2 Ruthenium Complexes

Other very efficient catalytic systems for cyclopropanation with diazoacetates described by Nishiyama et al. [26] are bis(oxazolinyl)pyridine (PyBOx)-ruthenium complexes (Scheme 11, structure **16**). The catalyst prepared in situ from **16** and [$RuCl_2$(*p*-cymene)$_2$] was found to be highly active and selective in the model test reaction between styrene and ethyldiazoacetate, achieving *trans*/*cis* selectivities up to 98/2 and 97% ee. Many structural variations of the ligand have been investigated. The same group [27] reported the synthesis of enantiopure bis(oxazolinyl)pyridine ligands (with only one substituted oxazoline, see **17** in Scheme 11) and their activity in the model test reaction. Enantioselectivities up to 94% were observed, confirming the author's assumption that C_2-symmetric ligands were not necessary in this system to achieve high selectivities.

These ligands were used in protic and biphasic media by modifying their structure using hydroxyalkyl groups [28, 29]. The solubility of the corresponding Ru(II) complexes was significantly increased in protic solvents. Hence, by performing the reaction in mixtures of toluene and water or al-

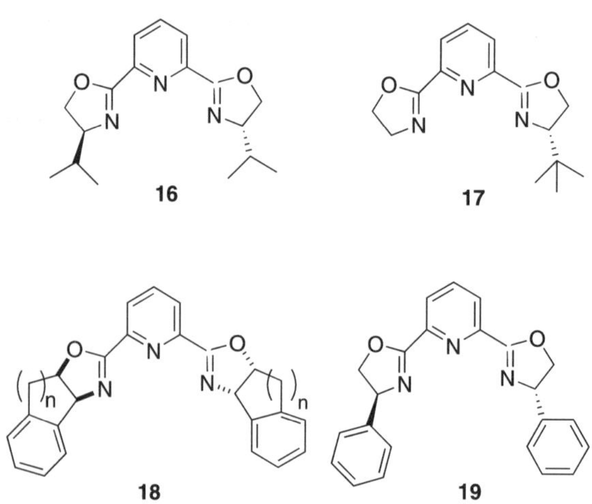

16 17 18 19

Scheme 11

cohols, the authors observed enhanced activities and selectivities for the cyclopropanation (up to 97% ee, and 97/3 *trans*/*cis* stereoselectivity, by using the hydroxymethyl derivative of PyBOx and ruthenium). The authors proposed that appropriate solvation of water or alcohols around the hydroxy group of the ligand may cause a more favorable stereochemical environment around the active site. Interestingly, the catalyst dissolved in the water phase could be reused, at least four times, but with a decrease in the yield and selectivity.

Davies [30] studied the PyBOx-induced conformational effects by testing several ligands sterically hindered on the oxazoline moieties (Scheme 11, structures **18** and **19**). However, these new ligands gave poorer results in terms of yields and enantioselectivities than ligand **16** for the Ru-catalyzed cyclopropanation reaction, indicating unfavorable steric interactions between styrene and the carbene complex.

Among the very few examples reported on the use of catalytic asymmetric C – C bond formation at a practical scale, the work of Deshpande et al. [31] has to be mentioned since this group described an application of asymmetric cyclopropanation. Using a styrene derivative as a limiting reagent and catalytic Ru(^{i}Pr-PyBOx), they achieved asymmetric cyclopropanation with high diastereo- (92%) and enantioselectivity (84%). The reaction gave the cyclopropyl acid derivative depicted in Scheme 12, towards the synthesis of several melatonin agonits. This synthesis was successfully performed in a pilot plant at a 50 kg styrene scale.

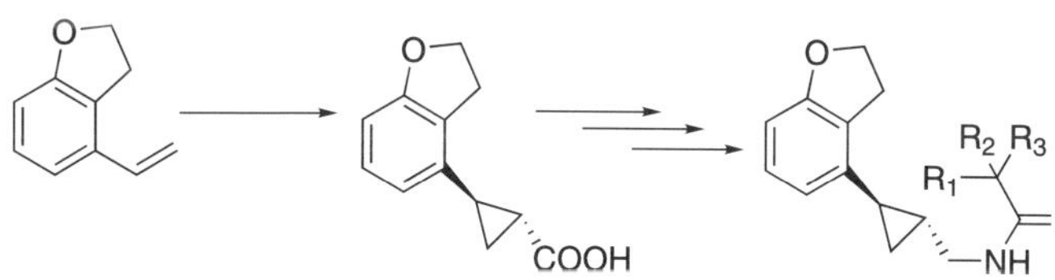

Scheme 12

2.1.3 Rhodium Complexes

Rhodium complexes with chelating bis(oxazoline) ligands have been described to a lesser extent for the cyclopropanation of olefins. For example, Bergman, Tilley et al. [32] have prepared a family of bis(oxazoline) complexes of coordinatively unsaturated monomeric rhodium(II) (see **20** in Scheme 13). Interestingly, the use of complex **20** in the cyclopropanation reaction of styrene afforded mainly the *cis* cyclopropane (*cis*/*trans* = 63/37), with 74% ee and not the thermodynamically favored *trans* isomer. No mechanistic suggestions are proposed by the authors to explain this unusual selectivity.

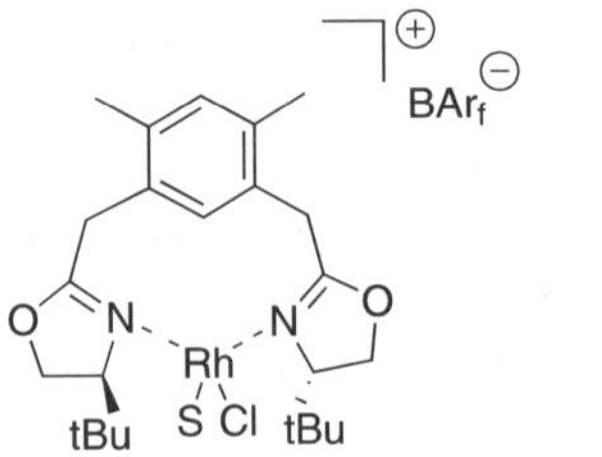

20

Scheme 13

2.2 Bipyridine Copper and Rhodium Complexes

Pyridine-based *N*-containing ligands have been tested in order to extend the scope of the copper-catalyzed cyclopropanation reaction of olefins. Chelucci et al. [33] have carefully examined and reviewed [34] the efficiency of a number of chiral pyridine derivatives as bidentate ligands (mainly 2,2′-bipyridines, 2,2′ : 6′,2′-terpyridines, phenanthrolines and aminopyridine) in the copper-catalyzed cyclopropanation of styrene by ethyl diazoacetate. The corresponding copper complexes proved to be only moderately active and enantioselective (ee up to 32% for a C_2-symmetric bipyridine). The same authors prepared other chiral ligands with nitrogen donors such as 2,2′-bipyridines **21**, 5,6-dihydro-1,10-phenanthrolines **22**, and 1,10-phenanthrolines **23** (see Scheme 14) [35].

Only poor enantioselectivities were obtained when no alkyl groups were attached next to the heterocyclic ring of these ligands. Increasing the steric bulk in this position by alkyl groups (Me and especially Bn substituents) led to up to 68% ee for ligand **21**. The authors could not explain, from these examples, the dependence of the stereoselectivity on the rigidity of the ligand backbone.

Wilson and Lyle [36] recently reported the synthesis and use of a new C_2-symmetric 2,2′-bipyridyl ligand that answered these steric demands, since, associated to copper(I)salts, high diastereoselectivities and enantioselectivities (*trans*/*cis* = 80/20 and 82% ee) could be achieved in the benchmark

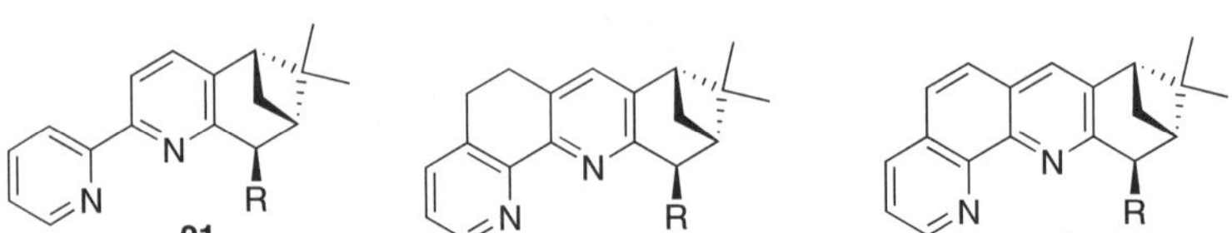

Scheme 14

reaction. Interestingly, the synthesis of this ligand (**24** in Scheme 15) was based on a catalytic asymmetric dihydroxylation reaction of a pyrindine, as the key step.

Malkov et al. [37] prepared a series of C_2-symmetric bipyridine-type ligands, the chiral moieties arising from the isoprenoid chiral pool (β-pinene, 3-carene, 2-carene, or α-pinene, for example). Some representative examples are drawn in Scheme 16 (see **25, 26, 27**) and were used as copper ligands of a copper(I) species obtained by an in-situ reduction of $Cu(OTf)_2$ with phenylhydrazine. The use of the resulting catalysts in enantioselective cyclopropanation proceeded with up to 76% ee (for ligand **27**) and high diastereoselectivity (up to 99 : 1).

Chan et al. [38] prepared optically active atropoisomeric 2,2′-bipyridine by nickel(0)-catalyzed homo-coupling of 2-bromopyridylphenol derivatives (structure **28** in Scheme 16). Tested in the model test reaction, the copper catalyst led to *trans*-cyclopropanes as major products with up to 86% ee.

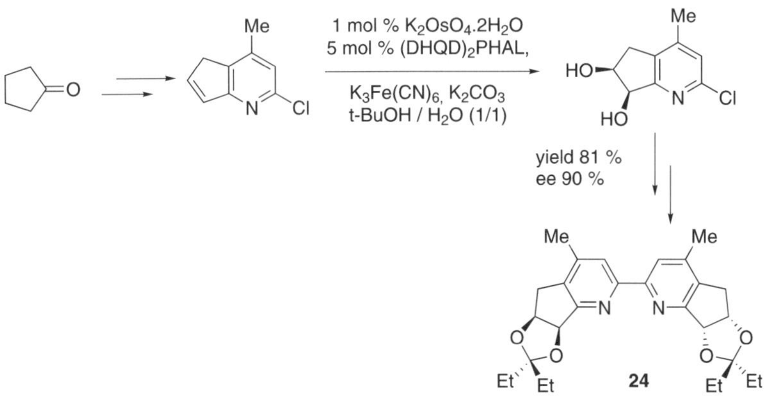

Scheme 15

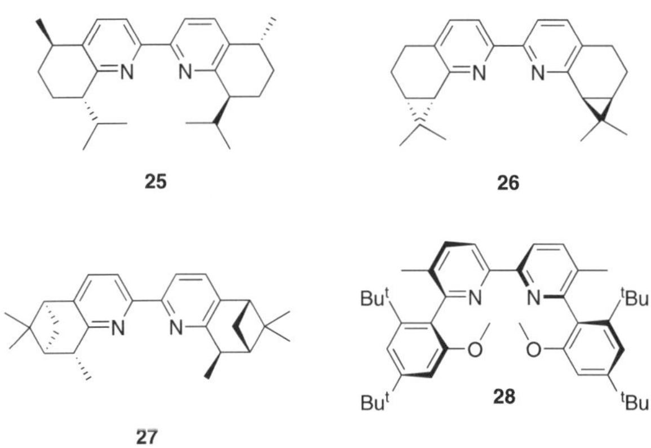

Scheme 16

Kwong and Lee [39] prepared various chiral 2,2′ : 6′,2″-terpyridines and tested them as copper ligands for the cyclopropanation of alkenes. High enantioselectivities were obtained, the presence of bulky alkyl groups at the 8-position of the tetrahydroquinoline ring being crucial (structure **29** in Scheme 17). Thus when $R = {}^{n}$Bu, up to 90% ee for the *trans* and 94% for the *cis* isomer were obtained by performing the reaction at 0 °C (*trans*/*cis* = 69/31).

These results prompted the authors to prepare other terpyridine ligands (structures **30** to **32** in Scheme 17) from the chiral pool by Kröhnke condensation of α,β-unsaturated ketones [40]. Active catalysts for the cyclopropanation reaction were prepared from these ligands and $Cu(OTf)_2$ or $RhCl_3$ in the presence of AgOTf. The copper catalysts yielded enantioselectivities between 10 and 82% (for ligand **32**) whereas the rhodium catalysts remained less selective. Interestingly, the products obtained from the copper catalysts or from the corresponding rhodium catalysts have opposite configurations. Moreover, contrary to the copper-catalysis that favored the *trans* isomers, *cis*-cyclopropanes were the major products in some of the rhodium-catalyzed reactions. The authors explained these results by proposing different models for the corresponding complexes. Concerning the copper complex, they propose that the copper carbene be located in the same plane as the terpyridine ligand, the ester group and the hydrogen group pointing above and below this

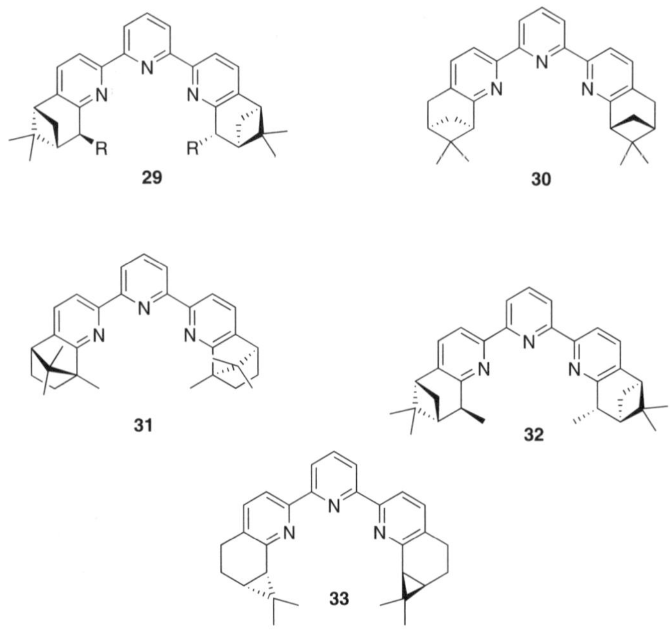

Scheme 17

plane. The absolute configuration of the *trans*-cyclopropane is explained by an approach of styrene minimizing the steric hindrance. The proposed model for rhodium involves an octahedral rhodium, the metal-carbene occupying the axial position, and bisecting the N – Rh – N bond angle. The styrene approach for minimizing the steric interaction afforded the product with the opposite configuration.

Chelucci et al. [41] synthesized further chiral terpyridines derived from (–)-β-pinene, (+)-camphor, and (+)-2-carene and tested their ability to chelate copper or rhodium for the asymmetric cyclopropanation of styrene. The copper catalysts were poorly efficient and selective in this reaction. The corresponding rhodium complexes led to the best result (64% ee) with the ligand derived from (+)-2-carene (ligand **33** in Scheme 17).

In conclusion, many chiral pyridine-based ligands have been prepared from the chiral pool and have been successfully tested as ligands for the copper- or rhodium-catalyzed cyclopropanation of olefins. Although efficient systems have been described, sometimes leading interestingly to the major *cis* isomer, the enantioselectivities usually remained lower than those obtained with the copper-bis(oxazoline) system.

2.3 Miscellaneous *N,N*-Containing Copper Complexes

Other types of new *N*-containing ligands have been described as effective chiral inductors for copper-catalyzed asymmetric cyclopropanation. Hence, Fu and Lo [42] prepared a new planar-chiral ligand, namely the C_2-symmetric bisazaferrocene (structure **34** in Scheme 18), which was found to be efficient for the cyclopropanation of various olefins with large diastereomeric excesses and ee values up to 95%.

These authors further described the synthesis and resolution (by chiral HPLC) of a new C_2-symmetric planar-chiral bipyridine ligand [43] (see structure **35** in Scheme 18). They obtained an X-ray crystal structure of the corresponding copper complex proving a bidentate complexation. This system led to high diastereo- (up to 94%) and enantioselectivity (up to 94%) in the

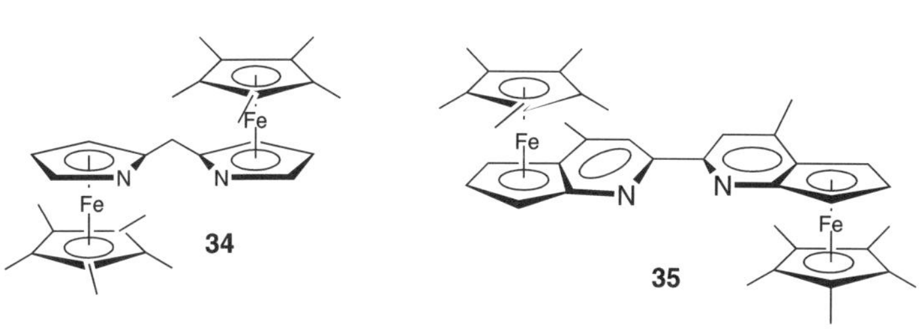

Scheme 18

asymmetric cyclopropanation of various olefins with sterically hindered aryl acetates.

Suga and Ibata [44] prepared binaphtyldiimine derivatives **36** (Scheme 19) affording 98% ee as best selectivity for the transformation of 1,1-diphenylethylene with *l*-menthyl diazoacetate. The authors performed PM3 calculations and proposed an optimized structure of the copper complex to explain the high enantioselectivity observed with 1,1-disubstituted olefins.

Scott et al. [45] prepared diimine derivatives of 2,2′-diamino-6,6′-dimethylbiphenyl (as structure **37** in Scheme 19) as copper chelates for the catalyzed cyclopropanation reaction. All catalysts were active in this reaction but enantioselectivities varied importantly according to the substitution pattern of the imine aryl group: only *ortho*-substituted ligands (by chloride or methyl groups) led to products with measurable enantioselectivity for the model test reaction (up to 57% ee with **37**).

Zhou and coworkers [46] prepared chiral ligands bearing both amne and pyridyl (quinolinyl) moieties (structures **38** and **39** in Scheme 19). Their copper complexes led to the cyclopropane carboxylates in up to 87% ee, for the quinolinyl derivative. The same group prepared ligands of type **40** in which the quinoline part of ligand **39** was changed to a chiral oxazoline [47]. The match of chiralities between binaphthyl and oxazoline was crucial in ligand **40**, which was found to be more selective than **39** (90% ee for the *cis* product and 54% de in favor of the *trans* isomer with structure **40**, *R* = Bn). However, the authors proved that the absolute configurations of cyclopropanation products were mainly controlled by the axial chirality of the ligands.

Chiral nitrogen chelates derived from sugars were prepared by Ruffo [48], introducing diimines and diamines functionalities on inexpensive monoses, α-D-glucose and α-D-mannose.

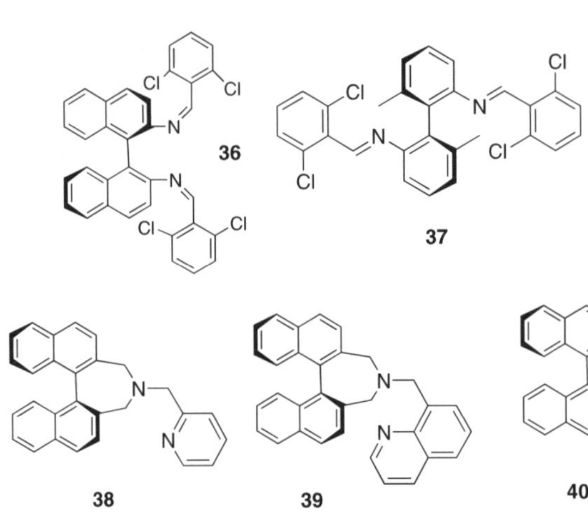

Scheme 19

Up to 55% ee was obtained with copper complexes from these ligands, the best results arising from the use of the α-D-glucose derivative **42** (Scheme 20), with Ar = mesityl. The products obtained using the mannose series were almost racemic, probably due to a local unfavorable *meso* configuration of the sugar at the nitrogen chelating atoms. The nature of both the sugars and the chelates (amines or imines) determined the sense of the enantioselection. Interestingly, diimines **41** derived from α-D-glucose were found to afford the opposite enantiomers with respect to diamines **42** derived from the same sugar.

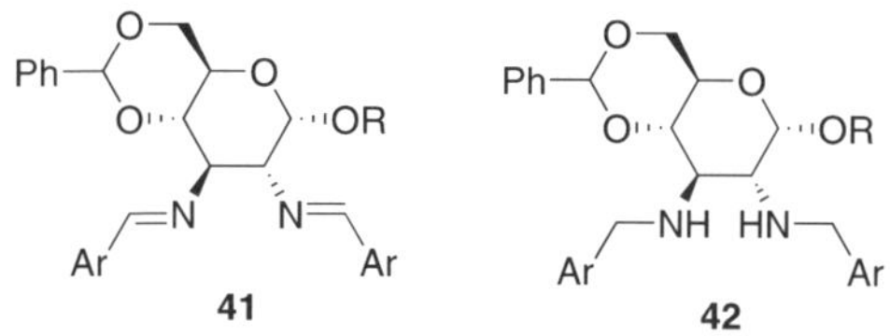

Scheme 20

2.4
Miscellaneous *N*,*N*-Containing Ruthenium Complexes

Zhang et al. [49] prepared a chiral ruthenium complex coordinated by a pyridine-bis(imine) ligand (structure **43** in Scheme 21).

In dichloromethane, this complex with a low catalyst loading (1 mol %) achieved the cyclopropanation of styrene with ethyldiazoacetate in high diastereoselectivity (*trans*/*cis* = 90/10) and enantioselectivity (up to 97% ee for the major isomer).

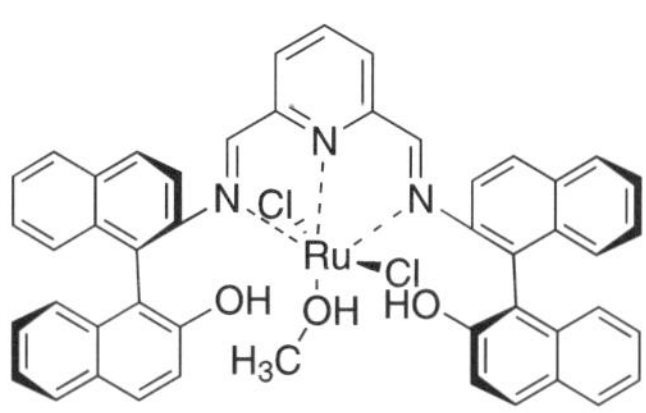

43

Scheme 21

2.5
Porphyrine-Containing Complexes

Metalloporphyrins have proved their efficiency as ruthenium carbonyl ligands for the enantioselective cyclopropanation of styrene [50, 51].

At a catalyst loading of only 0.15 mol %, quantitative transformation of olefins was obtained using **44** (Scheme 22) as ligand for an extremely active ruthenium complex. Moreover, excellent diastereoselectivities (*trans/cis* = 96/4) and high enantiomeric excesses (up to 91%) were obtained in the cyclopropanation of styrene with ethyl diazoacetate. However, Simonneaux et al. [52] reported that the homochiral porphyrin ruthenium(II) complex (structure **45** in Scheme 22) catalyzed the cyclopropanation of styrene derivatives in good yields but with moderate enantiomeric excesses (up to 52%). Zhang [53] developed analogous cobalt(II)porphyrin complexes (structures **46** and **47** in Scheme 22), a catalytic system that operates with alkenes as limiting reagents and does not require any slow addition of diazo compounds. Structure **46** led to the *cis* isomer as major product with 77% ee. By using catalyst **47**, near same amounts of *cis* and *trans* isomers were obtained with a maximum of 31% ee.

Woo et al. [54] prepared new chiral tetraaza macrocyclic ligands (**48** in Scheme 23) and their corresponding iron(II) complexes and tested them, as well as chiral iron(II) porphyrin complexes such as $Fe^{II}(D_4-TpAP)$ **49**, in asymmetric cyclopropanation of styrene.

Ligands of type **48** were synthesized by the cyclization reaction of diamines with dithioaldehydes. Iron complexes formed with those structures led, however, to active but weakly enantioselective catalysts. The best results were

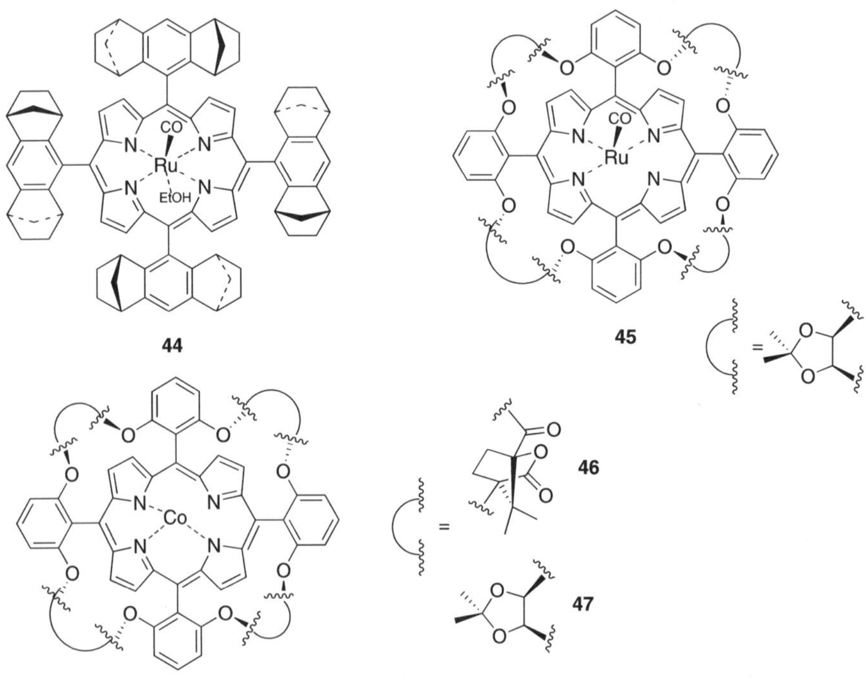

Scheme 22

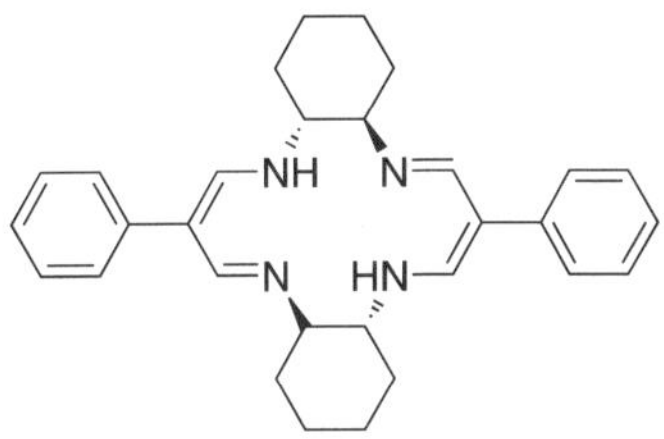

48

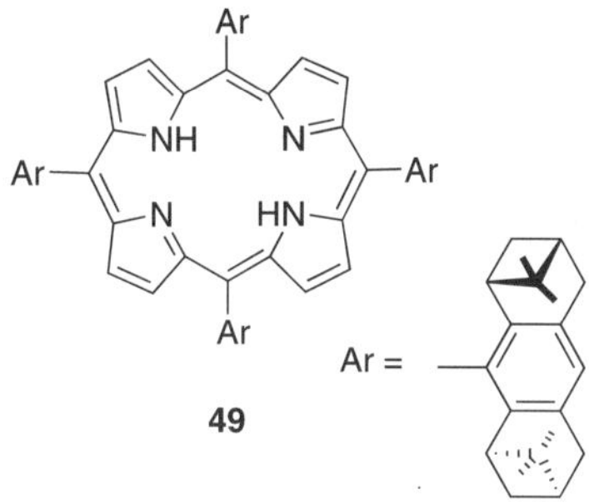

49

Scheme 23

obtained with ligand **48** (up to 79% ee using menthyldiazoacetate) and ligand **49** (78% under the same conditions but with a lower diastereoselectivity).

2.6 Heterogeneous Catalysis

Numerous examples have been published dealing with the heterogeneization of copper complexes, as immobilized catalysts for the asymmetric cyclopropanation of alkenes. Some of them have already been mentioned in the text for a direct comparison with their homogeneous counterparts. Other reusable catalytic systems have been developed and will be described as follows.

Corma et al. [55] have prepared chiral Cu(I) complexes with substituted pyrrolidine ligands bearing a triethoxysilyl group (Scheme 24).

These *N*,*N*-chelating ligands were then covalently grafted on a modified Y-zeolite containing "supermicropores". The selectivities observed were low (up to 11% ee) but led to similar values when comparing the unsupported and zeolite-supported Cu complexes. Interestingly, however, the zeolite catalysts could be recovered and reused several times with no loss of activity.

Fraile et al. [56] used cationic bis(oxazoline)-Cu(II) complexes, intercalated into lamellar clays by electrostatic interactions, as catalysts for C–C bond formation reactions. Interestingly, the heterogeneous catalysts led to higher conversions and selectivities than their homogeneous counterparts,

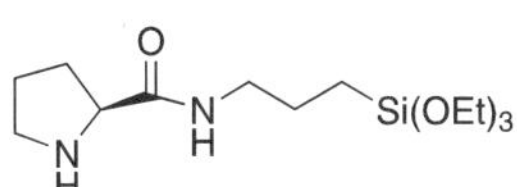

50

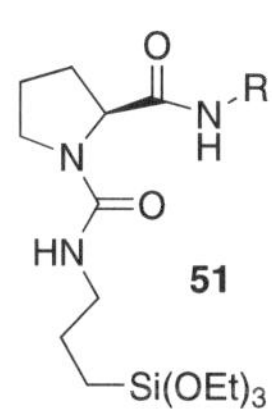

51

Scheme 24

due to changes in the microenvironment of the bis(oxazoline)-Cu(II) complexes. Furthermore, one of the solids, namely a laponite exchanged with a copper-bis(oxazoline) complex, was recovered and reused a second time with only a slight decrease in the catalytic performance. Fraile et al. [57] reported promising results when laponite was used as the support. Immobilization of the catalyst on this support induced major modifications in the stereochemical course of the reaction since the preference for the more stable *trans*-cyclopropanes was reduced, in contrast to what happened in the homogeneous phase. The authors proposed explanations for this phenomenon based on the modification of the relative energies of the four diastereomeric transition states by interaction with the support. Possible isolation of the catalytic sites was also mentioned. Some of the catalysts could be recovered and reused, but leaching of complexes and subsequent partial loss of catalytic performance were observed. By modifying chiral bis(oxazolines) in the methylene bridge with two allyl or vinylbenzyl groups, Burguete et al. [58] grafted these chiral ligands covalently onto mercaptopropyl silica. The catalytic performance of the corresponding solid copper catalysts was tested in the cyclopropanation of styrene and compared to the homogeneous analogs. The observed catalytic activity was in general similar to that obtained with the homogeneous catalysts, but the enantioselectivities decreased importantly, particularly with catalysts in which the complex was formed in solution prior to the immobilization process.

Clarke and Shannon also supported copper bis(oxazoline) complexes onto the surfaces of inorganic mesoporous materials, such as MCM-41 and MCM-48, through the covalent binding of the ligand, modified by alkoxysilane functionalities [59]. The immobilized catalysts allowed the cyclopropanation of styrene with ethyldiazoacetate to be performed as for the corresponding homogeneous case, and were reused once with almost no loss of activity or selectivity.

Fraile et al. developed a new system for immobilization of bis(oxazoline)-copper catalysts by electrostatic interactions with anionic organic supports, such as nafion [60]. Using the phenyl-substituted bis(oxazoline) as chiral ligand, the authors obtained results similar to those reported for homogeneous catalysis. No catalyst leaching was observed and recycling occurred successfully. Similar results were not obtained with the ^{t}Bu-substituted bis(oxazoline), probably because of important steric interactions between the chiral catalyst and the support leading to low electrostatic interactions. Regarding the covalent anchoring to insoluble organic materials, Mayoral et al. [58, 61] further modified bis(oxazolines) by functionalization of the central methylene bridge with *p*-vinylbenzyl groups (structure **52** in Scheme 25) and subsequent radical homo- or copolymerization. Though the results depended on the polymerization method, the activities and selectivities of all polymeric catalysts were similar or even better than those obtained with the homogeneous analog. The authors further proved that the functionalization in the bridge methylene pos-

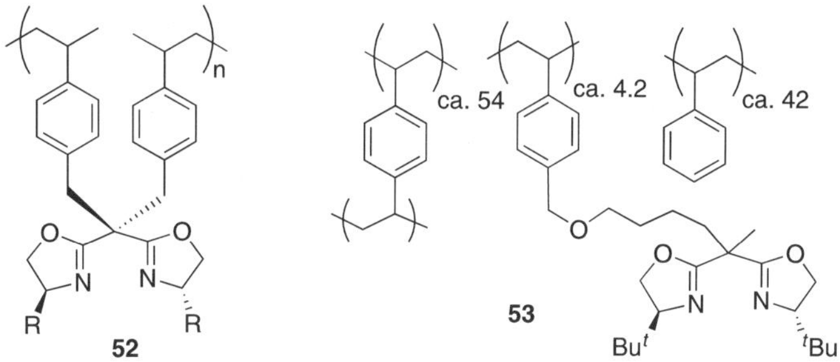

Scheme 25

itions (and not the immobilization process) strongly influenced the *trans*/*cis* selectivity in the product formation.

Salvadori et al. [62] tested the same strategy but derived the bis(oxazoline) ligands in such a way that they minimized the steric hindrance at the bridging methylene carbon (structure **53** in Scheme 25). The polymer was used affording enantiomeric excesses superior to 90% and was reused at least five times with almost no loss in enantioselectivity or activity.

Other original methodologies have been developed to efficiently recycle chiral homogeneous catalysts. Mayoral and Vaultier [63] studied the cyclopropanation of styrene by bis(oxazoline) complexes in ionic liquids. Without modifying the structure of the ligand, they were able to prepare copper complexes (associated with two different anions, Cl^- or OTf^-) soluble in ionic liquids. The authors proved the importance of the copper counterion by performing a reaction with $CuCl_2$ both in CH_2Cl_2 and in [EMIM][NTf_2] (EMIM for 1-ethyl-3-methylimidazolium). Though in the first case almost no enantioselectivity could be observed, reaction in the ionic liquid led to high selectivities, indicating that the most abundant copper species bore the counterion of the solvent. They were furthermore able to reuse the dissolved complex twice without loss of activity or selectivity. Davies et al. [64] also studied this strategy for immobilization. They performed the test reaction at 10 °C as a biphasic system, styrene not being soluble in the ionic liquids. The catalyst remained in the ionic liquid phase and could be reused several times (for [BMIM][BF_4], BMIM for 1-butyl-3-methylimidazolium) with only little loss of activity and selectivity. The observed performance of the catalyst was comparable to that obtained in a homogeneous reaction, using chloroform as solvent.

Cornejo et al. [65] reported the first immobilization of pyridine-bis(oxazoline) chiral ligands and the use of the corresponding solid ruthenium complex in the model cyclopropanation test. They synthesized vinyl-PyBOx, the vinyl functionality being introduced in the fourth position of the pyridine ring. This monomer was further homo- or copolymerized in the presence of styrene and divinylbenzene. The corresponding ruthenium catalysts proved their efficiency in the catalytic test reaction, leading to comparable results with the homogeneous case (with a 1/1 Ru/ligand ratio). It should be noted that the enantioselectivity for the *cis* product is dramatically reduced (up to 42% ee as compared with 71% ee in the homogeneous case). The polymeric catalysts were reused with, however, markedly decreased efficiency. According to the authors, this deactivation probably arose from partial coordination of by-products formed in the reaction.

2.7
Conclusion

N,*N*-containing ligands with various structures are effective for the enantioselective cyclopropanation of prochiral olefins with diazoacetates. The diastereo- and enantioselectivities obtained are not yet well understood and depend on both steric and electronic factors. The requirement of high diastereoselectivity is very important when one considers that potential useful target molecules often possess several substituents on the cyclopropane ring, with precise spatial orientation. It has to be noticed, however, that asymmetric cyclopropanation has been performed with excellent diastereo- and enantiocontrol when copper catalysts were used, especially those coordinated with bis(oxazoline) derivatives. This catalytic organic transformation is a particular example in which such good selectivities could not be obtained with *P*-containing ligands as chiral auxiliaries. Efficient methodologies for the reuse of the chiral catalysts have been described that allow, in some cases a lowering of the catalyst to substrate ratio, and an easy purification of functionalized cyclopropane rings.

3
Asymmetric Diels–Alder Reaction

The Diels–Alder reaction is one of the most powerful synthetic methods that fits the modern concept of atom economy [66, 67], to control the simultaneous formation of two C – C bonds. For this reason, the recent development of highly enantioselective catalytic Diels–Alder reactions represents a great advance in synthetic chemistry [68]. A large number of metals, ligands and dienophiles have been studied with most of the successful catalysts containing, until recently, chelating oxygen ligands [69]. Some recent success was obtained using diphosphine ligands such as BINAP but most of the studies are now focused on the use of optically active *N*-containing ligands, such as bis(oxazoline) derivatives, due to their better stability. The most recent results will be reported here and classified according to the nature of the ligand used for the preparation of the catalytic *N*-liganding complex.

3.1
Bis(oxazoline) Ligands

3.1.1
Copper Complexes

Ghosh et al. [70] reviewed a few years ago the utility of C_2-symmetric chiral bis(oxazoline)-metal complexes for catalytic asymmetric synthesis, and they reserved an important place for Diels–Alder and related transformations. Bis(oxazoline) copper(II)triflate derivatives have been indeed described by Evans et al. as effective catalysts for the asymmetric Diels–Alder reaction [71]. The bis(oxazoline) ligand **54** allowed the Diels–Alder transformation of two-point binding *N*-acylimide dienophiles with good yields, good diastereoselectivities (in favor of the *endo* diastereoisomer) and excellent ee values (up to 99%) [72]. These substrates represent the standard test for new catalysts development. To widen the use of Lewis acidic chiral Cu(II) complexes, Evans et al. prepared and tested bis(oxazolinyl)pyridine (PyBOx, structure **55**, Scheme 26) as ligand [73].

The tridentate PyBOx ligands were far more widely effective for the transformation of numerous carbonyl-derived dienophiles possessing a single accessible coordination site (high *exo:endo* selectivities and up to 96% ee). Evans group found that the nature of the counterion involved in the catalytic cycle is of major importance regarding the enantioselectivity, with SbF_6^- leading to the best results. High enantioselectivities are assumed to be obtained through a square-planar catalyst-substrate complex (**56** in Scheme 27) with a counterion possessing weak coordinating properties such as SbF_6^-. A less organized one-point complex (**57**) may explain the lower enantioselectivity observed in the case of triflate complexes.

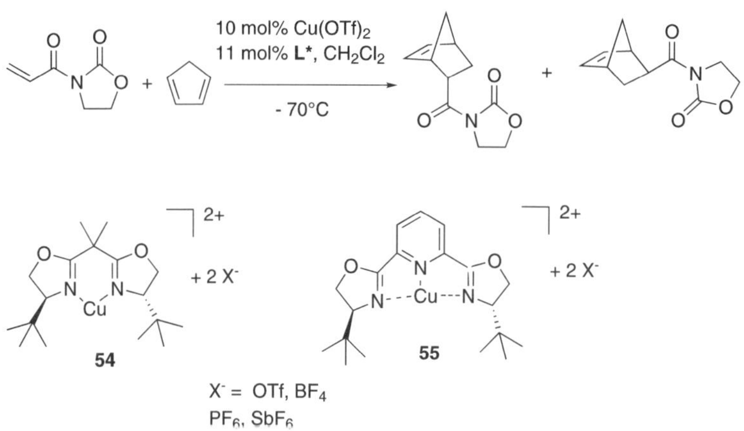

Scheme 26

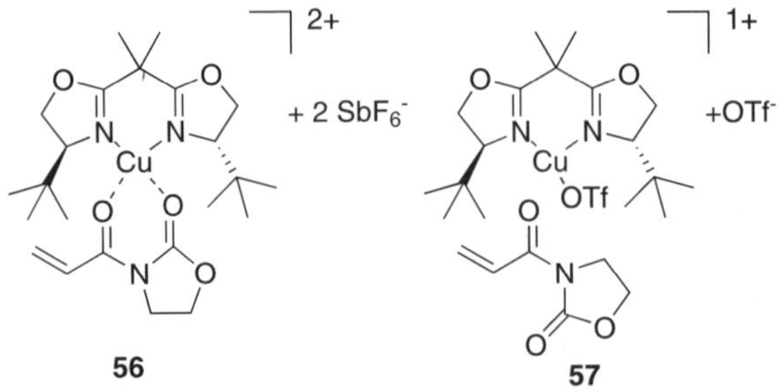

Scheme 27

Numerous articles have been devoted to the synthesis of structurally modified bis(oxazoline) ligands and to their ability to promote enantioselective Diels–Alder transformations. For example, Davies et al. [74] synthesized and tested several "Evans-type" auxiliaries, i.e., bis(oxazolines) or pyridine-bis(oxazolines), bearing various sterically-hindering substituents. The best results were obtained according to the conditions presented in Scheme 26, and afforded the *endo* diastereomer with 95% ee by using ligand **58** (Scheme 28).

By developing a single force field for modeling these copper(II)-complexes, the authors were able to confirm an important role of the ligand bite-angle on the enantioselectivity and to propose explanations for the influence of C_4-oxazoline substituents on the selectivity [75]. Ghosh et al. [76] prepared a cationic aqua complex of inexpensive $Cu(ClO_4)_2.6H_2O$ and constrained bidentate bis(oxazoline) (**59** in Scheme 28) to provide a highly effective system for enantioselective Diels–Alder reactions. For the test reaction depicted in Scheme 26, 98% ee (endo/exo > 99/1) could be obtained at low tempera-

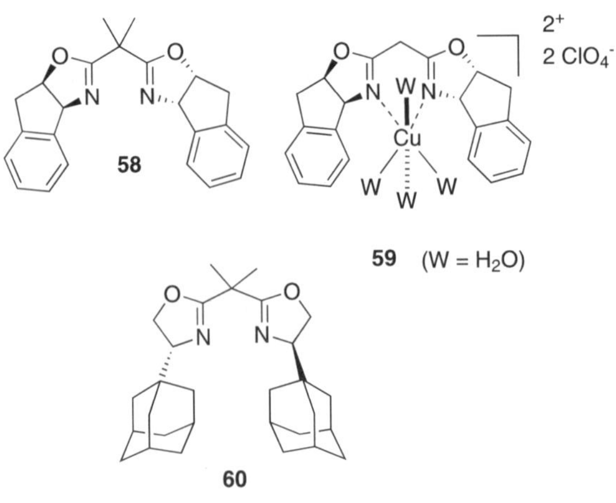

Scheme 28

ture (– 30 °C). Moreno-Mañas et al. prepared a bis(oxazoline) bearing the bulky adamantyl substituent [77] (see **60** in Scheme 28). The required (*R*)-2-(1-adamantyl)-2-aminoethanol was prepared by enzymatic resolution for the synthesis of the corresponding bis(oxazoline) and its copper complex was as enantioselective and active as ^{t}Bu-Box in the benchmark reaction. This new ligand allowed an easy access to highly pure enantiomers with the opposite configuration compared to the ones normally obtained in the literature from (*S*, *S*)-*t*-Bu-Box.

Tang synthesized pseudo-C_3-symmetric tris(oxazolines) in two steps from the corresponding triester and the amino alcohol [78]. They tested them as copper ligand for enantioselective Diels–Alder reaction [79].

These type of ligands provided air- and water-stable catalysts by reacting with $Cu(ClO_4)_2 \cdot 6H_2O$ and led to up to 82% ee (ligand **62** in Scheme 29) for the reaction of cyclopentadiene and acryloyl-2-oxazolidinone. Interestingly, in the reaction of cyclopentadiene with ketoesters (see Scheme 29), better enantio- and diastereoselectivities were obtained with these structures than those reported with the usual bis(oxazolines) or PyBOX. For example, ligand **62** led to the preparation of the cycloadduct with 71% ee (and 94% de) whereas the corresponding common *tert*-butylbis(oxazoline) provided the same product with only 53% ee.

Bis(oxazolines) ligands have been so widely used for the Diels–Alder reaction between *N*-2-alkenoyl-1,3-oxazolidine-2-one and cyclopentadiene that Lipkowitz and Pradhan developed a QSAR (quantitative structure-activity relationship) using Comparative Molecular Field Analysis (CoMFA) for a set of 23 copper-catalysts containing mainly bis(oxazoline) ligands. The generated

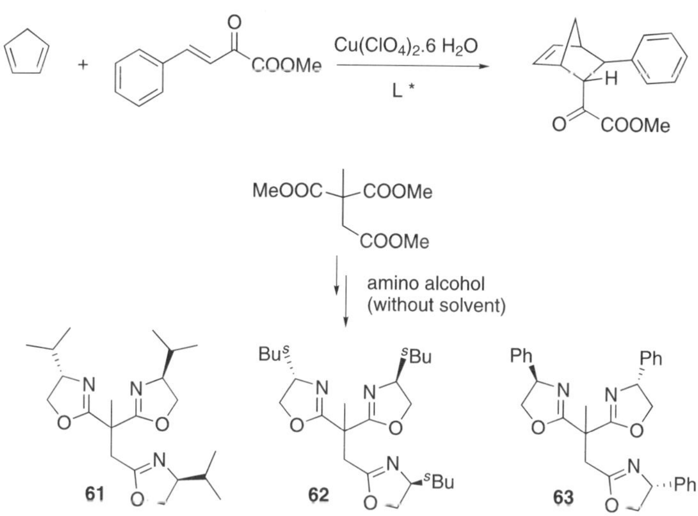

Scheme 29

models were in very good agreement (both statistically significant and predictive) with the catalytic systems and showed that approximately 70% of the variance arose from the steric field while the remaining 30% was electrostatic. The authors assumed these calculations to be meaningful for computer-aided catalyst design [80].

The use of chiral bis(oxazoline) copper catalysts has also been often reported as an efficient and economic way to perform asymmetric hetero-Diels–Alder reactions of carbonyl compounds and imines with conjugated dienes [81], with the main focus on the application of this methodology towards the preparation of biologically valuable synthons [82]. Only some representative examples are listed below. For example, the copper complex **54** (Scheme 26) has been successfully involved in the catalytic hetero Diels–Alder reaction of a substituted cyclohexadiene with ethyl glyoxylate [83], a key step in the total synthesis of (*R*)-dihydroactinidiolide (Scheme 30).

Jørgensen et al. [84] studied how solvent effects could influence the course of Diels–Alder reactions catalyzed by copper(II)-bisoxazoline. They assumed that the use of polar solvents (generally nitroalkanes) improved the activity and selectivity of the cationic copper-Lewis acid used in the hetero Diels–Alder reaction of alkylglyoxylates with dienes (Scheme 31, reaction 1). The explanation, close to that given by Evans regarding the crucial role of the counterion, is a stabilization of the dissociated ion, leading to a more defined complex conformation. They also used this reaction for the synthesis of a precursor for highly valuable sesquiterpene lactones with an enantiomeric excess superior to 99%.

Similar transformations have been performed with Danishefsky's diene and glyoxylate esters [85] catalyzed by bis (oxazoline)-metal complexes to afford the hetero Diels–Alder product in 70% isolated yield and up to 72% ee. Jørgensen [86, 87] reported a highly enantioselective, catalytic hetero Diels–Alder reaction of ketones and similar chiral copper(II) complexes leading to enantiomeric excesses up to 99% (Scheme 31, reaction 2). They also described [88] a highly diastereo- and enantioselective catalytic hetero Diels–Alder reaction of β,γ-unsaturated α-ketoesters with electron-rich alkenes

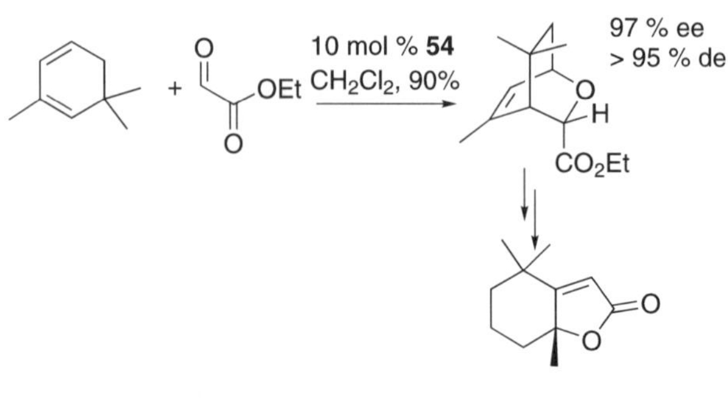

Scheme 30

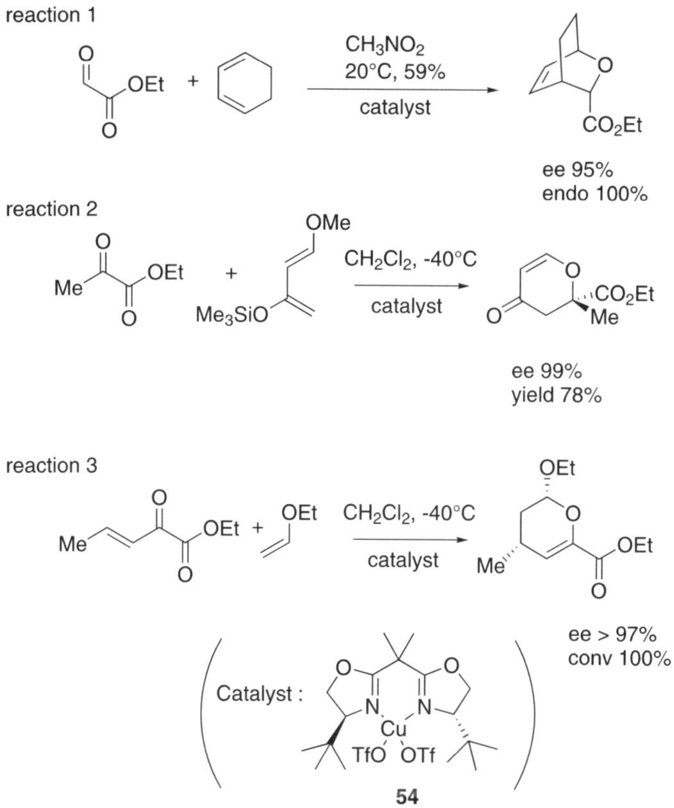

Scheme 31

(Scheme 31, reaction 3) to give dihydropyran adducts (more than 97% ee, total conversion). These compounds are highly valuable synthons for the synthesis of carbohydrates and natural products. Evans et al. [89] proposed a simplified procedure for this last reaction including particularly low catalyst loading and the use of recyclable aqua complexes. The cycloaddition was performed in hexane in the presence of florisil as an adsorbent, resulting in the formation of an insoluble catalyst. After decantation of the product solution at the end of the reaction, the catalyst was reused without significant loss of yield and selectivity. Evans et al. [90] reported that α,β-unsaturated acyl phosphonates undergo enantioselective hetero Diels–Alder reactions with enol ethers. This reaction was particularly efficient with chiral Cu(II) complexes such as **54** and afforded cyclic enol phosphonates (Scheme 32), efficient synthons for asymmetric synthesis.

α'-hydroxy enones have been studied by Palomo et al. [91] as substrates allowing 1,4-metal binding complexation for bis(oxazoline) copper-catalyzed Diels–Alder reactions of various dienes. The authors obtained excellent re-

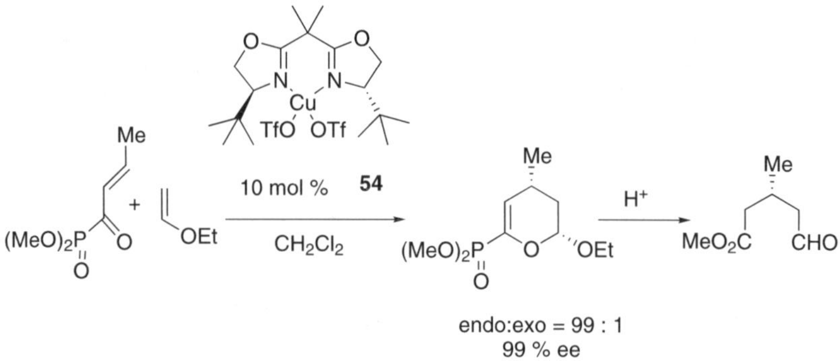

Scheme 32

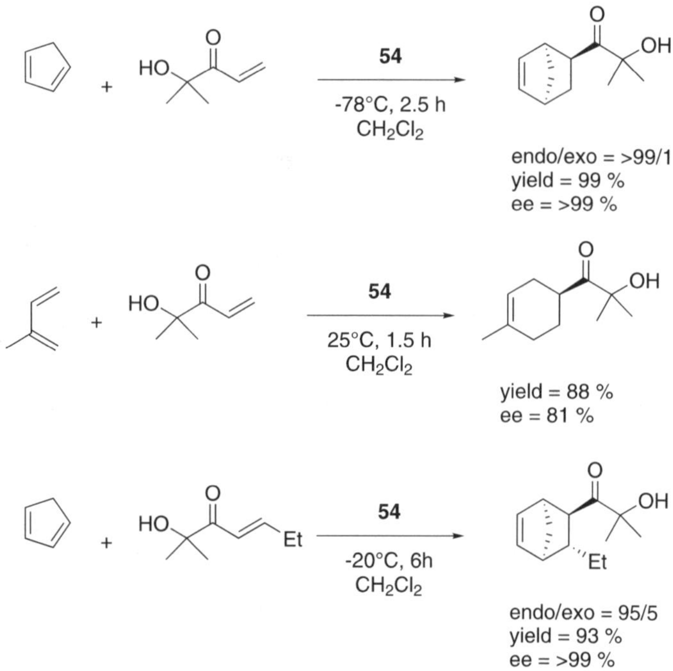

Scheme 33

sults in terms of enantio- and diastereoselectivity by using reactive cyclic dienes (cyclopentadiene or even cyclohexadiene) and also β-substituted enones (see some examples in Scheme 33).

From the several (not exhaustive) examples listed above, it is obvious that bis(oxazolines) associated to copper salts are efficient catalysts to perform Diels–Alder cycloadditions of numerous substrates, leading to highly valuable products with high diastereo- and enantioselectivity. Efforts have been moreover drawn towards the preparation of moisture- and air-stable systems allowing the use of simpler procedures. It may however be noticed that catalyst loading remains quite high to ensure reasonable yields in the expected products. Some interesting examples of catalyst recycling have been thus reported and will be described at the end of this section.

3.1.2 Zinc Complexes

Zinc(II) salts were also described as useful precursors for Lewis acid catalyzed transformations in organic synthesis. Takacs et al. [92] prepared and tested various chiral bis(oxazoline) ligands as zinc chelates, differing in the length of the chain binding the two oxazoline moieties, and in the nature of the substituent at the stereogenic center. They examined their ability to perform enantioselective Diels–Alder reactions of *N*-crotonyloxazolidinone with cyclopentadiene. The best $Zn(OTf)_2$ catalyst was derived from a 1,4-bis(oxazoline) ligand (structure **64**, Scheme 34) and led to the expected product with quantitative yield and 78% ee. These results prompted the authors to diversify the structure of this *N*-containing ligand [93].

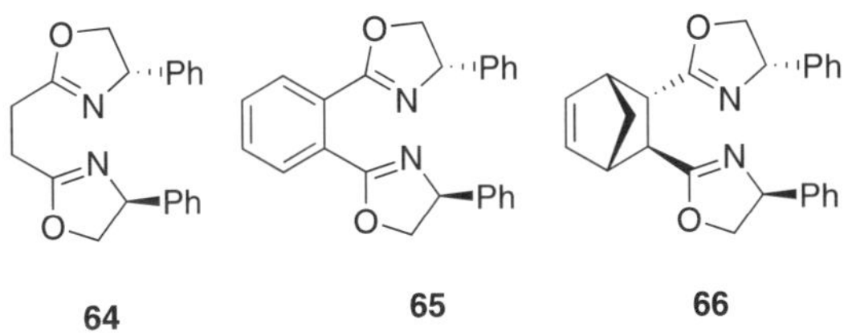

Scheme 34

Thus, they prepared ligand **65** that failed to give good ee, whereas ligand **66** afforded significant ee values (up to 78% ee). The authors observed poor enantioselectivities when magnesium was used instead of Zn (around 20% ee). Jørgensen [94] also studied the zinc(II) catalyzed hetero Diels–Alder reaction of different conjugated dienes such as 2,3-dimethylbuta-1,3-diene and cyclohexa-1,3-diene with ethyl glyoxylate in the presence of different C_2-symmetric bisoxazolines (Scheme 35).

The reaction involving the non-cyclic diene gave both the hetero Diels–Alder and -ene products, the former being the major compound with an enantiomeric excess of 87% using ligand **67**. With the cyclic diene, only 65% ee was achieved with ligand **68**, the selectivity being largely dependent on the solvent (i.e., dichloromethane or nitromethane). The authors noticed that these bis(oxazoline)zinc(II) catalysts were less active but more chemoselective than their copper counterparts for hetero Diels–Alder type reactions and they performed theoretical calculations to get insight into the geometry of the intermediates that failed, however, to give clear explanations.

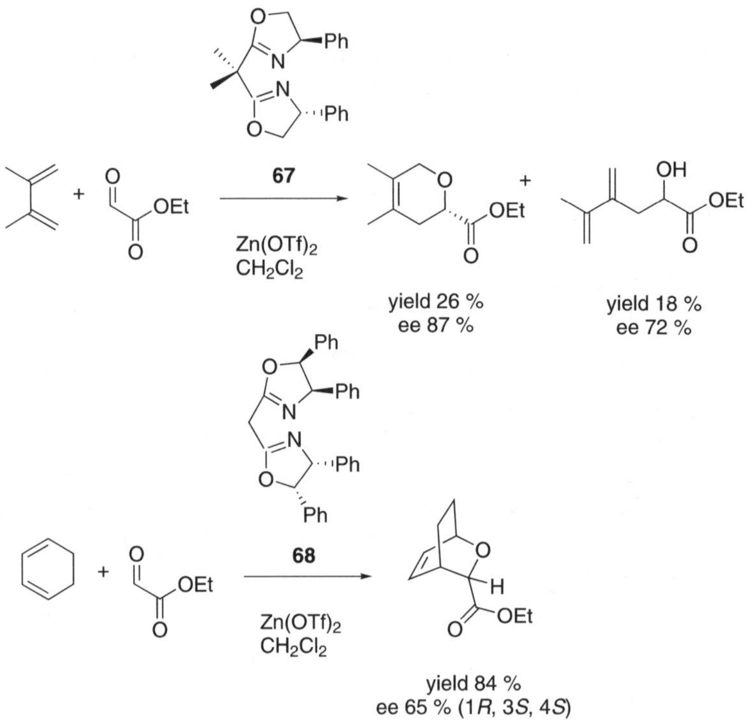

Scheme 35

3.1.3
Ruthenium Complexes

Ruthenium complexes have also been reported as active species for enantioselective Diels–Alder reactions. Faller et al. prepared a catalyst by treatment of (–)-[(η^6-cymene)RuCl(L)]SbF_6 with $AgSbF_6$ resulting in the formation of a dication by chloride abstraction [95]. The ligand was (+)-IndaBOx **69** (Scheme 36) and the corresponding complex allowed the condensation of methacrolein with cyclopentadiene in 95% conversion and 91% ee. As another example, Davies [96] prepared the complex [Ru(H_2O)L*(η^6-mes)] $[SbF_6]_2$ (with **70** as L* in Scheme 36), and tested its activity in the same reaction leading to the expected product with similar activity and lower enantioselectivity (70%).

The same group [97] studied the corresponding rhodium oxazoline complexes that led to very similar results in terms of activity (81% yield) and enantioselectivity (68%), however.

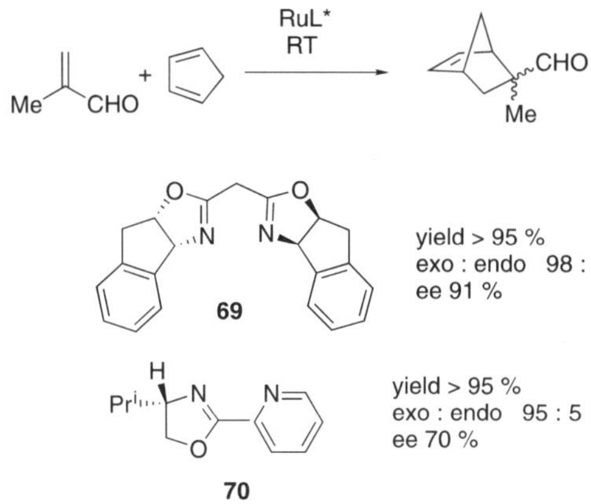

Scheme 36

3.1.4 Pyridine Bis(oxazoline) and Rare Earth Complexes

Several groups have reported the use of rare earth complexes as catalysts for asymmetric Diels–Alder reaction. Qian and Wang described thus the preparation and use of Yb complexes chelated by ^{i}Pr-PyBOx to successfully achieve the hetero-Diels–Alder reaction of methyl glyoxylate with Danishefsky's diene in 77% ee and 73% yield (Scheme 37) [98].

Fukuzawa et al. [99] found analogous scandium(III)triflate/^{i}Pr-PyBOx complex as efficient catalyst for the asymmetric Diels–Alder reaction between cyclopentadiene or acyclic dienes and acyl-1,3-oxazolidin-2-ones with up to 90% ee. They latter described the same reaction in super critical CO_2 in the presence of MS4Å [100] that proceeded more rapidly than in CH_2Cl_2 leading to the expected product with analogous selectivity.

Desimoni et al. [101] further investigated the influence of the variation of the lanthanide and of the PyBOx ligand (bearing a ^{i}Pr- or Ph-substituent with the same configuration) on the enantioselectivity of the Diels–Alder test

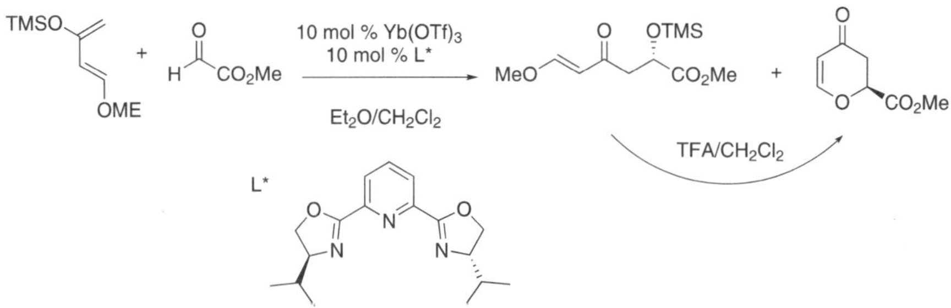

Scheme 37

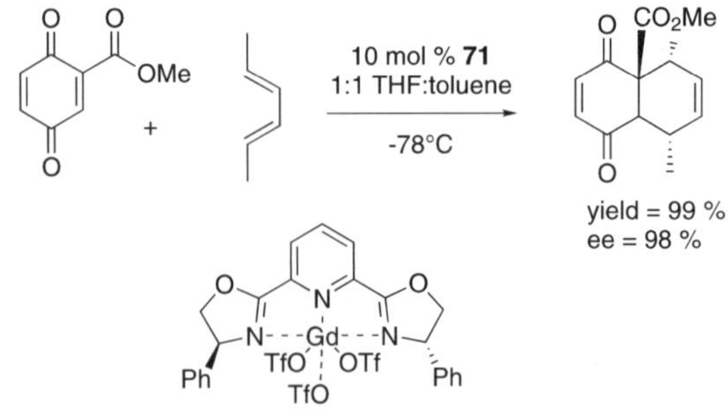

71

Scheme 38

reaction between cyclopentadiene and acryloyl-1,3-oxazolidin-2-one. They observed a regular variation of the enantioselectivity (also influenced by the molecular sieves) as a function of the cation ionic radius, both ligands reversing the absolute sense of the stereoselection on going from scandium to lanthanum. Furthermore, a different sense of induction was observed with the same cation, using either ^{i}Pr-PyBOx or Ph-PyBOx with the same configuration. Hence, ^{i}Pr-PyBOx led to the best results with Sc(III) whereas Ph-PyBOx had to be used with La(III) cation to afford the highest enantiomeric excesses (up to 84% ee and up to 95% ee, respectively). The authors proposed that two competitive reacting complexes with different coordination number could be formed, which favor the attack on the opposite heterotopic faces of the coordinated dienophile. As an explanation for these results, they assumed the contribution to the overall process to be maximum with Sc and La, respectively.

Evans and Wu have prepared complexes derived from PyBOx ligands and samarium or gadolinium triflates that were efficient for the Diels–Alder reaction between various quinones and dienes [102] (see Scheme 38 for an example).

3.2
Miscellaneous *N,N*-Containing Copper Complexes

Copper-complexes prepared with other type of *N*-chelating ligands have been also prepared and evaluated as catalysts for the Diels–Alder reaction. Engberts et al. [103] studied enantioselective Diels–Alder reaction of 3-phenyl-1-(2-pyridyl)-2-propen-1-one with cyclopentadiene in water (Scheme 39). By using coordinating chiral, commercially available α-amino-acids and their derivatives with copper salts as catalysts, they obtained the desired product with yields generally exceeding 90%. With L-abrine (**72** in Scheme 39) as chiral moiety, an enantiomeric excess of 74% could be achieved. Moreover, the catalyst solution was reused with no loss of enantioselectivity.

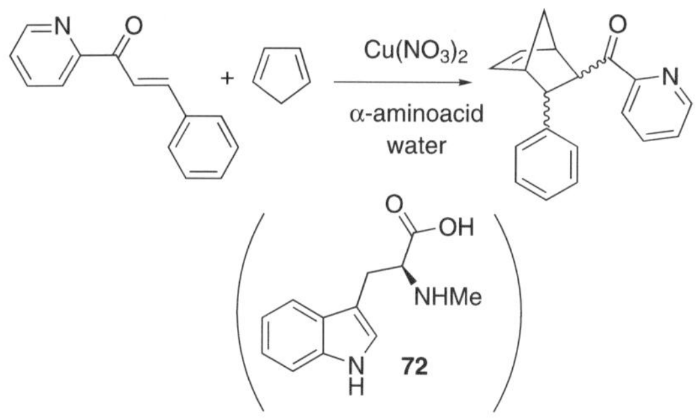

Scheme 39

Significant enantioselectivities were however obtained exclusively with ligands containing an aromatic side group. Therefore, the authors assumed that π-stacking must be of major importance in achieving high selectivities since aryl-aryl interactions are indeed more efficient in water than in organic solvents.

Lassaletta et al. designed new ligands for the copper-catalyzed Diels–Alder reaction by introduction of C_2-symmetric dialkylamino substructures in glyoxal bis-hydrazones [104] (**73** in Scheme 40). This ligand can advantageously be prepared in its both enantiomeric forms on a multigram scale from 1,4-diphenylbutanedione.

In the presence of the corresponding copper(II) catalyst, *N*-acryloyloxazolidinone reacted with various conjugated dienes (cyclic and acyclic) with good enantioselectivities in all cases, competing with results obtained for classical catalysts in the case of flexible dienes. Thus the cycloadducts obtained from isoprene or 2,3-dimethyl-1,3-butadiene were synthesized in high yields and good enantioselectivities (92% ee in both cases).

Sulfinyl imines ligands have been also largely used as copper-chelates for the asymmetric Diels–Alder reaction. Ellman et al. [105] reported their use as ligands, the chirality being solely introduced by the presence of sulfoxide moieties. They thus prepared novel sulfinyl imines **74** and **75** in analogy to the

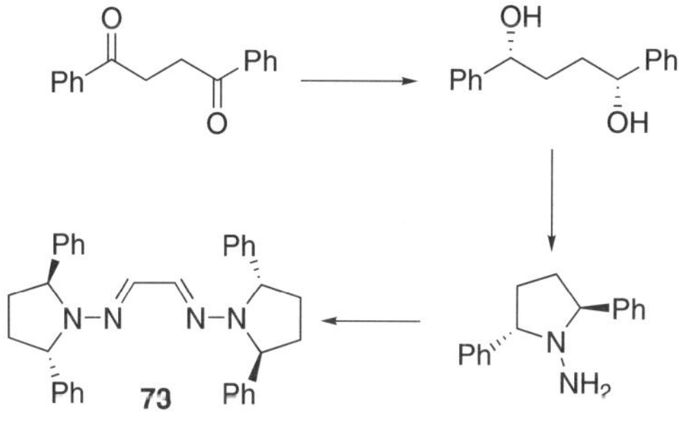

Scheme 40

bis(oxazoline) and PyBOx ligands, and a bis(sulfinyl)-iminoamidine **76** from commercially available and optically stable chiral sulfinamides (Scheme 41).

These various donor ligands were involved in the model Diels–Alder reaction and proved to display good catalytic activity but moderate enantioselectivity (up to 72% for **75**). To improve the enantioselectivity, the authors prepared **76**, a bis(sulfinyl)iminoamidine containing more basic donor atoms that gave, indeed, associated with $Cu(SbF_6)_2$, the desired endo product with high yield as well as high enantio- (98%) and diastereoselectivity ($> 99:1$). However, according to X-ray analyses, the authors demonstrated that this potentially *N*,*S*- or *O*-coordinating ligand led in fact to a complex existing as a M_2L_4 quadruple-stranded helicate in which both Cu atoms were coordinated to the sulfinyl oxygen in a square pyramidal array. Infra-red data obtained in solution proved also a binding mode via oxygen and not via nitrogen atoms.

Bolm et al. [106] have carefully studied the synthesis and the liganding ability of salen-like bis(sulfoximines). The chirality which is indeed generally introduced via the use of chiral diamines in the salen series, is in sulfoximines present via the sulfur atom. They investigated the Diels–Alder cycloaddition between cyclopentadiene and acryloyl-2-oxazolidinones with various bis(sulfoximines) (see Scheme 42) and $Cu(OTf)_2$ as the copper source [107].

The authors observed that the reaction could be run using a 1/1 mixture of bis(sulfoximine) and $Cu(OTf)_2$, at a 10 mol % ratio to give the expected product in excellent yield and high enantioselectivity (ratio *endo*/*exo* 94/6 and 83% ee for **77** and up to 93% ee for **78**). The authors studied next exhaustively how the other reaction parameters do influence the selectivity. They

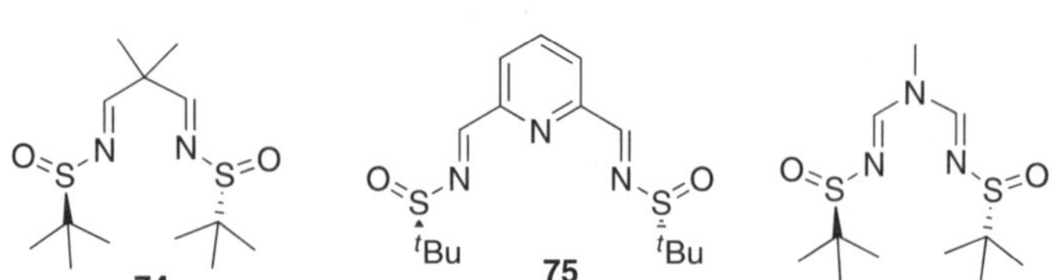

74 **75** **76**

Scheme 41

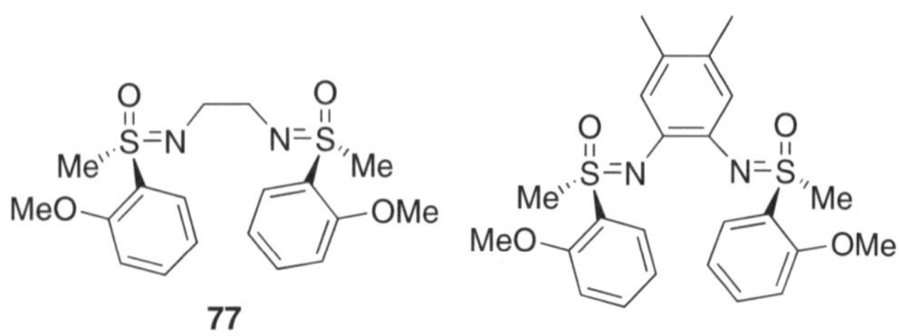

77 **78**

Scheme 42

thus proved that by using less coordinating counterions (for instance perchlorate) and chloroform as solvent, the ee raised up dramatically.

Bolm et al. [108] prepared a C_2-symmetric bis (sulfoximine) as ligand for the copper-catalyzed hetero-Diels–Alder reaction. The stereogenic sulfur atom being located near the *N*-coordinating atom, these structures were assumed to be promising for asymmetric catalysis. Their ligand (**79** in Scheme 43) was synthesized by palladium-catalyzed *N*-aryl imination from 1,2-dibromobenzene and (*S*)-S-methyl-S-phenylsulfoximine with Pd_2dba_3 in 70% yield.

The reaction of 1,3-cyclohexadiene and ethylglyoxylate catalyzed by the corresponding copper (II) triflate complex led to the expected product in high yield (81%) and high enantio- and diastereoselectivity. At – 40 °C, the reaction between cyclohexadiene and diethyl mesoxalate afforded similarly the expected product in high yield and up to 98% ee.

Spectroscopic investigations were conducted by different techniques including EXAFS, ESR and UV-Vis-spectroscopy [109]. The authors concluded that the chiral bis(sulfoximine) ligand linked to the Cu(II) atom via the imine nitrogens and the dienophile via the carbonyl oxygen atoms in a tetragonally distorted complex displaying a non-symmetric square pyramidal geometry. Since the two coordinating sulfoximine nitrogens were non-equivalent, the authors further studied the efficiency of monosulfoximine ligands to perform similar hetero Diels–Alder reactions as C_1-symmetric chelates [110]. The targeted ligands were obtained by palladium-catalyzed *N*-arylations of

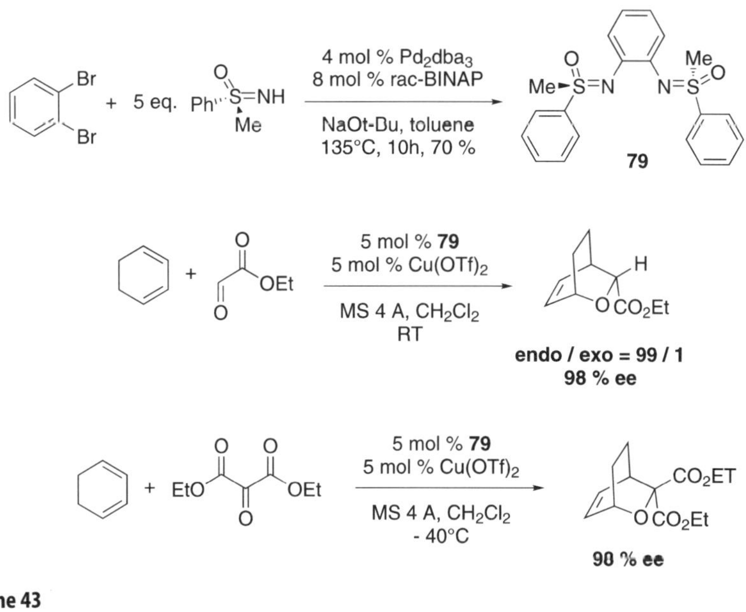

Scheme 43

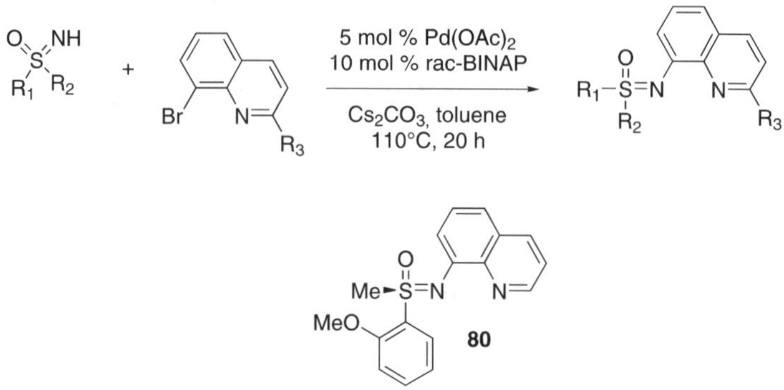

Scheme 44

enantiopure sulfoximines with 8-bromo-quinoline derivatives affording the quinoline-based C_1-symmetric sulfoximines in good yield. A large variety of ligands (see for example **80** in Scheme 44) with different alkyl and aryl substituents at the sulfoximine moiety were thus obtained. They were successfully tested in the hetero Diels–Alder reaction between 1,3-cyclohexadiene and ethylglyoxylate or diethyl mesoxalate reaching in general high values both in terms of activity and enantioselectivity.

The authors could obtain the desired product in up to 96% ee and 98% de by using ligand **80** as copper chelate and performing the reaction at – 10 °C with 1 mol % of complex.

From all these results, optically active sulfoximines, with their nitrogen-coordinating site located at the close proximity to the stereogenic sulfur atom, have thus proven their efficiency as copper-ligands for asymmetric Diels–Alder and hetero Diels–Alder reactions.

3.3 Miscellaneous *N*,*N*-Containing Magnesium Complexes

Fujisawa et al. [111] have reported that the magnesium complex prepared from chiral 2-[2-[(tolylsulfonyl)amino]phenyl]-4-phenyl-1,3-oxazoline **81** and methyl-magnesium iodide was efficient, in a stoechiometric amount, for promoting the enantioselective Diels–Alder reaction of 3-alkenoyl-1,3-oxazolidin-2-one with cyclopentadiene (Scheme 45) leading exclusively to the endo adducts in up to 92% ee. The use of 10 mol % of the complex led to an important decrease in enantioselectivity of the product (51% ee).

It was found, furthermore, that the substituent on the sulfonamide group of the chiral ligand strongly influenced the enantiofacial selectivity. Hence, ligand **81** bearing a tosyl substituent delivered the *endo*-(2*R*)-cycloadduct, whereas a trifluoromethanesulfonamide group afforded its enantiomer. The authors proposed that the latter substituent should increase the Lewis acid-

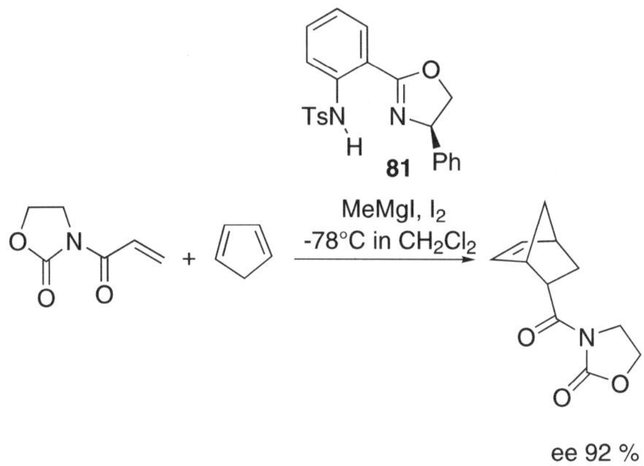

Scheme 45

ity of the metal inducing a probable coordination of the fluorine or oxygen atom of the sulfonyl group, leading thus to a stable intermediate providing the other configuration of the product. The presence of iodine as an additive was proven to be a determining factor for enantioselectivity by dissociating the iodide anion from the magnesium cation.

3.4 Miscellaneous *N*,*N*-Containing Rare Earth Complexes

Zeijden [112] used chiral *N*-functionalized cyclopentadiene ligands to prepare a series of transition metal complexes. The zirconium derivative (**82** in Scheme 46), as a moderate Lewis acid, catalyzed the Diels–Alder reaction between methacroleine and cyclopentadiene, with 72% de but no measurable enantiomeric excess. Nakagawa [113] reported 1,1′-(2,2′-bis-acylamino)binaphthalene (**83** in Scheme 46) to be effective in the ytterbium-catalyzed asymmetric Diels–Alder reaction between cyclopentadiene and crotonyl-1,3-oxazolidin-2-one. The adduct was obtained with high yield and enantioselectivity (97% yield, endo/exo = 91/9, > 98% ee for the endo adduct). The addition of diisopropylethylamine was necessary to afford high enantioselectivities, since without this additive, the product was essentially

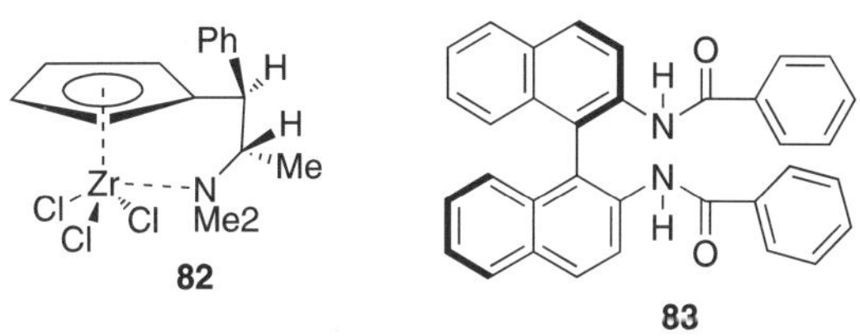

Scheme 46

racemic. The authors attributed the structure $Yb(OTf)_3$ **83**.$[iPr_2NEt]_2$ to the active species, thanks to NMR studies.

3.5 Heterogeneous Catalysis

The examples presented above are quite convincing to propose bis(oxazoline) copper catalysts as some of the most powerful complexes in terms of activity and selectivity, to perform enantioselective Diels–Alder transformations. However, the catalyst loading that is often used remains quite high and not attractive for a possible industrial development even at a pilot scale. In the very recent years, many efforts have therefore been undertaken towards the tuning of heterogeneously modified catalytic systems for a convenient regeneration and recycling [114]. The immobilization of such complexes was for instance performed on inorganic solids by non-covalent bonding such as electrostatic interactions. Mayoral et al. described cationic exchanges of bis(oxazoline) complexes (Cu, Mg and Zn) with two types of anionic solids (a laponite clay and a nafion-silica nanocomposite) [115]. The authors assumed the immobilization to take place quantitatively with square-planar complexes (i.e., copper complexes) whereas the exchange of tetrahedral bis(oxazoline)-M(II) complexes was not so efficient, probably due to stronger steric interactions between the chiral ligand and the support. As a consequence, when the clays were tested as support for catalysts in the benchmark Diels–Alder reaction, poorer enantioselectivities than in the homogeneous case were observed. Nevertheless, the Cu-Laponite catalyst could be reused with no loss of enantioselectivity. Very recently, Klein Gebbink et al. [116] reported an easy immobilization of chiral triflate copper(II) bis(oxazoline) complexes on silica via electrostatic interactions corresponding to hydrogen bonds between the surface hydroxyl functionalities and the triflate anion. Two chiral copper(II)-bis(oxazoline) complexes were tested but they led to modest enantioselectivities in the Diels–Alder reaction between acryloyloxazolidinone and cyclopentadiene (up to 57% ee with immobilized copper(II) *t*Bu-bis(oxazoline)). Amazingly, the immobilized copper(Ph-bis(oxazoline))$(OTf)_2$ led to a major compound with the configuration opposite to that obtained with the same catalyst in a homogeneous reaction (33% ee for the (*S*) product and 20% ee for the (*R*) isomer, respectively). The nature of the interaction of this complex and the achiral support seemed important enough to modify its reactivity sufficiently for producing the opposite enantiomer in excess.

Rechavi and Lemaire reported the heterogeneization of copper bis(oxazoline) catalysts through covalent grafting onto silica via functionalization of indaBOX ligand by triethoxysilane groups [117] (**84** in Scheme 47).

In the presence of copper perchlorate, the heterogenized catalyst was successfully used for the test Diels–Alder reaction and proved to be similarly active and enantioselective than its homogeneous counterpart. It was further-

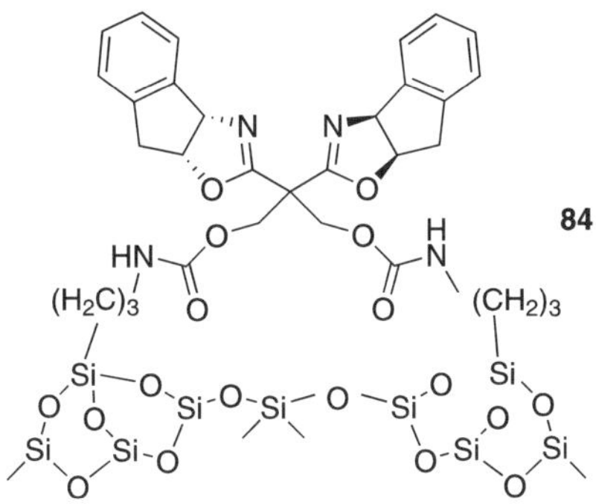

84

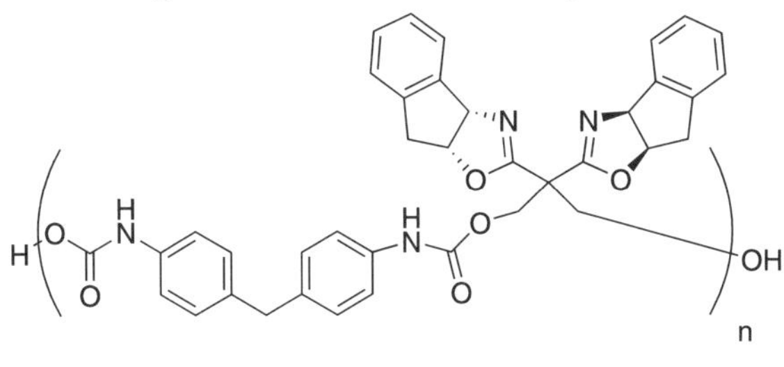

85

Scheme 47

more reused at least three times without noticeable loss of enantioselectivity or activity. By passivation of the silica surface (protection of the free silanol groups by reaction with *N*-trimethylsilylimidazole), the authors were able to obtain better results in terms of enantioselectivities (up to 92% by performing the reaction at – 78 °C).

The same authors developed another heterogeneization technique by polymerization of the indaBOX ligand as part of the main chain of a polyurethane backbone (**85** in Scheme 47). High conversion and diastereoselectivity (about 90%) towards the endo enantiomer were obtained when the polymer was treated with copper triflate and used as catalyst for the same Diels–Alder transformation [118]. Disappointingly, the catalyst proved however not to be as stable since it could be only reused three times (55% ee), the fourth cycle showing a complete loss of enantioselectivity. Cozzi and coworkers [119] supported chiral bis(oxazolines) on poly(ethylene glycol) allowing the catalytic reaction to be performed under homogeneous conditions, due to the solubility properties of PEG. The ligand could be recovered (and recycled) after precipitation and filtration, but the enantioselectivity obtained in the Diels–Alder reaction remained unsatisfactory (up to 45% ee, **86**, *R* = *t*Bu in Scheme 48).

Hallman and Moberg described the preparation of a chiral bis(oxazoline) grafted on ArgoGel [120] (**87**, in Scheme 48). In the zinc-catalyzed Diels–Alder reaction, polymer **87** was found to be less active than the analogous

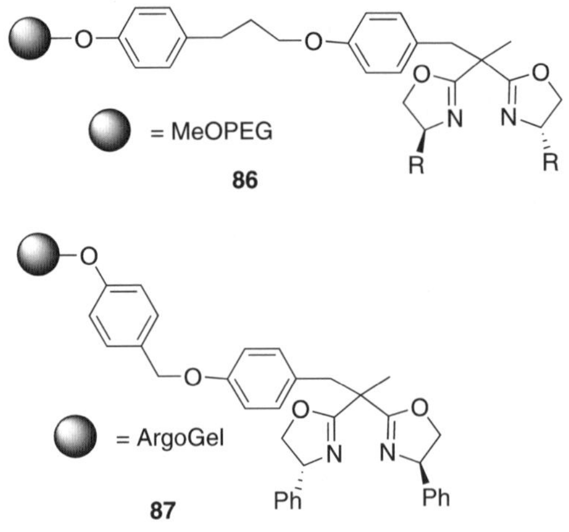

Scheme 48

homogeneous ligand. Moreover, the polymer-supported catalyst gave racemic product. The authors proposed the rigidity of the polymer (and its low swollen ability) to be responsible of the inhibition of the reaction. Meracz and Oh [121] have performed Diels–Alder reactions in ionic liquids, especially in DiBuIm (1,3-dibutylimidazolium) and observed enhanced activities and enantioselectivities with bis(oxazoline) copper triflate complexes compared to the use of dicholoromethane as solvent, when the reaction was performed at room temperature. No recycling tests were described in this article.

3.6 Conclusion

The great versatility of the Diels–Alder reaction has certainly been demonstrated here by numerous examples covering very different catalytic systems. *N*-containing ligands are highly efficient for the preparation of chiral complexes for asymmetric Diels–Alder cycloadditions. In particular, bis(oxazolines) associated with copper salts have been reported to lead to selective reactions with various substrates. Indeed, the most successful system associated a bidentate ligand chelating the metal, and a dienophile (which acts as a two-point binder to the ligand-metal complex). The use of other metal ions is one of the new trends for achieving this reaction and interesting examples have been described with zinc, magnesium, and even ruthenium or rhodium. Although bis(oxazolines) are often successful chelates for this transformation, the development of other types of ligands (such as amines or sulfoximines) seems to have considerable potential. Heterogeneous catalysis has recently been developed in this field, and some pertinent examples have been described in terms of selectivity and activity. However, the stability of the immobilized catalysts over several uses has to be further optimized.

4
Asymmetric Nucleophilic Allylic Substitution

Activated allylic substrates are easily available and are very useful for creating C – C bonds by substitution. Pd-catalyzed nucleophilic allylic substitutions, i.e., Tsuji–Trost reactions [122], have been particularly studied over the last few years for adjusting the structure of the catalytic species to allow good control of regio-, diastereo-, and enantioselectivity, depending on the substrate. The selectivity of new chiral ligands is mainly evaluated by transformation of the 1,3-diphenylallyl system with various nucleophiles, but especially dimethyl malonate (Scheme 49). This reaction allows a facile comparison and analysis of the results since this substrate, despite its poor practical interest, usually gives rise to rather high ee values and good yields, as compared to the analogous 1,3-dimethylallyl system, for example.

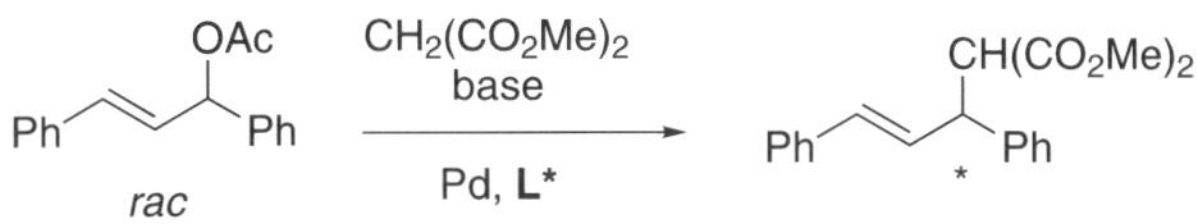

Scheme 49

Thus, new catalytic systems allowing the nucleophilic substitution of cyclopentenyl, cyclohexenyl derivatives, or other aliphatic-activated allylic substrates are less developed, whereas these transformations could potentially lead to valuable synthons. Only catalysts allowing good conversions and ee values with numerous allyllic substrates are of general interest. Chiral phosphine-containing ligands [123] have been demonstrated to be valuable ligands. Furthermore, thanks to the recent use of *N,P*- or N_2,P_2-containing ligands, high enantioselectivities could be obtained for the Tsuji–Trost transformation of many useful substrates. Reports on the use of optically active *N,N*-ligands have appeared only very recently for efficient allylic nucleophilic substitution. Indeed, nitrogen donors are rather poor ligands for the stabilization of Pd(0) species involved in the catalytic cycle but some examples have been reported in which the chelation is performed via two nitrogen atoms, leading to high enantioselectivities. These results will be classified here according to the structure of the *N,N*-chelate and the efficiency of the corresponding metallic complex in what we will from now on call the test reaction, i.e., the alkylation of 1,3-diphenyl-2-propenyl acetate.

4.1 Bipyridines, Terpyridines, Phenanthrolines and Related Ligands

Transition metal complexes with sp^2-nitrogens as chelating atoms have proven to be very efficient for asymmetric catalysis, especially those complexes containing chiral oxazolines as ligands. Pyridine-based ligands are other potentially promising chelates since they can be rendered chiral by appending additional substituents, often arising from the chiral pool. These types of *N,N*-chelates have been the most studied for coordination to palladium and used as enantioselective complexes for asymmetric nucleophilic allylic substitution. For instance, Chelucci et al. prepared chiral nitrogen-containing ligands with pyridine [124] and phenanthroline derivatives [125] bearing the 6,6-dimethyl-norpynan-2-yl group as chiral moiety. Some relevant examples of structures are listed in Scheme 50.

These ligands were active for allylic substitutions but the process was not enantioselective in the benchmark reaction (**88**, in Scheme 49). More structurally constrained chelates led, however, to measurable enantioselectivities: 40% ee for **89**, 50% ee for **90**, and 64% ee for **91** in the test reaction. By further modifications in the structure of these bipyridine-type ligands (see **92** in Scheme 51, a chiral C_1-symmetric 2,2′-bipyridine) [126], enantioselectivities up to 89% were obtained.

Ligand **92** was readily prepared by reaction of (+)-pinocarvone with 1-phenacylpyridinium iodide. The authors similarly prepared corresponding 5,6-dihydro-1,10-phenanthrolines derived from (+)-pinocarvone and a tetrahydroquinolone (structure **93**, [127]) and obtained up to 81% in the palladium-catalyzed test reaction. Chelucci et al. [128] reported the synthesis of chiral C_1-symmetric 1,10-phenanthrolines incorporated in a steroid backbone. Structure **94** derived from 5α-cholestan-4-one in Scheme 51, allowed very high yield and up to 96% ee using BSA and tetrabutylammonium fluoride to generate the malonate anion.

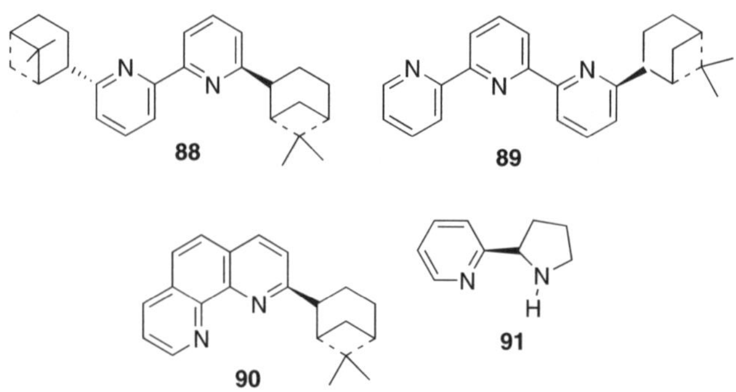

Scheme 50

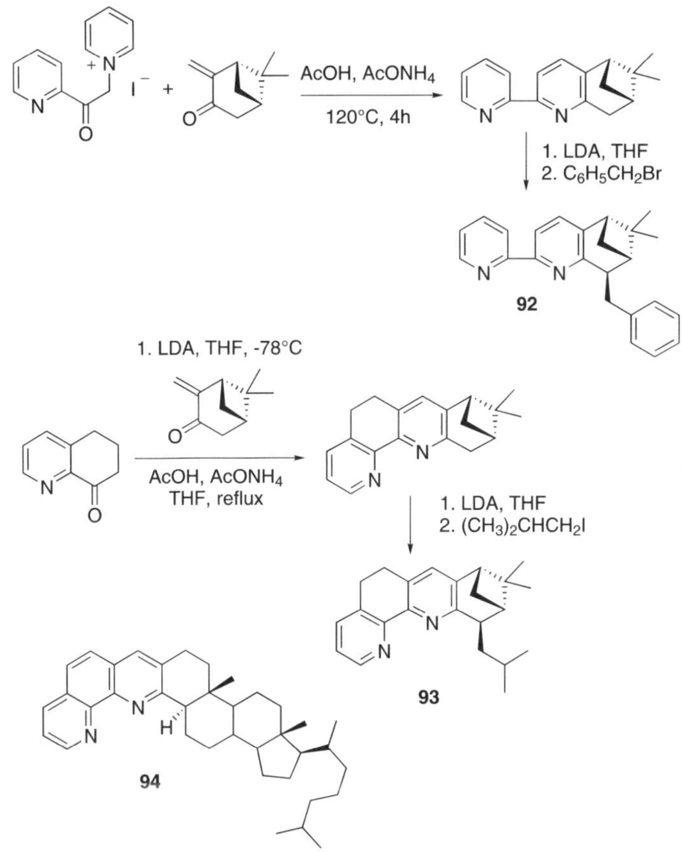

Scheme 51

Helquist et al. [129] have reported molecular mechanics calculations to predict the suitability of a number of chiral-substituted phenanthrolines and their corresponding palladium-complexes for use in asymmetric nucleophilic substitutions of allylic acetates. Good correlation was obtained with experimental results, the highest levels of asymmetric induction being predicted and obtained with a readily available 2-(2-bornyl)-phenanthroline ligand (**90** in Scheme 50). Kocovsky et al. [130] prepared a series of chiral bipyridines, also derived from monoterpene (namely pinocarvone or myrtenal). They synthesized and characterized corresponding Mo complexes, which were found to be moderately enantioselective in allylic substitution (up to 22%).

4.2
Oxazolinylpyridines and Related Ligands

Another class of chiral dinitrogen ligands, combining chelation by a pyridine nitrogen atom and an oxazoline nitrogen atom, was described by Chelucci et al. [131]. An enantioselectivity of up to 91% was obtained with ligand **95** (Scheme 52) for the test reaction. According to this type of structure, several groups developed new ligands by varying the size and nature of the linker between the pyridine and the oxazoline ring. Chelucci et al. [132] synthesized pyridinylmethyl-oxazolines of type **96** (R = H) in Scheme 52, leading however to poor enantioselectivities. Analogous ligands but bridged by disubstituted methylene, prepared by Zhou et al. [133] provided higher values, with ee up to 88% (structure **96** in Scheme 52, with R = Me and R' = *t*Bu).

Chelucci et al. further prepared quinolylmethyloxazolines and acridinyloxazolines as depicted in structures **97** and **98** of Scheme 52 [134]. These ligands proved effective with up to 78% enantioselectivity for the quinolylderivatives (**97**, R_1 = Ph and R_2 = H) while the acridininyl compounds afforded low enantiomeric excess under severe reaction conditions. Interestingly, those ligands gave the product with the opposite configuration compared to that obtained with the corresponding ligands pyridylmethyloxazolines and quinolyloxazolines, lacking a benzo-fused ring on the pyridine framework.

Moberg described similar *N*,*N*-ligands (**99** and **100**, Scheme 53) and prepared [(hydroxyalkyl)-pyridinooxazoline]- and [(alkoxyalkyl)pyridinooxazoline]-palladium complexes possessing chiral substituents, both on the pyridine and on the oxazoline rings. These catalysts were highly selective for the test reaction [135, 136], the results being dependent on the nature of the substituents and on the relative configuration of both stereogenic centers present in the ligand.

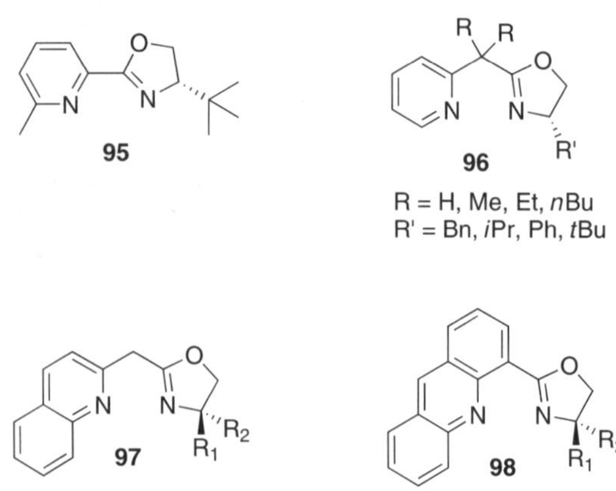

Scheme 52

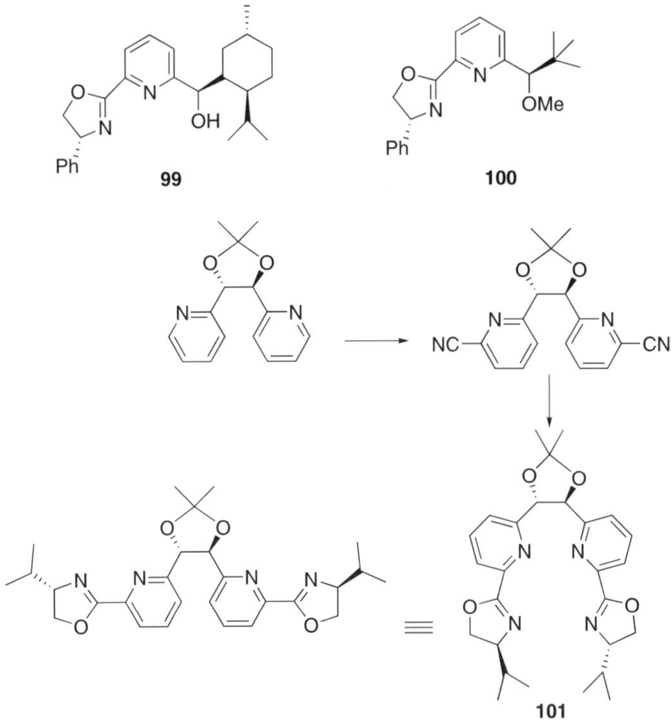

Scheme 53

Better than 99% ee was obtained with ligands **99** and **100**. The authors later reported the use of this type of ligand for asymmetric transition metal-catalyzed reactions promoted by microwave flash-heating [137]. The very fast allylic alkylation reactions were accomplished in high yields and with fair enantioselectivities, but somewhat lower than those obtained under classic conditions at room temperature. Chelucci prepared new chiral C_2-symmetric bis(oxazolinyl-pyridinyl)dioxolane ligands such as **101** [138] as possible bis-bidentate (by coordination with a nitrogen atom from the pyridine core and one nitrogen atom from the oxazoline moiety) or tetradentate chelates. Such ligands hence led to a high selectivity (ee > 98%) in the classic palladium-assisted allylic substitution. Since such high enantioselectivity was also reached with an analog without any stereogenic center on the oxazoline ring, the authors assumed the selectivity to arise from the presence of the dioxolane backbone. Furthermore a bis-bidentate coordination was proposed since poor activity and enantioselectivity were observed using the bis(pyridinyl)-1,3-dioxolane precursor as ligand.

4.3
Bis(oxazoline) Ligands

Aït-Haddou and Balavoine prepared an homochiral bis(oxazoline), with stereogenic centers on the oxazoline rings and on the side chains (**102** in Scheme 54). By using the corresponding palladium complex for the nucleophilic allylic substitution, the desired product was produced in up to 90% ee [21]. Similar ligands were studied by Pericas and Muller [139], who developed a family of bis(oxazolines) bearing various substituents at different positions of the five-membered rings (**103** and**104** in Scheme 54). They performed various NMR studies on palladium intermediates and proved the presence of *syn*/*syn*- and *syn*/*anti*-allyl isomers in solution. The influence of all types of oxazoline substituents on the activity of the resulting palladium complexes was studied and it was found that *cis* substitution on the oxazoline ring (i.e., **104** in Scheme 54) led to inactive complexes.

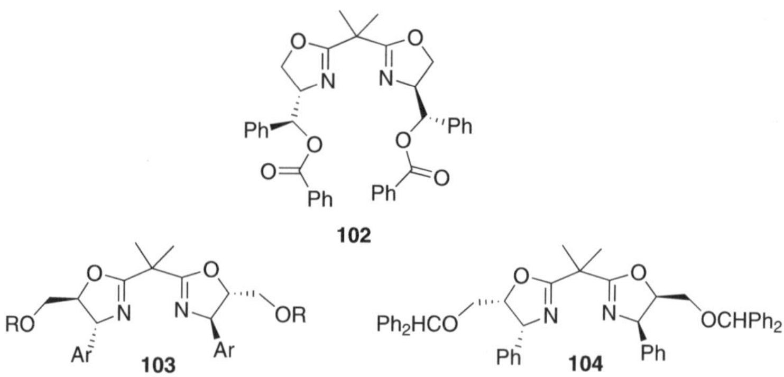

Scheme 54

Up to 96% ee were obtained with *trans*-substituted oxazoline ligands of type **103**.

4.4
Bis(pyridylamide) Ligands

Trost and Hachiya [140] studied asymmetric molybdenum-catalyzed alkylations. Interestingly, they noticed that the regioselectivity of this transformation performed with a non-symmetric allylic substrate varied according to the nature of the metal: Pd-catalyzed substitutions on aryl-substituted allyl systems led to attack at the less substituted carbon, whereas molybdenum catalysis afforded the more substituted product. They prepared the bis(pyridylamide) ligand **105** (Scheme 55) and synthesized the corresponding Mo-complex from $(C_2H_5-CN)_3Mo(CO)_3$. With such a catalyst, the allylic

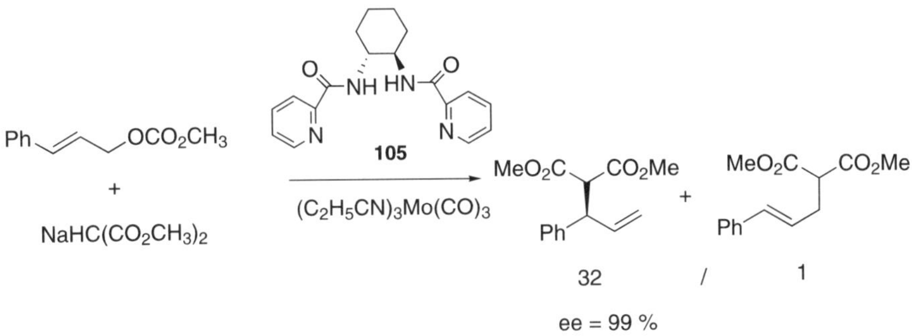

Scheme 55

substitution occurred at the most substituted position (with a 32/1 ratio) and afforded the expected product in 99% ee.

Therefore, these molybdenum-catalyzed processes are efficient and complementary to the palladium-catalyzed processes, thus widening the scope of product synthesis.

4.5
Miscellaneous *N*,*N*-Containing Copper Complexes

The copper(I)-catalyzed enantioselective substitution of allyl chlorides with diorganozinc compounds has been described by Knochel [141] using ferrocenyl amines as chiral ligand (**106** in Scheme 56). This reaction proceeded with high S_N2' regioselectivity allowing the efficient transformation of unsymmetrical allylic substrates. Under the conditions depicted in Scheme 56, the chiral substitution product was obtained in high yield and enantioselectivity (up to 87% ee), starting from substituted cinnamyl chloride.

Hoveyda and coworkers [142] developed the Cu-catalyzed allylic substitutions of phosphonate derivatives with pyridinyl peptide structures as efficient ligands. The structure of the ligands was chosen through synthesis, and analysis of libraries. Optimized compounds were used as ligands for the

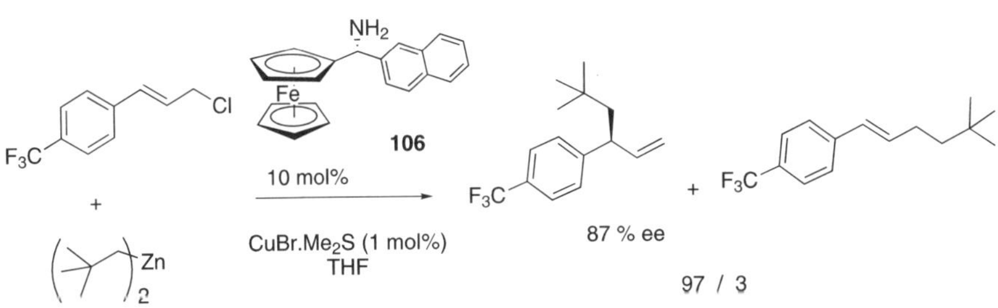

Scheme 56

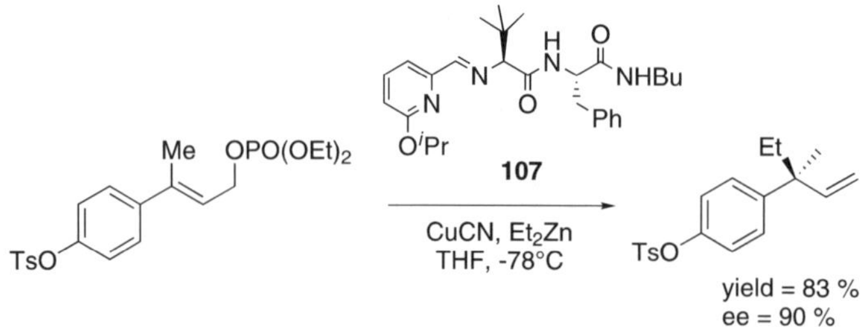

Scheme 57

enantio- and regioselective formation of quaternary carbons (see an example in Scheme 57).

4.6
Heterogeneous Catalysis

Asymmetric nucleophilic allylic substitution has rarely been studied in its heterogeneous version, probably because of the difficulties encountered in properly stabilizing and recycling Pd(0) species. Nevertheless, some promising examples have been published. Lemaire et al. [143] studied the activity and enantioselectivity of various chiral C_2-diamines for the asymmetric Pd-catalyzed transformation of various allyl acetates. The structures tested are represented in Scheme 58.

Diamine **108** led to 95% ee for the alkylation of 1,3-diphenyl-2-propenyl acetate with 90% yield. By polycondensation with a diacid chloride or polyaddition with a diisocyanate, this ligand led, respectively, to an insoluble poly(amide) **109** or poly(urea) **110** with excellent yields. Poly(amide) **109** gave a better ee (80%) than poly(urea) **110** (38%), albeit with a lower conversion (respectively, 38 and 72%), when they were used as palladium ligands

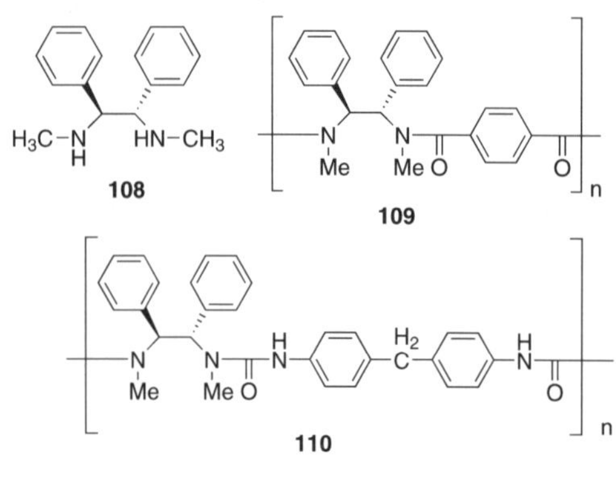

Scheme 58

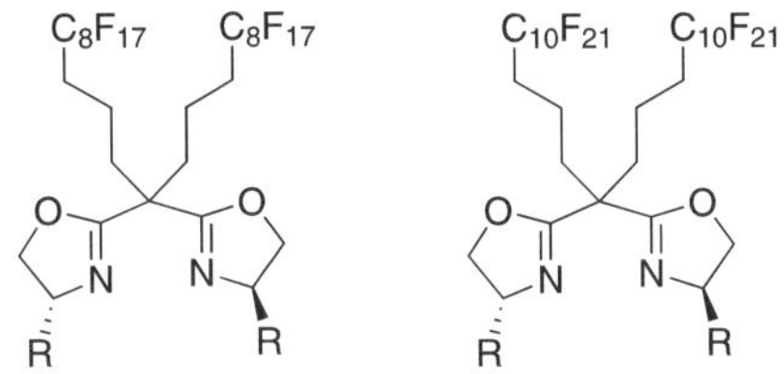

111 112

Scheme 59

in the same transformation tested with the diamine **108**. However, efforts to reuse these heterogeneous catalysts after filtration remained in vain: the palladium turned black and the recovered catalyst was no longer active. This is one of the very first examples of carbon–carbon bond formation in heterogeneous phase with enantioselective control. Another methodology aiming for the easy recovery of chiral palladium complexes was published more recently by Sinou [144]. Enantiopure fluorous bis(oxazolines) were prepared (Scheme 59) and tested as new ligands in the test palladium-catalyzed nucleophilic reaction.

All these ligands proved to give highly active and enantioselective catalysts for the reaction to be performed either in classical organic solvents (i.e., CH_2Cl_2) or in fluorinated solvents (benzotrifluoride). Recycling of the catalyst remained unsuccessful using the two-phase system FC72/CH_2Cl_2, probably due to a low partition coefficient of the palladium catalyst or to its deactivation (formation of palladium black). Recovery of the fluorinated bis(oxazolines) (and their efficient reuse as Pd-ligands) was, however, possible either by liquid–liquid extraction using a fluorous solvent or by solid–liquid extraction via a fluorous silicagel.

Another approach described by Moberg et al. [145] uses the anchoring of chiral pyridinooxazoline and bis(oxazoline) compounds on first to fourth generation dentritic substituents based on 2,2-bis(hydroxymethyl)propionic acid and (1*R*, 2*S*, 5*R*)-menthoxyacetic acid. This modification had little effect on the pyridinooxazoline series concerning the activity or the enantioselectivity of the corresponding catalyst. In the bis(oxazoline) series, in contrast, the introduction of dendrons led to higher enantioselectivities, but accompanied by loss of activity, particularly when the fourth generation dendrons were used. The introduction of a chiral dendritic substituent in this series also yielded higher stereoselective catalysts. Nothing is written in this article concerning the recycling of such dendritic oxazoline ligands.

Moberg et al. [146] modified further the bis(pyridylamide) ligand described by Trost for the preparation of a polymer-supported pyridylamide (**113** in Scheme 60) for the microwave-accelerated molybdenum-catalyzed allylic alkylation. TentaGel resin was tested in the presence of high concentrations of reactants and gave, after a 30 min reaction, total conversion in the

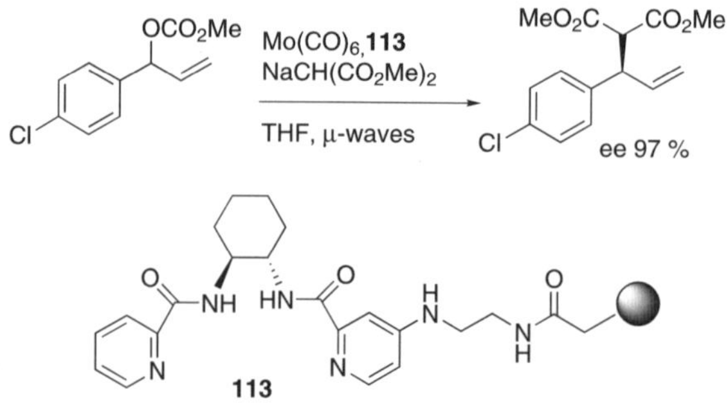

Scheme 60

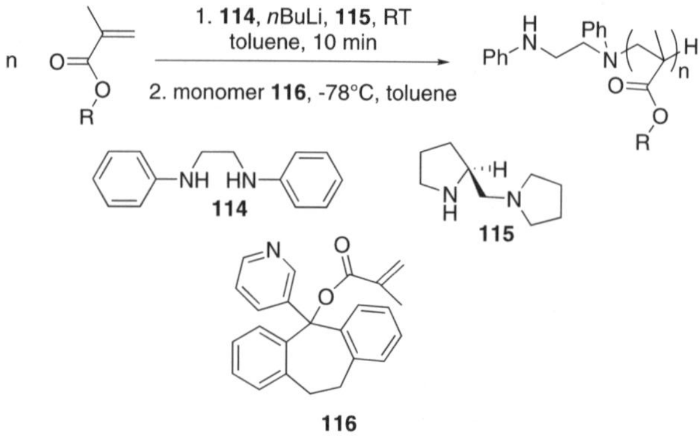

Scheme 61

product exhibiting a 35 : 1 branched-to-linear ratio and an enantiomeric excess of 97%. The resin-supported ligand was reused at least seven times after filtration and washings with no reported loss in enantioselectivity nor in reactivity for the corresponding catalyst.

Interestingly, Reggelin et al. [147] prepared helical chiral polymers by helix-sense selective anionic polymerization of methacrylates, using an asymmetric base mixture as initiator (Scheme 61).

These polymers, as helical chiral ligand lacking any other elements of chirality, were tested as chiral ligands for the palladium-catalyzed allylic substitution, and in the test reaction, enantiomeric excesses of up to 33% were obtained (higher than those reached with monomer **115** as unique ligand).

4.7 Conclusion

Pd-catalyzed nucleophilic allylic substitutions perform efficiently in an enantioselective way by using chiral phosphorous-containing ligands. Due to their high stability and convenient handling, *N*-containing ligands have recently been tested for such a catalytic reaction. Particularly, chiral pyridine derivatives (mono-, bi-, or terpyridines and their condensed analogs) proved efficient for the preparation of active and selective catalysts. Associated to oxazoline moieties, these ligands allowed the preparation of the substituted product in the test reaction with nearly perfect asymmetric induction. Asymmetric molybdenum-catalyzed alkylations, especially of non-symmetric allylic derivatives as substrates, have been very efficiently performed with bis(pyridylamide) ligands. Since the molybdenum catalysis afforded the more substituted product, this process represents a complementary method to the palladium-catalyzed process for widening the scope of product synthesis.

Efforts have been made to propose a heterogeneous version of this reaction by polymerization or support-anchoring of these *N*-containing ligands. In most cases, however, even if success was obtained by using these heterogeneous catalysts, their recycling remained non-efficient, mainly due to the poor stability of the active Pd(0) species.

5 Conclusion

Even though this article is not exhaustive, one can be convinced by the efficient use of *N*,*N*-containing ligands for asymmetric catalysis. It is now obvious that, for performing selective catalytic C – C bond formations and ensuring excellent enantio- and diastereocontrol, *N*,*N*-containing ligands are the structure of choice, particularly for cyclopropanations or Diels–Alder reactions. This remark has, however, to be moderated if one considers the nucleophilic allylic substitution for which ligands containing other coordinating heteroatoms still efficiently compete (in terms of activity and selectivity of the corresponding catalyst) with purely *N*,*N*-containing ligands. Indeed, the most useful catalyst remained the system developed by Trost, with a ligand possessing two amides and two phosphine-coordinating groups in a chiral pocket.

Bis(oxazoline) structures have been the most studied, probably due to their straightforward synthesis from the chiral pool, their easy handling, and their high stability. Their structures have been altered, introducing further chiral centers (in the side chain, or as an additional planar chirality) in order to prepare tailor-made ligands for specific substrates. Cyclopropanations and Diels–Alder reactions performed with these chiral inductors allowed the preparation of targeted product with nearly perfect asymmetric induction. Chiral bipyridines, terpyridines, phenanthrolines and analogs, when associated with copper, proved efficient for performing asymmetric cyclopropanation, or as palladium chelates for nucleophilic substitutions. An association of both structures (as exemplified in the PyBOx ligand) led to efficient catalysts for these reactions in the presence of copper, ruthenium, palladium, or even lanthanides.

The above-described structures are the main representatives of the family of nitrogen ligands, which cover a wide spectrum of activity and efficiency for catalytic C – C bond formations. To a lesser extent, amines or imines, associated with copper salts, and metalloporphyrins led to good catalysts for cyclopropanation. Interestingly, sulfinylimine ligands, with the chirality provided solely by the sulfoxide moieties, have been also used as copper-chelates for the asymmetric Diels–Alder reaction. Amide derivatives (or pyridylamides) also proved their efficiency for the Tsuji–Trost reaction.

All these C – C bond formations are, however, often performed with a high loading of catalyst, leading to expensive processes. Many efforts have been made towards the development of new methods for easily separating and recycling the catalysts. Thus, the heterogeneization of nitrogen-containing ligands was performed for most of the efficient catalytic systems. Nevertheless, these methodologies have to be optimized further in order to enhance the activity, and stability of the "solid" catalyst.

In this article, special attention has been paid to cyclopropanations, Diels–Alder reactions, and nucleophilic substitutions, for which numerous works have been devoted to the use of *N*,*N*-containing ligands. Other classical reactions allowing the formation of a new C – C bond have been omitted here (e.g., Michael-type additions or aldol reactions) where they have also been, to a lesser extent, efficiently performed using nitrogen-containing ligands.

The importance and high potential of nitrogen-containing structures as chiral ligands for asymmetric catalysis is clearly demonstrated. This research area is rapidly growing, always promising new challenges, e.g., to the best of our knowledge, a powerful transformation, the Heck-type reaction, has not yet been described in the presence of *N*,*N*-containing chiral ligands.

References

1. Jacobsen EN, Pfaltz A, Yamamoto H (1999) Comprehensive asymmetric catalysis. Springer, Berlin, Heidelberg, New York
2. Blaser HU, Spindler F, Studer M (2001) Appl Catal A: General 221:119
3. Fache F, Schulz E, Tommasino LM, Lemaire M (2000) Chem Rev 100:2159
4. Lebel H, Marcoux JF, Molinaro C, Charrette AB (2003) Chem Rev 103:977
5. Fukuda T, Katsuki T (1997) Tetrahedron 53:7201
6. Dakovic S, Liscic-Tumir L, Kirin SI, Vinkovic V, Razr Z, Suste A, Sunjic V (1997) J Mol Cat A: Chem 118:27
7. Brunner H, Berghofer JJ (1995) Organomet Chem 501:161
8. Wu XY, Li XH, Zhou QL (1998) Tetrahedron: Asymmetry 9:4143
9. Wu XY, Shen YY, Ma B, Zhou QL, Chan ASC (2000) J Mol Cat A: Chem 157:59
10. Chelucci G, Sanna MG, Gladiali S (2000) Tetrahedron 56:2889
11. Fritschi H, Leutenegger U, Pfaltz A (1986) Angew Chem, Int Ed Engl 25:1005
12. Pfaltz A (1996) Acta Chem Scand 50:189

13. Evans DA, Woerpel KA, Hinman MM, Faul MM (1991) J Am Chem Soc 113:726
14. Boulch R, Scheurer A, Mosset P, Saalfrank RW (2000) Tetrahedron Lett 41:1023
15. Bedekar AV, Andersson PG (1996) Tetrahedron Lett 37:4073
16. Bedekar AV, Koroleva EB, Andersson PG (1997) J Org Chem 62:2518
17. Harm AM, Knight JG, Stemp G (1996) Tetrahedron Lett 37:6189
18. Uozumi Y, Kyota H, Kishi E, Kitayama K, Hayashi T (1996) Tetrahedron: Asymmetry 7:1603
19. Gant TG, Noe MC, Corey EJ (1995) Tetrahedron Lett 36:8745
20. Kim SG, Cho CW, Ahn KH (1997) Tetrahedron: Asymmetry 8:1023
21. Hoarau O, Aït-Haddou H, Castro M, Balavoine GGA (1997) Tetrahedron: Asymmetry 8:3755
22. Schinnerl M, Böhm C, Seitz M, Reiser O (2003) Tetrahedron: Asymmetry 14:765
23. Glos M, Reiser O (2000) Org Lett 14:2045
24. Charrette AB, Janes MK, Lebel H (2003) Tetrahedron: Asymmetry 14:867
25. France MB, Milojevich AK, Stitt TA, Kim AJ (2003) Tetrahedron Lett 44:9287
26. Nishiyama H, Itoh Y, Matsumoto H, Park SB, Itoh K (1994) J Am Chem Soc 116:2223
27. Nishiyama H, Soeda N, Naito T, Motoyama Y (1998) Tetrahedron: Asymmetry 9:2865
28. Isawa S, Takezawa F, Tuchiya Y, Nishiyama H (2001) Chem Commun 59
29. Iwasa S, Tsushima S, Nishiyama K, Tsuchiya Y, Takezawa F, Nishiyama H (2003) Tetrahedron: Asymmetry 14:855
30. Davies IW, Gerena L, Cai D, Larsen RD, Verhoeven TR, Reider PJ (1997) Tetrahedron Lett 38:1145
31. Simpson JH, Godfrey J, Fox R, Kotnis A, Kacsur D, Hamm J, Totelben M, Rosso V, Mueller R, Delaney E, Deshpande RP (2003) Tetrahedron: Asymmetry 14:3569
32. Krumper JR, Gerisch M, Suh JM, Bergman RG, Tilley TD (2003) J Org Chem 68:9705
33. Chelucci G, Cabras MA, Saba A (1995) J Mol Cat A: Chem 95:L7
34. Chelucci G, Thummel RP (2002) Chem Rev 102:3129
35. Chelucci G, Gladiali S, Sanna MG, Brunner H (2000) Tetrahedron: Asymmetry 11:3419
36. Lyle MPA, Wilson PD (2004) Org Lett 6:855
37. Malkov AV, Pernazza D, Bell M, Bella M, Massa A, Teply F, Meghani P, Kocovsky P (2003) J Org Chem 68:4727
38. Wong HL, Tian Y, Chan KS (2000) Tetrahedron Lett 41:7723
39. Kwong HL, Lee WS (2000) Tetrahedron: Asymmetry 11:2299
40. Kwong HL, Wong WL, Lee WS, Cheng LS, Wong WT (2001) Tetrahedron: Asymmetry 12:2683
41. Chelucci G, Saba A, Soccolini F, Vignola D (2002) J Mol Cat A: Chem 178:27
42. Lo MMC, Fu GC (1998) J Am Chem Soc 120:10270
43. Rios R, Liang J, Lo MMC, Fu GC (2000) Chem Commun 377
44. Suga H, Kakehi A, Ito S, Ibata T, Fudo T, Watanabe Y, Kinoshita Y (2003) Bull Chem Soc Jpn 76:189
45. Sanders CJ, Gillespie KM, Scott P (2001) Tetrahedron: Asymmetry 12:1055
46. Ma JA, Wang LX, Zhang W, Zhou QL (2001) Tetrahedron: Asymmetry 12:2801
47. Ma JA, Wan JH, Zhou YB, Wang LX, Zhang W, Zhou QL (2003) J Mol Cat A: Chem 196:109
48. Borriello C, Cucciolito ME, Panunzi A, Ruffo F (2001) Tetrahedron: Asymmetry 12:2467
49. Tang W, Hu X, Zhang X (2002) Tetrahedron Lett 43:3075
50. Frauenkron M, Berkessel A (1997) Tetrahedron Lett 38:7175
51. Lo WC, Che CM, Cheng KF, Mak CW (1997) J Chem Soc, Chem Commun 1205

52. Galardon E, Roué S, Le Maux P, Simonneaux G (1998) Tetrahedron Lett 39:2333
53. Huang L, Chen Y, Gao GY, Zhang XP (2003) J Org Chem 68:8179
54. Du G, Andrioletti B, Rose E, Woo LK (2002) Organomet 21:4490
55. Carmona A, Corma A, Iglesias M, Sanchez F (1996) Inorg Chim Acta 244:79
56. Fraile JM, García JI, Mayoral JA, Tarnai T (1997) Tetrahedron: Asymmetry 8:2089
57. Fraile JM, García JI, Mayoral JA, Tarnai T (1998) Tetrahedron: Asymmetry 9:3997
58. Burguete MI, Fraile JM, García JI, García-Verdugo E, Herrerías CI, Luis SV, Mayoral JA (2001) J Org Chem 66:8893
59. Clarke RJ, Shannon IJ (2001) Chem Commun 1936
60. Fraile JM, García JI, Mayoral JA, Tarnai T, Harmer MA (1999) J Catal 186:214
61. Burguete MI, Fraile JM, García JI, García-Verdugo E, Luis SV, Mayoral JA (2000) Org Lett 2:3905
62. Mandoli A, Orlandi S, Pini D, Salvadori P (2003) Chem Commun 2466
63. Fraile JM, García JI, Herrerías CI, Mayoral JA, Carrié D, Vaultier M (2001) Tetrahedron: Asymmetry 12:1891
64. Davies DL, Kandola SK, Patel RK (2004) Tetrahedron: Asymmetry 15:77
65. Cornejo A, Fraile JM, García JI, García-Verdugo E, Gil MJ, Legarreta G, Luis SV, Martínez-Merino V, Mayoral JA (2002) Org Lett 4:3927
66. Trost BM (1991) Science 1471
67. Sheldon RA (1992) Chem Ind 903
68. Corey EJ (2002) Angew Chem Int Ed 41:1650
69. Kagan HB, Riant O (1992) Chem Rev 92:1007
70. Ghosh AK, Mathivanan P, Cappiello J (1998) Tetrahedron: Asymmetry 9:1
71. Johnson JS, Evans DA (2000) Acc Chem Res 33:325
72. Evans DA, Miller SJ, Lectka T (1993) J Am Chem Soc 115:6460
73. Evans DA, Murry JA, von Matt P, Norcross RD, Miller SJ (1995) Angew Chem Int Ed Engl 34:798
74. Davies IW, Gerena L, Cai D, Larsen RD, Verhoeven TR, Reider P (1997) Tetrahedron Lett 38:1145
75. Davies IW, Deeth RJ, Larsen RD, Reider PJ (1999) Tetrahedron Lett 40:1233
76. Ghosh AK, Cho H, Cappiello J (1998) Tetrahedron: Asymmetry 9:3687
77. Clariana J, Comelles J, Moreno-Mañas M, Vallribera A (2002) Tetrahedron: Asymmetry 13:1551
78. Zhou J, Tang Y (2002) J Am Chem Soc 124:9030
79. Zhou J, Tang Y (2004) Org Biomol Chem 2:429
80. Lipkowitz KB, Pradhan M (2003) J Org Chem 68:4648
81. Jorgensen KA (2000) Angew Chem Int Ed 39:3558
82. Jorgensen KA (2004) Eur J Org Chem 2093
83. Yao S, Johannsen M, Hazell RG, Jørgensen KA (1998) J Org Chem 63:118
84. Johannsen M, Jørgensen KA (1996) Tetrahedron 52:7321
85. Ghosh AK, Mathivanan P, Cappiello J, Krishnan K (1996) Tetrahedron: Asymmetry 7:2165
86. Johannsen M, Yao S, Jørgensen KA (1997) J Chem Soc, Chem Commun 2169
87. Yao S, Johannsen M, Audrain H, Hazell RG, Jørgensen KA (1998) J Am Chem Soc 120:8599
88. Thorhauge J, Johannsen M, Jørgensen KA (1998) Angew Chem Int Ed 37:2404
89. Evans DA, Olhava EJ, Johnson JS, Janey JM (1998) Angew Chem Int Ed 37:3372
90. Evans DA, Johnson JS (1998) J Am Chem Soc 120:4895
91. Palomo C, Oiarbide M, Garcia JM, Gonzalez A, Arceo E (2003) J Am Chem Soc 125:13942

92. Takacs JM, Lawson EC, Reno MJ, Youngman MA, Quincy DA (1997) Tetrahedron: Asymmetry 8:3073
93. Takacs JM, Quincy DA, Shay W, Jones BE, Ross CR II (1997) Tetrahedron: Asymmetry 8:3079
94. Yao S, Johannsen M, Jorgensen KA (1997) J Chem Soc, Perkin Trans 1:2345
95. Faller JW, Lavoie A (2001) J Organomet Chem 630:17
96. Davies DL, Fawcett J, Garratt SA, Russell DR (1997) Chem Commun 1351
97. Davenport AJ, Davies DL, Fawcett J, Garratt SA, Lad L, Russell DR (1997) Chem Commun 2347
98. Qian C, Wang L (2000) Tetrahedron Lett 41:2203
99. Fukuzawa SI, Matsuzawa H, Metoki K (2001) Synlett 709
100. Fukuzawa SI, Metoki K, Esumi SI (2003) Tetrahedron 59:10445
101. Desimoni G, Faita G, Guala M, Pratelli C (2003) J Org Chem 68:7862
102. Evans DA, Wu J (2003) J Am Chem Soc 125:10162
103. Otto S, Boccaletti G, Engberts JBFN (1998) J Am Chem Soc 120:4238
104. Lassaletta JM, Alcarazo M, Fernández R (2004) Chem Commun 298
105. Owens TD, Hollander FJ, Oliver AG, Ellman JA (2001) J Am Chem Soc 123:1539
106. Bolm C, Bienewald F, Harms K (1996) Synlett 775
107. Bolm C, Martin M, Simic O, Verrucci M (2003) Org Lett 5:427
108. Bolm C, Simic O (2001) J Am Chem Soc 123:3830
109. Bolm C, Martin M, Gescheidt G, Palivan C, Neshchadin D, Bertagnolli H, Feth M, Schweiger A, Mitrikas G, Harmer J (2003) J Am Chem Soc 125:6222
110. Bolm C, Verrucci M, Simic O, Cozzi PG, Raabe G, Okamura H (2003) Chem Commun 2826
111. Ichiyanagi T, Shimizu M, Fujisawa T (1997) J Org Chem 62:7937
112. van der Zeidjen AAH (1996) J Organomet Chem 518:147
113. Nishida A, Yamanaka M, Nakagawa M (1999) Tetrahedron Lett 40:1555
114. Rechavi D, Lemaire M (2002) Chem Rev 102:3467
115. Fraile JM, García JI, Harmer MA, Herrerías CI, Mayoral JA (2001) J Mol Cat A: Chem 165:211
116. O'Leary P, Krosveld NP, De Jong KP, van Koten G, Klein Gebbink RJM (2004) Tetrahedron Lett 45:3177
117. Rechavi D, Lemaire M (2001) Org Lett 3:2493
118. Rechavi D, Lemaire M (2002) J Mol Cat A: Chem 182–183:239
119. Annunziata R, Benaglia M, Cinquini M, Cozzi F, Pitillo M (2001) J Org Chem 66:3160
120. Hallman K, Moberg C (2001) Tetrahedron: Asymmetry 12:1475
121. Meracz I, Oh T (2003) Tetrahedron Lett 44:6465
122. Trost BM, Van Vranken DL (1996) Chem Rev 96:395
123. Consiglio G, Waymouth RM (1989) Chem Rev 89:257
124. Chelucci G, Caria V, Saba A (1998) J Mol Cat A: Chem 130:51
125. Chelucci G, Saba A (1998) Tetrahedron: Asymmetry 9:2575
126. Chelucci G, Pinna GA, Saba A (1998) Tetrahedron: Asymmetry 9:531
127. Chelucci G, Saba A, Sanna G, Soccolini F (2000) Tetrahedron: Asymmetry 11:3427
128. Chelucci G, Pinna GA, Saba A, Sanna G (2000) J Mol Cat A: Chem 159:423
129. Peña-Cabrera E, Norrby PO, Sjörgren M, Vitagliano A, De Felice V, Oslob J, Ishii S, O'Neil D, Åkermark B, Helquist P (1996) J Am Chem Soc 118:4299
130. Malkov AV, Baxendale IR, Bella M, Langer V, Fawcett J, Russell DR, Mansfield DJ, Valko M, Kocovsky P (2001) Organomet 20:673
131. Chelucci G, Medici S, Saba A (1997) Tetrahedron: Asymmetry 8:3183
132. Chelucci G, Gladiali S, Saba A (1999) Tetrahedron: Asymmetry 10:1393

133. Li ZP, Tang FY, Xu HD, Wu XY, Zhou QL, Chan ASC (2003) J Mol Cat A: Chem 193:89
134. Chelucci G, Pinna GA, Saba A, Valenti R (2000) Tetrahedron: Asymmetry 11:4027
135. Nordström K, Macedo E, Moberg C (1997) J Org Chem 62:1604
136. Bremberg U, Rahm F, Moberg C (1998) Tetrahedron: Asymmetry 9:3437
137. Bremberg U, Larhed M, Moberg C, Hallberg A (1999) J Org Chem 64:1082
138. Chelucci G (1997) Tetrahedron: Asymmetry 8:2667
139. Pericas MA, Puigjaner C, Riera A, Vidal-Ferran A, Gomez M, Jimenez F, Muller G, Rocamora M (2002) Chem Eur J 8:4164
140. Trost BM, Hachiya I (1998) J Am Chem Soc 120:1104
141. Dübner F, Knochel P (1999) Angew Chem Int Ed Engl 38:379
142. Luchaco-Cullis CA, Mizutani H, Murphy KE, Hoveyda AH (2001) Angew Chem Int Ed 40:1456
143. Gamez P, Dunjic B, Fache F, Lemaire M (1995) Tetrahedron: Asymmetry 6:1109
144. Bayardon J, Sinou D (2004) J Org Chem 69:3121
145. Malkoch M, Hallman K, Lutsenko S, Hult A, Malmström E, Moberg C (2002) J Org Chem 67:8197
146. Belda O, Lundgren S, Moberg C (2003) Org Lett 5:2275
147. Reggelin M, Schultz M, Holbach M (2002) Angew Chem Int Ed 41:1614

Top Organomet Chem (2005) 15: 149–190
DOI 10.1007/b136350
© Springer-Verlag Berlin Heidelberg 2005
Published online: 10 September 2005

Non-covalent Immobilization of Catalysts Based on Chiral Diazaligands

José M. Fraile · José I. García · José A. Mayoral (✉)

Departamento de Química Orgánica, Instituto de Ciencia de Materiales de Aragón,
Instituto Universitario de Catálisis Homogénea, Facultad de Ciencias,
Universidad de Zaragoza–C.S.I.C., 50009 Zaragoza, Spain
mayoral@unizar.es

Abstract The immobilization of chiral catalysts through non-covalent methods, as opposed to covalent immobilization, allows an easier preparation of chiral heterogeneous catalysts with, in principle, less influence of the support on the conformational preferences of the catalytic complex. In this review the different possibilities for immobilization without forming a covalent bond between the chiral diazaligand and the support, which can be either solid or liquid, are presented.

Keywords Asymmetric catalysis · Salen · Biphasic liquid phase · Bis(oxazoline) · Immobilization

1 Introduction

Heterogeneous catalysts, in the general sense of catalysts placed in a phase different from that of the reagents and products, present clear advantages from a practical point of view, including ease of recovery and potential recycling and reuse. The latter point is especially important when the catalyst cost is high, as is the case for chiral catalysts [1].

In the last 20 years a great deal of effort has been focused towards the immobilization of chiral catalysts [2] and disparate results have been obtained. In order to ensure the retention of the valuable chiral ligand, the most commonly used immobilization method has been the creation of a covalent bond between the ligand and the support, which is usually a solid. In many cases this strategy requires additional functionalization of the chiral ligand, and this change – together with the presence of the very bulky support – may produce unpredictable effects on the conformational preferences of the catalytic complex. This in turn affects the transition-state structures and thus the enantioselectivity of the process.

On the basis of these results it is desirable to develop immobilization methods that do not require modification of the chiral ligand, in order to preclude these effects on enantioselectivity. This goal can be achieved by employing non-covalent immobilization, although in certain instances modification of the ligand is required in order to facilitate the immobilization. In this review we present the possibilities developed for non-covalent immobilization, in the general sense, of chiral catalysts based on diazaligands. It must be stressed that there is a self-imposed limit to the definition of diazaligands: We have considered only ligands that use two or more nitrogen atoms *to coordinate* with the metal center. We have expressly excluded those ligands where the nitrogens are used as auxiliary groups for immobilization or in which the presence of nitrogens arises from the availability of the starting material, as in the case of cinchona alkaloids.

2 Methods for Non-covalent Immobilization

Bearing in mind that most asymmetric reactions take place in the liquid phase, we have considered two general types of heterogeneous systems: a liquid phase that is immiscible with the reaction phase and a solid phase.

2.1
Liquid–Liquid Systems

In these systems the retention of the catalyst in the liquid phase is determined by the relative solubility of the complex in the two solvents and most of the problems associated with this approach concern this point. In order to overcome these limitations, modifications of the chiral ligand are introduced that tend to increase the solubility of the complex in the new liquid phase.

The most important biphasic liquid systems are probably those that combine a "conventional" organic phase with another type of solvent, such as water, a fluorous organic solvent, or an ionic liquid [3]. In those cases the solvent can be considered as the "support" for the catalyst phase and we have therefore limited the examples in this review to those where the recycled liquid catalyst phase is recovered as a whole.

2.2
Solid–Liquid Systems

In the case of solid supports, the immobilization methods can be classified according to the support–complex interaction. However, this is not a simple classification given that more than one interaction can be present and it is often difficult to discern which interaction is the most significant in terms of immobilization. In this review, however, we have distinguished three categories of interaction [4] (Fig. 1):

Adsorption
This involves deposition of the catalyst on the accessible surface of the support. The complex is held on the surface by rather weak interactions, such as van der Waals' forces, hydrogen bonding, or a donating bond. The stability of the complex on the surface is determined by its solubility in the reaction solvent and/or the complexation of reagents and products.

Entrapment
The catalysts must be located in the inner part of the support and are retained because their size is larger than the pores of the solid. The complex

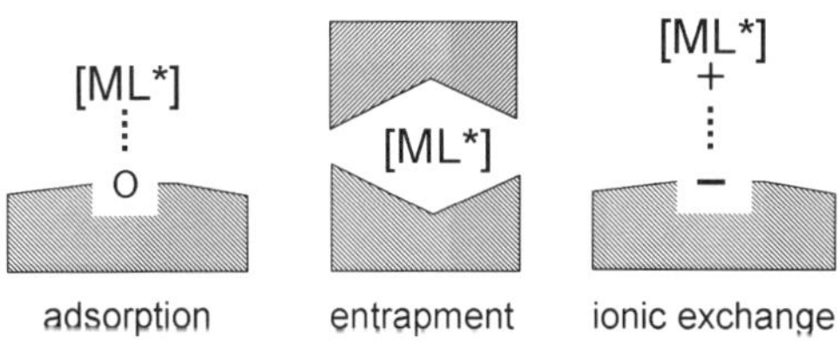

Fig. 1 Methods for immobilization on solid supports

is introduced "in small pieces" and built inside the support ("ship-in-bottle" method), or the support is prepared around the catalyst. Rigid supports, such as zeolites, have limitations with regard to the size of the catalytic complex and reagents. Flexible supports, such as polymers, do not have a permanent porosity and retention of the catalyst is therefore highly dependent on the solvent.

Electrostatic interaction
In principle this is the method that gives rise to the strongest support–complex interaction. We have considered in this category all the methods in which the support compensates for at least one of the charges of the complex, usually due to the metal, although without considering the exact nature of the metal–support bond, i.e., purely ionic or polarized covalent. In any case, the only possible covalent bond between support and complex would be established with the metal center, not with the chiral ligand.

3 Non-covalently Immobilized Catalysts Based on Chiral Salen Ligands

Chiral salen ligands are diimines of salicylaldehydes with chiral diamines, usually cyclohexane-1,2-diamine (salen ligands **1**) or 1,2-diphenylethylenediamine (salen ligands **2**). The most widely used salen ligand in homogeneous catalysis is probably Jacobsen's ligand (**1a**, Fig. 2), which is commercially available and hence has been used as reference to compare the results of im-

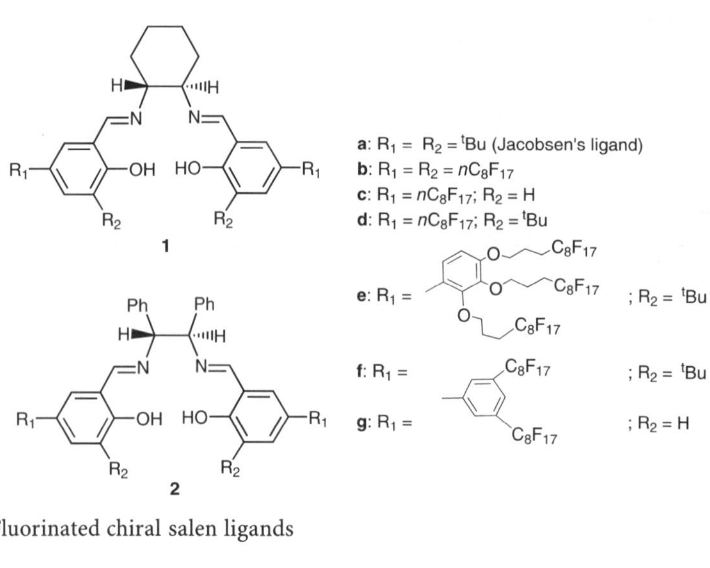

Fig. 2 Fluorinated chiral salen ligands

mobilization of complexes with analogous ligands. Examples can be found in the literature of liquid and solid phase immobilization of transition metal-salen complexes using all the available immobilization techniques, including the non-covalent methods.

3.1 Liquid Phase Immobilization

3.1.1 Fluorinated Solvents

Only a few years after the development of the homogeneous chiral Mn(salen) complexes by Jacobsen and Katsuki, several research groups began to study different immobilization methods in both liquid and solid phases. Fluorinated organic solvents were the first type of liquid supports studied for this purpose. The main problem in the application of this methodology is the low solubility of the catalytic complex in the fluorous phase. Several papers were published by Pozzi and coworkers, who prepared a variety of salen ligands with perfluorinated chains in positions 3 and 5 of the salicylidene moiety (Fig. 2).

Compounds **1b** and **2b** were the first fluorinated ligands tested in Mn-catalyzed alkene epoxidation [5, 6]. The biphasic liquid system perfluorooctane/dichloromethane led to excellent activity and enantioselectivity (90% ee) in the epoxidation of indene with oxygen and pivalaldehyde (Scheme 1, Table 1). In addition, the fluorous solution of the catalyst was reused once and showed the same activity and selectivity. This represents a considerable improvement over the behavior in the homogeneous phase, where the used catalyst was bleached and reuse was impossible. Unfortunately, indene was the only suitable substrate for this system, which failed to epoxidize other alkenes (such as styrene or 1,2-dihydronaphthalene) with high enantioselectivity. The system was also strongly dependent on the oxidant and only 71% ee was obtained in the epoxidation of indene with *m*CPBA at – 50 °C.

The limitations of the system with regard to substrates and oxidants was attributed to the strong electron-withdrawing character of the perfluorinated chains and the lower steric hindrance in the position adjacent to phenols, in marked contrast to the *tert*-butyl groups present in Jacobsen's catalyst. In view of this, a second generation of fluorinated salen ligands **1e** and **1f** was

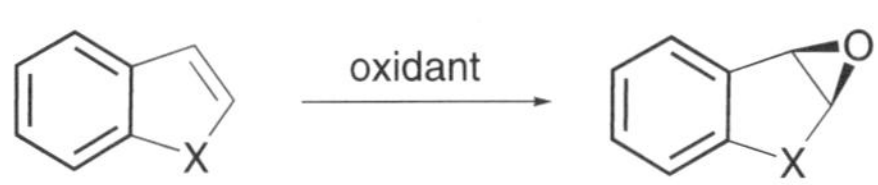

indene: X = $-CH_2-$
1,2-dihydronaphthalene: X = $-(CH_2)_2-$
2,2-dimethylchromene: X = $-OCMe_2-$

Scheme 1

Table 1 Results of the alkene epoxidation reactions with fluorinated (salen)Mn complexes under biphasic conditions

Catalyst	Substrate	Oxidant	T (°C)	Yield[a] (%)	ee[a] (%)
1b-MnCl	Indene	$O_2/^t$-BuCHO	20	83-73	92-89
2b-MnCl	Indene	$O_2/^t$-BuCHO	20	77-75	90-92
2b-MnCl	DHNapht[b]	$O_2/^t$BuCHO	20	70	13
1f-$MnOOCC_7F_{15}$	Indene	PhIO/PNO	100	98-95-93-76	92-92-93-79
1f-$MnOOCC_7F_{15}$	DHNapht[b]	PhIO/PNO	100	77-67-63-19	50-45-46-40
1f-$MnOOCC_7F_{15}$	TPhEt[c]	PhIO/PNO	100	98-96-92-80	87-85-83-71
1f-$MnOOCC_7F_{15}$	MeChex[d]	PhIO/PNO	100	91-87-76-45	58-53-40-25

[a] Numbers separated by dashes indicate results in successive reuses
[b] 1,2-Dihydronaphthalene
[c] Triphenylethylene
[d] 1-Methylcyclohexene

prepared [7,8]. These ligands bore the bulky *tert*-butyl groups in position 3 and required the inclusion of additional perfluorinated chains in position 5 in order to increase the fluorinated/organic partition coefficient. This aspect was also improved by exchanging the original chloride of the catalyst with a perfluorinated carboxylate.

All of the factors in the epoxidation reaction were optimized, from the solvent (perfluorooctane/acetonitrile) to the oxidant (iodosylbenzene and pyridine *N*-oxide). Under such conditions, the efficiency of the catalyst, both in terms of activity and selectivity, increased with reaction temperature, which is opposite to the behavior observed in the homogeneous phase. A plausible explanation is that, although remaining immiscible, the solubility of each solvent in the other is higher at higher temperature, thus improving the mass transfer between the two liquid phases. Enantioselectivities in the range of 50–92% ee were obtained with a wide variety of alkenes (Table 1), including indene, 1,2-dihydronaphthalene, triphenylethylene, and even non-aromatic alkenes such as 1-methylcyclohexene. Recycling was also improved with this method. Up to four reactions were possible, although in all cases a gradual decrease in activity and enantioselectivity was observed. As in other immobilization methods, this effect was due to partial degradation of the chiral ligand under the reaction conditions, with the extent of degradation seeming to be dependent on the nature of the alkene and the oxidant.

The application of this biphasic system to the asymmetric oxidation of arylmethylsulfides [9] did not lead to such successful results. Conversions (78–100%) and selectivities to sulfoxide (88–99%) were excellent, much better in general than in homogeneous phase, but enantioselectivities were always very low (up to 17% ee).

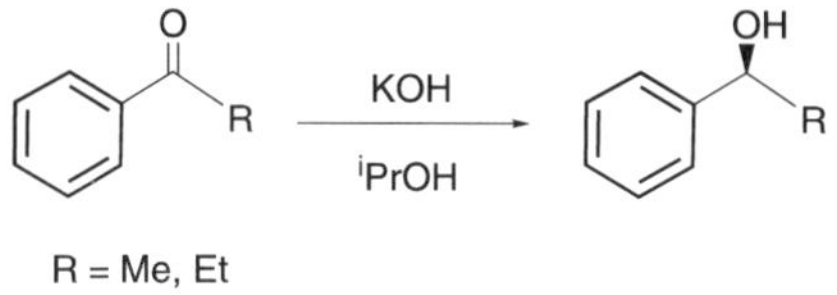

Scheme 2

The same authors tried the use of salen complexes with other transition metals in asymmetric reactions other than oxidation. The iridium complexes of fluorinated salen ligands **1b**, **2c**, **1d**, **2d**, and **1f** (Fig. 2) were active in the transfer hydrogenation of arylketones (Scheme 2) under biphasic conditions (perfluorooctane/isopropanol) [10]. Besides the modest enantioselectivity (e.g., 56% ee with acetophenone) the most significant problem concerned recycling of the catalyst (Table 2). Significant leaching of Ir to the organic phase was demonstrated by the high catalytic activity of that solution. Partial hydrolysis of the chiral ligand has been proposed as the reason for this leaching. In view of this situation, amine **4** – an analog of salen **1b** – and the corresponding imine **3** and amine **5** without phenol groups (Fig. 3) were prepared and tested in the same reaction [11]. The best ligand with regard to all the reaction parameters was amine **5** (Table 2). Not only did this system lead to the highest catalytic activity and enantioselectivity (69% ee), but it also showed the best behavior in terms of recycling, with a very low level of Ir leaching (4% after the first run) and good activity and enantioselectivity even in the fourth run. The higher stability of the ligand under the strongly basic conditions seems to be responsible for this improved recyclability.

Another application of salen ligands is the hydrolytic kinetic resolution of epoxides (Scheme 3). For this purpose cobalt complexes are efficient, and flu-

Table 2 Results of the transfer hydrogenation of ketones with fluorinated (salen)Ir complexes under biphasic conditions[a]

Ligand	Substrate	Run	Time (h)	Conversion (%)	ee (%)
2d	Acetophenone	1	24	84	56
		2	21	37	6
2d	Propiophenone	1	24	97	60
3	Acetophenone	1	24	93	47
4	Acetophenone	1	5	95	23
5	Acetophenone	1	0.5	92	69
		2	0.5	90	79
		3	1	86	59
		4	2	69	58

[a] Iridium precursor $[Ir(COD)Cl]_2$. T 70 °C. Phase separation at 0 °C

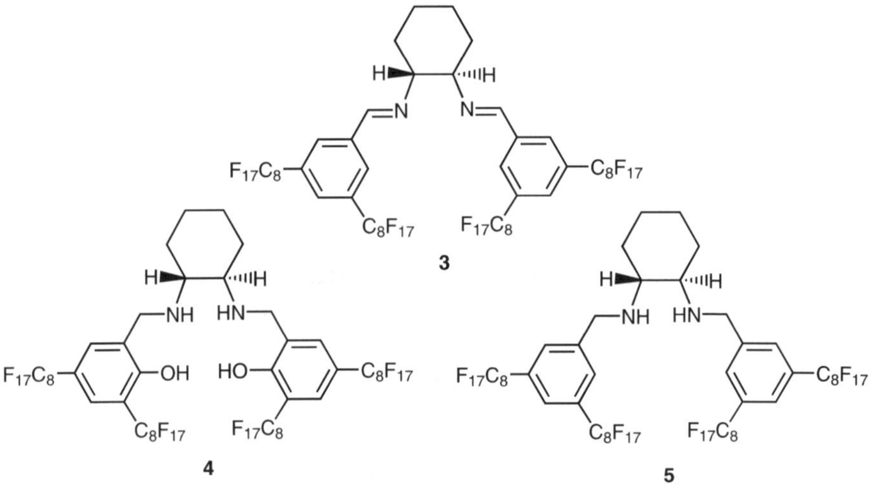

Fig. 3 Fluorinated chiral salen-analogous ligands

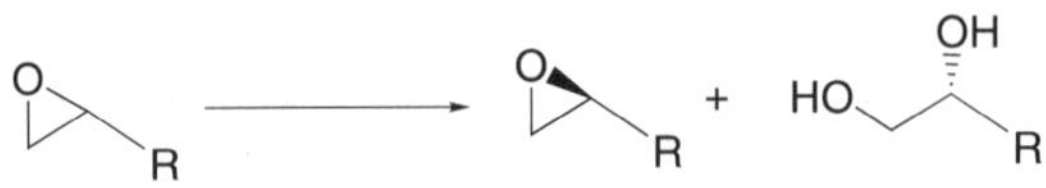

Scheme 3

orinated salen ligands were also tested. The Co^{II} complexes of **1b**, **1d**, and **2d** were oxidized with air in the presence of acetic acid but, unfortunately, the Co^{III} complexes were not soluble in perfluoroalkanes [12]. Even when the oxidation was carried out in the presence of $C_8F_{17}COOH$, the complexes were only sparingly soluble in perfluoroalkanes. However, the complexes were unexpectedly soluble in neat 1-hexene oxide, making them unsuitable for work under biphasic conditions. This problem was solved by using ligand **1f** (Fig. 2), the Co^{III} complex of which was prepared in the presence of $C_8F_{17}COOH$ and was found to be soluble in perfluorooctane but insoluble in terminal epoxides [13]. The results were excellent for aliphatic terminal epoxides, with total (50%) conversion and only one enantiomer detectable for both the diol and remaining epoxide. Conversion was not as good in the case of styrene oxide, which gave excellent enantioselectivity (97% ee) in the diol but only 31% ee in the epoxide. However, it proved impossible to recover the fluorous phase as the complex precipitated during the reaction as a Co^{II} – Co^{III} mixture. It was possible to filter and reoxidize the complex three more times, with slight loss of activity and, consequently, lower ee in the remaining epoxide.

3.1.2 Ionic Liquids

As outlined above, immobilization in a fluorinated liquid phase demands the functionalization of the ligand with perfluoroalkyl chains and, even then, the solubility is strongly influenced by the nature of the complex. Ionic liquids of the alkylmethylimidazolium type (Fig. 4) have been recently developed as alternative solvents for organometallic catalysis and have the practical advantage of using directly the commercially available chiral ligands and complexes.

The first application of ionic liquids for salen complexes dealt with the epoxidation of alkenes [14]. Jacobsen's Mn complex was immobilized in [bmim][PF_6] and different alkenes were epoxidized with aqueous NaOCl solution at 0 °C. As the ionic solvent solidified at this temperature, dichloromethane was used as a cosolvent. The recycling procedure consisted of washing with water, evaporation of dichloromethane, and product extraction with hexane. The results (Table 3) were excellent and only a slow decay in activity and enantioselectivity was detected after several cycles.

The analogous chromium complex was used in the asymmetric ring opening of *meso* epoxides with trimethylsilyl azide [15] (Scheme 4). In this case a strong dependence on the anion of the ionic liquid was observed. Anions leading to hydrophobic ionic liquids, such as PF_6^- and SbF_6^-, led to high

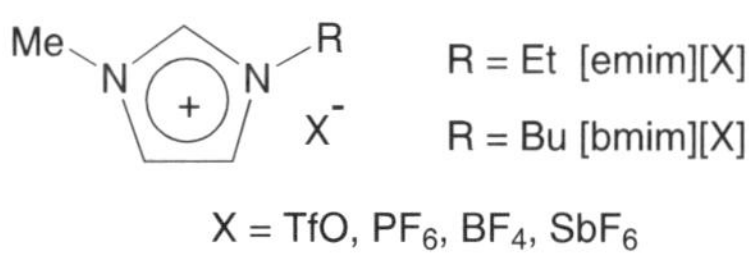

Fig. 4 Ionic liquid solvents

Table 3 Results of reactions with **1g**-metal complexes in ionic liquids

Catalyst	Substrate	Reaction	Yield[a] (%)	ee[a] (%)
1a-MnCl	2,2-Dimethylchromene	Epoxidation[b]	86-73-73-60-53	96-90-90-89-88
1a-CrCl	Cyclopentene oxide	Ring opening[c]	68-72-85-75-76	94-93-93-94-93
1a-Co	Epichlorohydrin	HKR[d]	53 (10 times)	>99 (10 times)
1a-VO	Benzaldehyde	Cyanation[e]	85-79-89-80-83	89-88-90-88-89

[a] Numbers separated by dashes indicate results in successive reuses
[b] In [bmim][PF_6] (CH_2Cl_2 cosolvent) at 0 °C Oxidant NaOCl
[c] In [bmim][PF_6] + [bmim][OTf] mixture at 20 °C. Reagent $TMSN_3$
[d] Hydrolytic kinetic resolution in [bmim][NTf_2] at 20 °C
[e] In [emim][PF_6] at room temperature. Reagent TMSCN

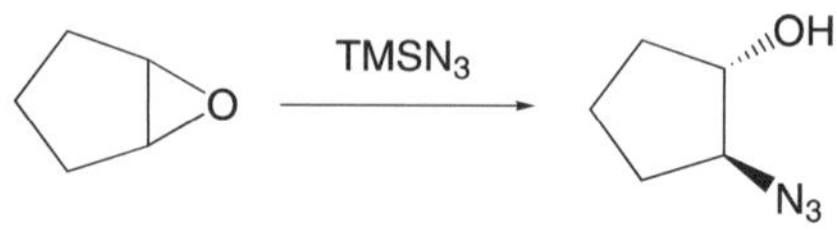

Scheme 4

yields (75%) and enantioselectivities (87–94% ee). In contrast, in hydrophilic ionic liquids, with anions such as BF_4^- and TfO^-, the reaction did not take place at all. However, the solubility of the catalyst was much higher in the latter system, which made recycling of the complex far easier. In view of these findings the authors decided to test a mixture of one hydrophobic and one hydrophilic ionic liquid as the reaction solvent and excellent results were obtained in five consecutive runs (Table 3).

The same research group explored the application of the analogous cobalt complexes in the hydrolytic kinetic resolution of terminal epoxides (Scheme 3) [16]. In the case of epichlorohydrin the diol was obtained with 91% ee, whereas the remaining epoxide was enantiopure. The main problem with this system is the ease of the $Co^{II} - Co^{III}$ redox reaction, which had already been detected in fluorinated solvents. However, ionic liquids are advantageous in this sense because both types of complex are soluble and in this way the recycling process is easier. Another advantage is that the Co^{II} complex, the catalyst precursor, is oxidized in the ionic liquid without addition of a carboxylic acid, which again contributes to the efficient recycling. Indeed the ionic liquid solution is more active after each recycling stage, probably a consequence of an increase in the active Co^{III} species caused by successive oxidations.

Finally, the vanadyl complex was tested in the trimethylsilylcyanation of aldehydes (Scheme 5) [17] and a significant effect of the anion in the solvent

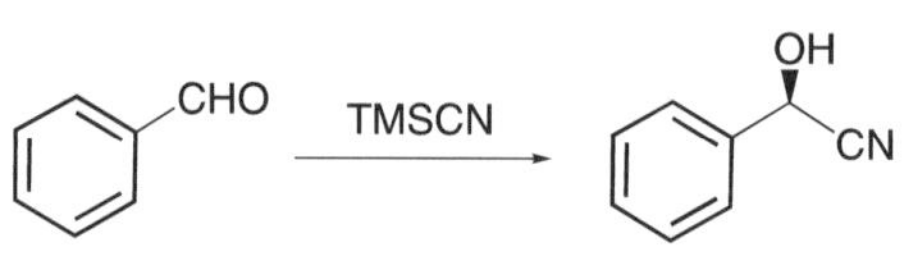

Scheme 5

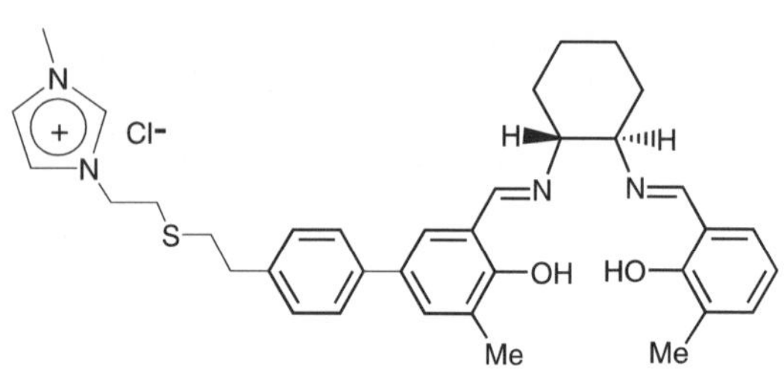

Fig. 5 Chiral salen ligand with improved solubility in ionic liquids

was found. Although the recycling was good (Table 3), some complex leaching to the hexane extraction phase was detected [18].

In an attempt to increase the ionic liquid/hexane partition coefficient, a new salen ligand appended with an imidazolium salt was developed (Fig. 5) [18]. Unfortunately, modification of the ligand caused a dramatic reduction in the enantioselectivity - down to 57% ee at most - although the reasons for this behavior remain unclear.

3.2 Solid Phase Immobilization

3.2.1 Entrapment

Entrapment into solid matrices was the first system employed to immobilize salen-based complexes as it was considered advantageous that it was unnecessary to modify the original Jacobsen's ligand (**1a**). The (salen)MnCl complex was occluded in a polysiloxane membrane during the cross-linking process in the membrane synthesis (Fig. 6) [19]. With this supported catalyst, the alkene epoxidation reaction was carried out in a countercurrent membrane reactor using an aqueous NaOCl solution as the oxidant. The best results were obtained in the styrene epoxidation, which gave 84% conversion with 80% selectivity to epoxide and 52% ee. The same results were obtained in two recycles and only low levels of Mn leaching from the membrane were found. However, these encouraging initial results were not supported in a subsequent paper [20], where Mn leaching in the range of 8–78% was reported and depended on the alkene and reaction solvent. This problem is due to the swelling of the polysiloxane membrane and the solubility of the (salen)MnCl complex in the reaction solvent. Chlorobenzene, diethyl ether, and dichloromethane (84–100% leaching) were not suitable solvents due

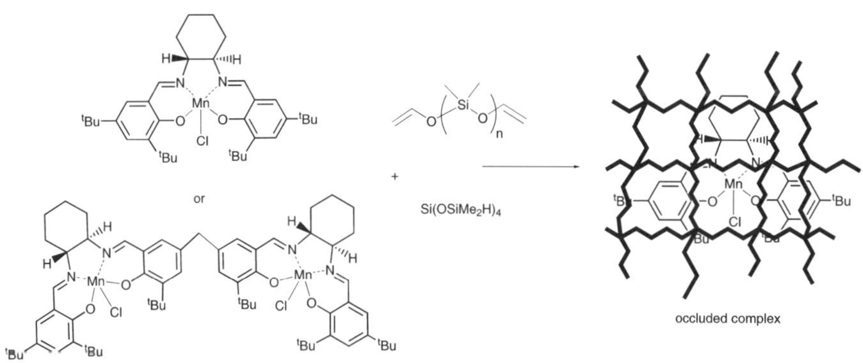

Fig. 6 Method of (salen)MnCl entrapment into polysiloxane membranes

to their high swelling ability. Acetone, acetonitrile, and methanol (54–62% leaching) did not swell the membrane to the same extent but the high solubility of the complex in these solvents made them unsuitable. Only heptane was an acceptable solvent (12% leaching) due to the very low complex solubility. A dimeric salen ligand (Fig. 6) was prepared in an attempt to prevent leaching on the basis of the greater size of the complex. Leaching was considerably reduced in acetone, acetonitrile, and methanol (16–56% leaching) but the problem was not completely solved. These results showed that solvent compatibility is the main problem associated with the occlusion approach.

The other strategy for entrapment is the "ship-in-a-bottle" synthesis of the chiral complex within the supercages of the pore system of a zeolite (Fig. 7). It was noted that the size of the commonly used Jacobsen's complex was too big to fit into the micropores of the zeolites and less bulky ligands were therefore prepared for immobilization (Fig. 8). In the first example [21] the chiral salen ligand **1h** was prepared within the pores of EMT zeolite by reaction between the chiral diamine and the corresponding salicylaldehyde. The resulting ligand was subsequently complexed with Mn^{II}, which was at the same time oxidized to Mn^{III} in air (method A). In a second example [22] the chiral ligand **1i** was prepared around the Mn^{II} ions, which had previously been exchanged in Y zeolite, also by reaction of the diamine and salicylaldehyde, and the complex was finally oxidized in air (method B). In a more recent

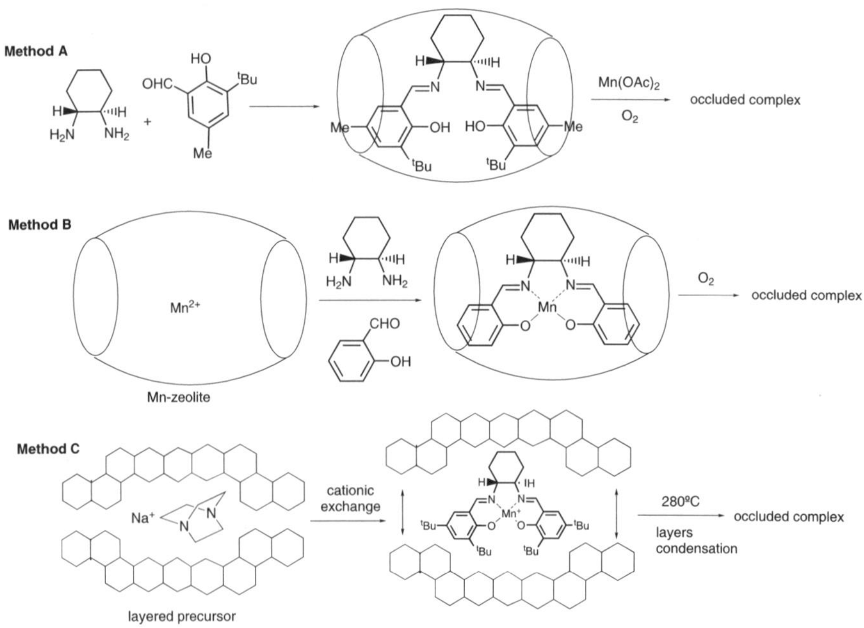

Fig. 7 Methods for "ship-in-bottle" synthesis of (salen)Mn complexes inside the pores of zeolites

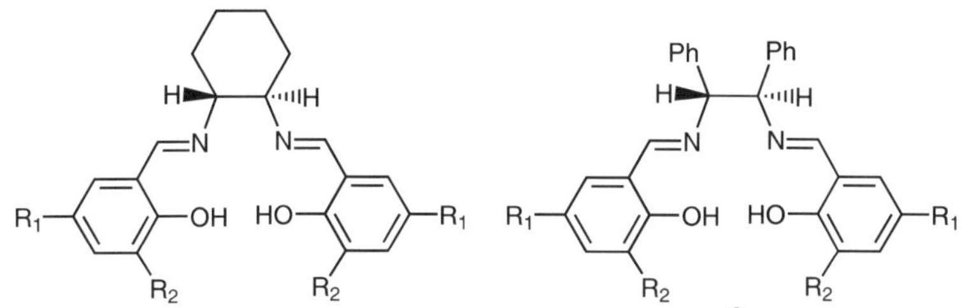

1 **2**

a: R_1 = R_2 = ^{t}Bu (Jacobsen's ligand)
h: R_1 = Me; R_2 = ^{t}Bu
i: R_1 = R_2 = H
j: R_1 = H; R_2 = ^{t}Bu
k: R_1 = H; R_2 = OMe
l: R_1 = Me; R_2 = CPh_3

Fig. 8 Symmetrical chiral salen ligands used in solid phase immobilization

study [23] Jacobsen's complex was entrapped in the final step of the zeolite synthesis (method C). This process was possible because MCM-22 zeolite is prepared by condensation of a layered precursor, which is exchangeable by the catalytic complex. Leaching of Mn was not observed in these systems, which is not unexpected bearing in mind that the complex is also bound to the zeolite structure through an electrostatic interaction.

The results obtained in reactions involving the two first examples showed a reduced catalytic activity compared to the homogeneous catalyst, a situation that may be due to diffusion problems. Enantioselectivity was similar or slightly lower than in solution, with 80% ee [21] and 58% ee [22] in the epoxidation of *cis*-β-methylstyrene with NaOCl providing the best results. Only in the last example was an improvement in enantioselectivity reported: from 51% to 91% ee in the epoxidation of α-methylstyrene. Recovery of the catalyst was only considered in one case [21] and a significant decrease in enantioselectivity was observed on reuse.

In one case, the insertion of the whole chiral ligand into a Co-exchanged zeolite by sublimation was described [24]. Only small ligands, such as **1i** and **2i**, can be efficiently introduced into the micropores of the Y zeolite, whereas the bulkier Jacobsen's ligand **1a** only remains on the external surface of the solid. Unfortunately, these occluded (salen)Co complexes led to very low enantioselectivities (up to 8% ee) in the reduction of acetophenone with $NaBH_4$.

Method B was also used in the preparation of occluded (salen)Cr^{III} complexes. Ligands **1h** and **1i** were prepared within the pores of Cr^{3+}-exchanged EMT and Y zeolites, respectively [25]. These complexes were tested as catalysts in the ring opening of *meso*-epoxides with trimethylsilyl azide (Scheme 4). The occluded complexes showed a dramatic decrease in catalytic activity, selectivity to reaction products and enantioselectivity. Up to 16% ee was obtained with the same complex leading to 68% ee in solution.

Electrochemical studies, in combination with EPR measurements, of the analogous non-chiral occluded (salen)Mn complex in Y zeolite showed that only a small proportion of the complex, i.e., that located on the outer part of the support, is accessible and takes part in the catalytic process [26]. Only this proportion (about 20%) is finally oxidized to Mn^{III} and hence the amount of catalyst is much lower than expected. This phenomenon explains the low catalytic activity of this system. We have considered other attempts at this approach using zeolites with larger pore sizes as examples of cationic exchange and these have been included in Sect. 3.2.3.

3.2.2 Adsorption

In principle the solid supports can act as additional ligands for the catalytic complexes, either with their own groups, such as siloxane or silanol in silica, or with groups tethered to the surface. The strength of this dative bond depends not only on the group and the complex, but also on the catalytic reaction, given that solvent, reagents, products, or by-products may also act as ligands, replacing the support and causing leaching.

In spite of these limitations, three examples of (salen)-metal complex adsorption have been described. In the first one, Jacobsen's complex (**1a**-MnCl) was adsorbed on Al-MCM-41 [27] by impregnation with a solution of the complex in dichloromethane, an approach that prevents the possible cationic exchange. The results in the epoxidation of 1,2-dihydronaphthalene with aqueous NaOCl were comparable to those obtained in solution, with only a slight reduction in enantioselectivity (55% ee instead of 60% ee). However, recycling of this catalyst was not described.

In the other two examples, (salen)Cr complexes were immobilized. Aminopropyl-MCM-41 was used as a support for the immobilization of a (salen)CrCl complex derived from $1,1'$-binaphthyl-$2,2'$-diamine (Fig. 9) [28]. The primary amino group was introduced on the support to act as a supplementary ligand for chromium. The supported complex was tested as a catalyst in the epoxidation of alkenes with iodosylbenzene. Good results were obtained in terms of catalytic activity, with TON in the range of 50–250, and moderate to good enantioselectivity (54–67% ee) was observed in the epoxidation of substituted styrenes and *cis*-β-methylstyrene. An improvement in enantioselectivity with respect to homogeneous catalysis was claimed, but the results were not directly comparable given the possible effect of axial nitrogen ligands on the performance of this type of catalyst. The catalyst was recycled several times (Fig. 10) but it progressively lost both activity and enantioselectivity – as reported in other cases. In spite of this problem, this catalyst

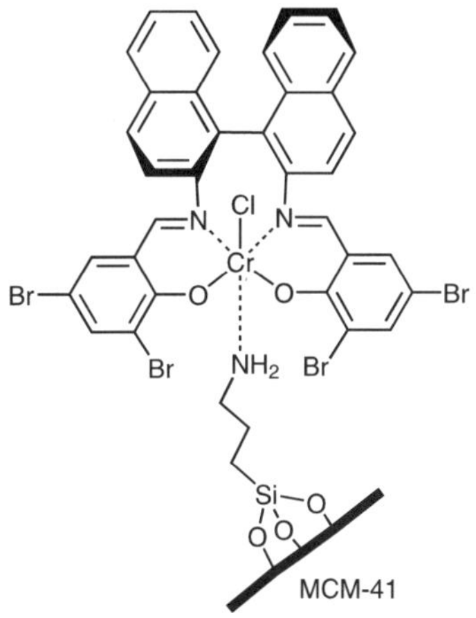

Fig. 9 (Salen)CrCl complex immobilized on aminopropyl-MCM-41

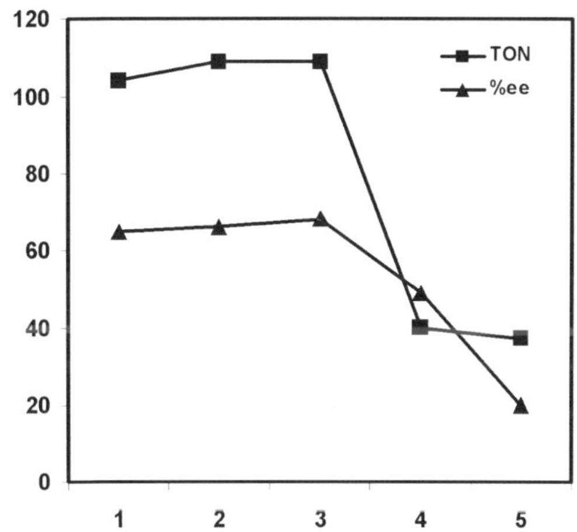

Fig. 10 Recycling results for the immobilized complex shown in Fig. 9 in the epoxidation of *cis*-β-methylstyrene

was able to afford more than 300 TON with 67% ee after three runs, which represents a very high productivity level for this type of reaction.

Very recently the complex **1a**-CrCl was directly adsorbed onto silica by impregnation [29]. This solid was used in the kinetic resolution of epoxides by ring opening with trimethylsilyl azide. In the case of 1,2-epoxyhexane, the catalyst led to excellent enantioselectivities for the remaining epoxide over ten cycles (82–93% ee), while maintaining high conversion and good enantioselectivity in the resulting azidoalcohol (60–71% ee). Cr leaching was observed in the first three cycles (0.59–1.83%) but it remained below 0.2% in the subsequent five cycles. Mechanical abrasion was proposed as the main reason for the increase in leaching observed in the last two cycles.

3.2.3
Ion Exchange

3.2.3.1
Complexes with the Charge on the Metal

As indicated above, the cationic nature of (salen)-metal complexes can be used to bind them onto the surface of anionic supports after a cationic exchange process. This method was first used with clays [30]. After several tests with a non-chiral (salen)Mn complex, synthetic laponite was chosen as the best support for this type of catalyst. Jacobsen's complex was immobilized in this way and the resulting catalyst led to moderate enantioselectivity (34% ee) in the epoxidation of 1,2-dihydronaphthalene (Scheme 1) with iodosylbenzene. This result represented a slight reduction from the value obtained in solution (46% ee) but was consistent with the use of an additional axial ligand, such as pyridine, and showed the possible role of the clay sheets not only as anions but also as ligands for the catalytic complex. Probably the most relevant finding in this paper was the demonstration of complex degradation under the reaction conditions – a process that is responsible for the loss of catalytic activity and enantioselectivity in this and probably other immobilized catalysts. The IR spectrum of the recycled catalyst showed that most of the bands corresponding to the initial complex were lost, whereas the N/Mn ratio remained constant, thus demonstrating the absence of ligand leaching.

Given the obvious size limitations for (salen)-metal complexes in the micropore systems of zeolites, dealuminated faujasites prepared by a combination of $SiCl_4$ treatment and steaming were used as supports [31–33]. It was claimed that the newly formed mesopores were completely surrounded by micropores, which retain the complex prepared by the ship-in-a-bottle method in the larger mesopore cavities. However, some uncertainty remains about the nature of the complex–support interaction. Firstly, several metals (e.g., V, Fe, Mn, Cr, Co, or Rh) were used but without reference to their oxidation state. As all of these metals were used in the epoxidation of terpenes with molecular oxygen, it is supposed that they were in the higher oxidation state, which means that the complexes, at least those of Co^{III}, Fe^{III}, Cr^{III}, and Mn^{III}, were in their cationic form. The complexes of ligands **1a**, **1h–1j**, **2a**, and **2h–2j** (Fig. 8) with all of the above metals were tested, both in homogeneous and heterogeneous phases, in the epoxidation of (*R*)-(+)-limonene and (–)-α-pinene (Scheme 6) with molecular oxygen (10–40 bar) and pivalaldehyde in fluorobenzene. The best results were obtained with Mn and Co complexes in the (–)-α-pinene epoxidation, with up to 100% conversion, 96% selectivity, and > 90% de. Recycling was proven to be possible, at least in the epoxidation of (*R*)-(+)-limonene, and metal leaching was not detected.

Although these examples show the possible immobilization on clays and mesoporous zeolites, the most widely used support for salen complexes has

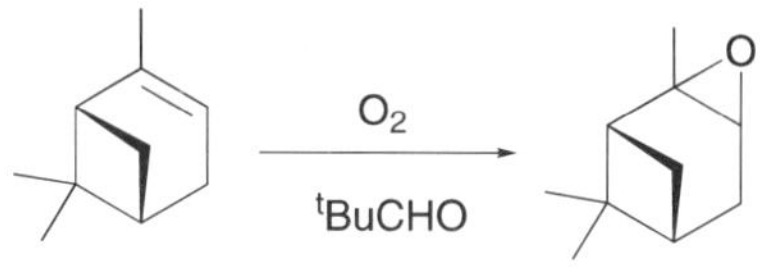

Scheme 6

been the mesoporous material MCM-41, either in the form of Al-MCM-41 to generate negative charges in the framework or modified with organic anionic groups.

The mesoporous character of MCM-41 overcomes the size limitations imposed by the use of zeolites and it is possible to prepare the complex by refluxing the chiral ligand in the presence of Mn^{2+}-exchanged Al-MCM-41 [34–36]. However, this method only gives 10% of Mn in the form of the complex, as shown by elemental analysis, and good results are only possible due to the very low catalytic activity of the uncomplexed Mn sites. The immobilized catalyst was used in the epoxidation of (*Z*)-stilbene with iodosylbenzene and this led to a mixture of *cis* (meso) and *trans* (chiral) epoxides. Enantioselectivity in the *trans* epoxides was up to 70%, which is close to the value obtained in solution (78% ee). However, this value was much lower when (*E*)-stilbene was used (25% ee). As occurred with other immobilized catalysts, reuse of the catalyst led to a significant loss in activity and, to a greater extent, in enantioselectivity.

The new ligands **2k** and **2l** (Fig. 8) were synthesized and immobilized in MCM-41 using two methods: direct exchange of the (salen)$MnPF_6$ complex or treatment of Mn^{II}-MCM-41 with the chiral ligand and subsequent oxidation of Mn [37]. The behavior of the two types of immobilized catalysts was not markedly different, although those prepared by direct exchange showed slightly higher activity and selectivity in the epoxidation of styrene and α-methylstyrene with *m*-CPBA in the presence of *N*-methylmorpholine *N*-oxide (Table 4). Better results were obtained with ligand **2l** and this is thought to be due to the bulkiness of the triphenylmethyl substituent in position 3 of the salicylaldehyde. It is worth noting that the immobilized catalyst led to better enantioselectivity than the homogeneous one, especially at 0 °C, and excellent enantioselectivity (86% ee) was obtained at – 80 °C. In spite of the fact that the results with α-methylstyrene were not as good, the effect of immobilization on enantioselectivity was again observed. Recycling was attempted with one catalyst and it was shown that three consecutive reactions with the same activity and enantioselectivity could be performed. The same authors described the synthesis of analogous non-symmetrical salen ligands (Fig. 11) [38]. As expected, ligand **1m**, which is an analog of Jacobsen's ligand **1a**, was less efficient in the epoxidation of styrene but, surprisingly, was better for α-methylstyrene (Table 4).

Jacobsen's complex (**1a**-MnCl) was also immobilized on MCM-41 modified with phenolic groups [39]. The surface phenoxide groups act as an-

Table 4 Results of the alkene epoxidation reactions with *m*-CPBA catalyzed by (salen)Mn complexes immobilized on MCM-41 by cationic exchange

Ligand	Catalysis	Substrate	T (°C)	Yield[a] (%)	ee[a] (%)
2l	Homogeneous	Styrene	0	95	56
	Heterogeneous[b]	Styrene	0	82	70
	Heterogeneous[b]	Styrene	−80	75	86
	Heterogeneous[c]	Styrene	0	76-78-75	66-64-64
	Heterogeneous[c]	α-Methylstyrene	−80	47	54
1a	Heterogeneous[b]	α-Methylstyrene	0	81	32
1m	Heterogeneous[b]	α-Methylstyrene	0	83	55

[a] Numbers separated by dashes indicate results in successive reuses
[b] Prepared by direct exchange
[c] Prepared by refluxing Mn-MCM-41 with the chiral ligand

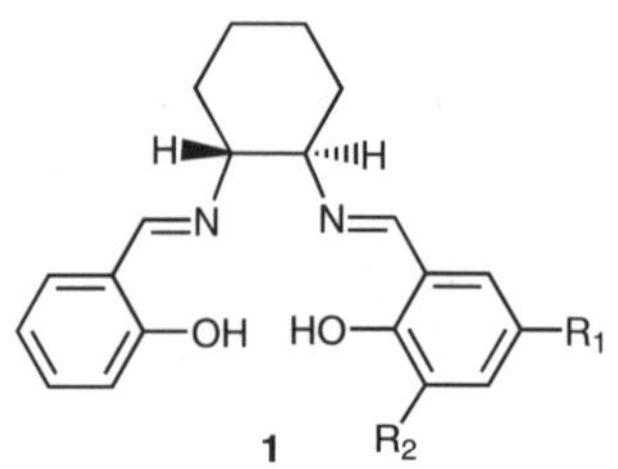

1

m: $R_1 = R_2 = {}^tBu$
n: $R_1 = H$; $R_2 = OMe$

Fig. 11 Non-symmetrical chiral salen ligands used in solid phase immobilization

ions to compensate the positive charge of the (salen)Mn complex (Fig. 12). The immobilization clearly proved to be positive in the epoxidation of α-methylstyrene with NaOCl because 67–72% ee was obtained in different solvents; the homogeneous catalyst on the other hand led to only 56% ee under the same conditions. Moreover, the catalyst was used three times without loss of activity or enantioselectivity. However, this solid was not active in the epoxidation of 1-phenylcyclohexene, probably as a consequence of steric restrictions around the active site located close to the support surface.

Salen ligands have also been used in the titanium-catalyzed trimethylsilylcyanation of benzaldehyde. The complexes were immobilized by substitution of a chloride with a surface silanol from the support. In the first study on this reaction [38], the most efficient ligand was the non-symmetrical salen **1m** (Fig. 11) (94% ee), whereas the selectivity obtained with the symmetrical ligand **1a** was significantly lower (72% ee). In a recent paper, the immobilization of different titanium species, including monomeric and dimeric systems with

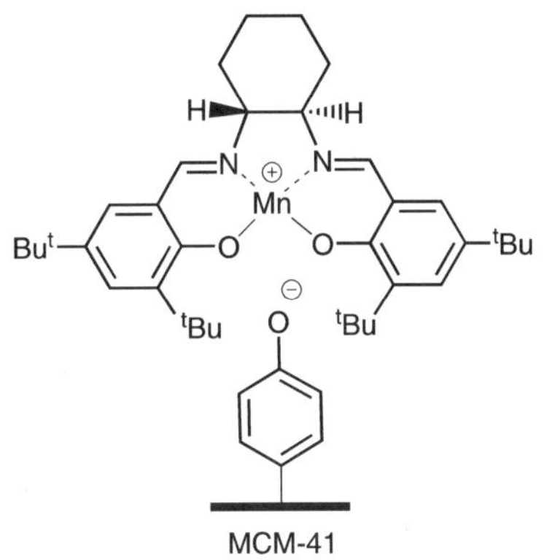

Fig. 12 (Salen)Mn complex immobilized on phenoxide-MCM-41

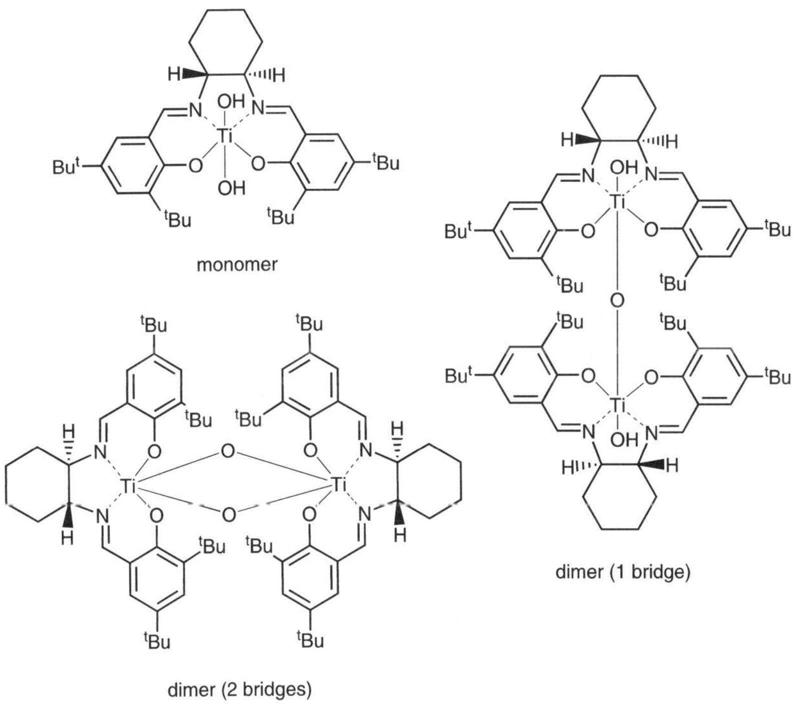

Fig. 13 Monomeric and dimeric (salen)Ti complexes

one or two oxygen bridges (Fig. 13), on MCM-41 and amorphous silica was described [40]. This type of immobilization led to a significant reduction of the enantioselectivity from values around 75% ee in solution to around 60% ee on MCM-41 and 50% ee on silica. The most marked reduction was observed in the case of the dimeric species with two oxygen bridges, probably due to the steric constraints in the environment close to the catalytic sites.

3.2.3.2
Complexes with the Charge on the Ligand

A different strategy has very recently been applied to the immobilization of (salen)Mn complexes by electrostatic interactions. In these examples the charges were placed on substituents of the aromatic rings in salen ligands (Fig. 14).

Positive charges were introduced in the form of quaternary ammonium salts in ligands **1o**, **1p**, **2o**, and **2p** [41, 42] and the complexes were exchanged on a montmorillonite clay. The catalyst loading was very low, which is consistent with the large size of the complexes, and exchange mainly took place on the outer surface of the clay. All these catalysts showed a very high activity in the epoxidation of alkenes with NaOCl using pyridine *N*-oxide as an additive. In the case of styrene, 70% ee was obtained and this represents an improvement on the result in homogeneous phase (up to 52% ee). Similar enantioselectivity was obtained in the epoxidation of indene, although in this case the immobilization was slightly detrimental. The best results, with up to 99% ee, were obtained in the epoxidation of 2,2-dimethyl-6-nitrochromene. The possibility of recycling the catalyst was tested in the epoxidation of styrene, with five consecutive reactions in which loss of enantioselectivity was not observed and only slow deactivation of the catalyst occurred.

A complementary approach has been reported very recently [43]. In this case negative charges were introduced into the salen ligand **1q** (Fig. 14) with the aim of exchanging it on cationic supports, such as a layered double (Zn, Al) hydroxide. The expansion in the basal spacing indicated intercalation, at least partially, of the **1g**-Mn complex between the layers of $[Zn_{2.15}Al_{0.86}(OH)_{6.02}]$. The complex was used in the epoxidation of (*R*)-limonene with molecular oxygen and pivalaldehyde. The use of *N*-

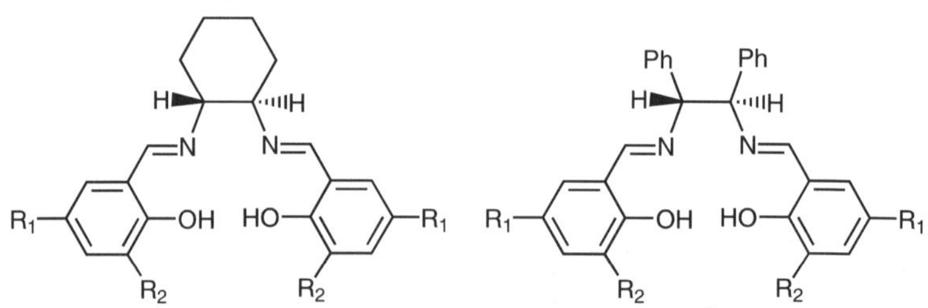

1 **2**

o: R_1 = $-CH_2\overset{\oplus}{N}(isoC_8H_{17})_3$ R_2 = ^tBu

p: R_1 = $-CH_2\overset{\oplus}{N}Et_3$ R_2 = ^tBu

q: R_1 = $-SO_3^{\ominus}$ R_2 = H

Fig. 14 Chiral salen ligands with charged substituents

methylimidazole as an additive in the epoxidation gave moderate stereoselectivity, 55% de, that compared favorably with the results obtained with zeolite-exchanged catalysts [32].

As can be seen, many methods are available for the immobilization of salen-metal complexes, although it is difficult to elucidate which one is the best. Even in the case of epoxidation, the differences in substrates, oxidants, and reaction conditions preclude any direct comparison.

4 Non-covalently Immobilized Catalysts Based on Chiral Bis(oxazoline) Ligands

Chiral bis(oxazoline) ligands are among the most successful and versatile ligands for enantioselective catalysis. They are readily obtained from chiral aminoalcohols, which come in many cases from the corresponding amino acid. Some of these ligands, such as 2,2′-isopropylidenebis[(4*S*)-phenyl-2-oxazoline] (**6a**) and 2,2′-isopropylidenebis[(4*S*)-*tert*-butyl-2-oxazoline] (**6b**) (Fig. 15) among others, are commercially available and, in a similar way to salen ligands, have been used as references to compare the results of immobilization of complexes with analogous ligands. Many examples have been described of liquid and solid phase immobilization of transition metal-bis(oxazoline) complexes using all of the available immobilization techniques, including the non-covalent methods.

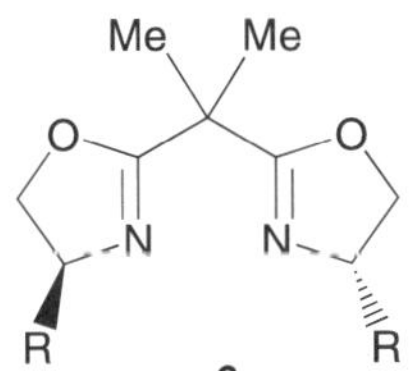

6

a: R = Ph
b: R = ^{t}Bu
c: R = CH_2Ph

Fig. 15 Chiral bis(oxazoline) ligands used for immobilization

4.1 Liquid Phase Immobilization

4.1.1 Fluorinated Solvents

Only one example of the functionalization of a bis(oxazoline) ligand with fluorinated chains has been described to date [44], but the use of these functionalized ligands does not meet our description of a "supported" catalyst

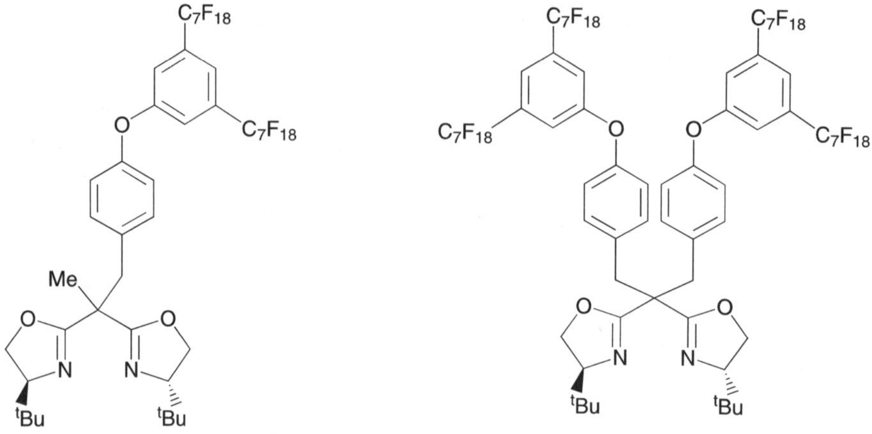

7 8

Fig. 16 Fluorinated chiral bis(oxazoline) ligands

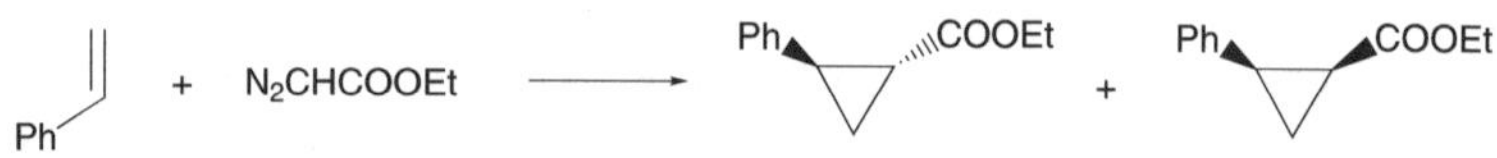

Scheme 7

Scheme 8

since the complex solution is not recovered as a whole after the reaction. These fluorinated bis(oxazolines) have the structure depicted in Fig. 16. Although the fluorine content is very similar in both cases, the solubility in conventional and fluorinated solvents is different, with ligand **7** being soluble in conventional organic solvents.

The copper complexes of these ligands were tested in the cyclopropanation of styrene with ethyl diazoacetate (Scheme 7) and the ene reaction between α-methylstyrene and ethyl glyoxylate (Scheme 8). In both cases moderate enantioselectivities were obtained but these were lower than those found with the parent ligand.

4.1.2 Ionic Liquids

In contrast with salen ligands, ionic liquids were used earlier than fluorinated solvents for biphasic liquid systems with bis(oxazoline)-based complexes. In spite of this, there are still very few applications of ionic liquids with this type of complex. The first example described involved the use of the copper complexes of ligands **6a** and **6b** for the enantioselective cyclopropanation of styrene with ethyl diazoacetate (Scheme 7) [45]. Two imidazolium ionic liquids, [emim][NTf_2] and [emim][BF_4], together with the aliphatic methyltrioctylammonium bis(trifluoromethylsulfonyl)amide ([Oct_3Me][NTf_2]) were used for the immobilization of the copper complexes. Some relevant results from this study are gathered in Table 5.

One important point to stress from these results is the possibility of using copper chloride instead of copper triflate to prepare the complexes. It is well known that in organic solvents there is a dramatic counteranion effect on the activity and enantioselectivity of these catalysts. On the other hand, the rapid anion exchange produced in the ionic liquid resulted in better performance of the complexes, as the bis(triflyl)imide behaves in a similar way to the triflate counteranion.

The activity and enantioselectivity of the complex **6a**-Cu were slightly lower than those observed in dichloromethane, but the catalysts were stable after two reuses. After each reaction, the products were extracted with hexane and the catalyst remained in the ionic liquid phase, ready for reuse. In the case of the complex **6b**-Cu, the enantioselectivities obtained in the former reactions (entry 3 in Table 5) were significantly lower than those observed in dichloromethane (~ 90% ee) under the same conditions. When the complex was prepared in situ in the ionic liquid, as opposed to dissolving the preformed complex, the enantioselectivity results improved considerably (entry 5) but decreased again after catalyst recovery.

As far as the influence of the ionic liquid is concerned, both the nature of the cation and the anion had a clear effect on the reaction. For this reason [emim][NTf_2] was the best solvent for this reaction.

Table 5 Results of cyclopropanation reactions with complexes **6a**-Cu and **6b**-Cu in ionic liquids[a,b]

Catalyst	Solvent	Yield[c] (%)	ee *trans*[c] (%)	ee *cis*[c] (%)
6a-$CuCl_2$	[emim][NTf_2]	34–32–33	55–53–53	47–45–44
6a-$CuCl_2$	[Oct_3NMe] [NTf_2]	18	49	41
6b-$Cu(OTf)_2$	[emim][NTf_2]	38–38–37	66–66–64	64–64–62
6b-$Cu(OTf)_2$	[Oct_3NMe] [NTf_2]	24	2	7
6b-$CuCl_2$	[emim][NTf_2]	50–42	86–55	85–56

[a] Using equimolecular amounts of styrene and ethyl diazoacetate and 1% catalyst
[b] In all cases a *trans*/*cis* selectivity around 7/3 is obtained
[c] Numbers separated by dashes indicate results in successive reuses

This preliminary study has recently been extended to these and other ionic liquids, using CuCl as the catalyst precursor [46]. Some relevant results are gathered in Table 6.

In this work, special attention was paid to those factors that could interfere with the catalytic activity, such as the presence of water arising from the hygroscopic nature of the ionic liquid, and bromine impurities arising from the preparation of the ionic liquid. The ionic liquids were therefore dried under vacuum in the presence of P_2O_5 prior to the reactions and, in some cases, the ionic liquids were prepared by alternative procedures that avoided the presence of bromine. The use of these methods, together with the use of CuCl, resulted in improved performance of the catalytic systems both with regard to the enantioselectivities and to the recoverability of the catalyst. A noticeable influence of the catalyst concentration was also noticed and when the ionic liquid/catalyst proportion was tripled, enantioselectivities were obtained that were essentially identical to those obtained in CH_2Cl_2 under the same conditions. The origin of this effect was not clear and the authors suggested that catalysis by residual amounts of **6b**-CuCl complex, which is much less enantioselective, could be responsible for this behavior.

In general, a decrease in enantioselectivity was observed after the second reuse. The origin of this effect lies in the partial extraction of the chiral ligand in the hexane phase after each reaction. This was demonstrated by adding a small amount of ligand after the fourth reuse. In all cases, the original enantioselectivity was fully recovered (entries 2 and 3 in Table 6).

Table 6 Results of cyclopropanation reactions with **6b**-CuCl in ionic liquids [a,b]

Solvent	Yield[c] (%)	ee *trans*[c] (%)	ee *cis*[c] (%)
[emim][OTf]	38–40–38–20–24	85–85–77–58–83	73–73–64–45–72
[emim][OTf][d]	51–56–45–51–42–39[f]	85–85–77–71–63–85[f]	78–78–71–63–55–78[f]
[bmim][OTf][d]	38–37–40–43–43–38[f]	79–79–69–64–61–86[f]	71–71–71–60–55–79[f]
[bmim][OTf][e]	34–37–39	92–90–87	88–84–79
[Oct_3Me][OTf]	15	64	54
[bmim][BF_4][g]	88–88–90–85–69–71–55–51	97–95–93–98–76–81–76–77	94–93–91–98–77–78–72–78

[a] Using equimolecular amounts of styrene and ethyl diazoacetate and 1% catalyst. Catalyst concentration 76 mM [46]
[b] In all cases a *trans*/*cis* selectivity of around 7/3 is obtained
[c] Numbers separated by dashes indicate results in successive reuses
[d] Bromine-free ionic liquid
[e] Catalyst concentration 25 mM
[f] Additional amount of chiral ligand was added to the recovered ionic liquid solution
[g] [47]

Similar conclusions concerning the effect of the anion and impurities in the ionic liquid on the cyclopropanation reaction have been drawn in a recently published study [47]. Ionic liquids with the formula [bmim][X], where $X =$ OTf, NTf_2, PF_6, and BF_4, were used. The catalyst used in this study was **6b**-Cu(OTf). In all cases, good enantioselectivities (89–97% ee) were obtained and these are similar to those obtained in chloroform. The influence of the presence of halogen anions was tested by the addition of 5% [bmim][Cl] or [bmim][Br] to [bmim][BF_4]. In both cases, a catalytically inactive solution was obtained, showing the detrimental effect of these anions on the reaction.

The possibility of recycling the catalyst was also studied. In order to decrease the amount of ethyl maleate and fumarate, the addition rate of ethyl diazoacetate was reduced and the reaction temperature was kept low during the addition. Furthermore, the catalyst concentration was reduced by doubling the volume of ionic liquid. Under these conditions the catalyst was reused seven times, although both the yield and the enantioselectivity decrease from the fourth reuse on (entry 6 in Table 6).

Apart from the cyclopropanation reaction, only one example has been published of the application of ionic liquids as reaction media for enantioselective catalysis with bis(oxazoline) ligands. In this case, the complex **6b**-$ZnCl_2$ was used as a catalyst for the Diels–Alder reaction between cyclopentadiene and *N*-crotonyloxazolidin-2-one in dibutylimidazolium tetrafluoroborate (Scheme 9) [48]. Compared with the same process in CH_2Cl_2, the reaction was faster and both the *endo*/*exo* selectivity and the enantioselectivity in the *endo* product were excellent. However, experiments aimed at recovering the catalysts were not carried out.

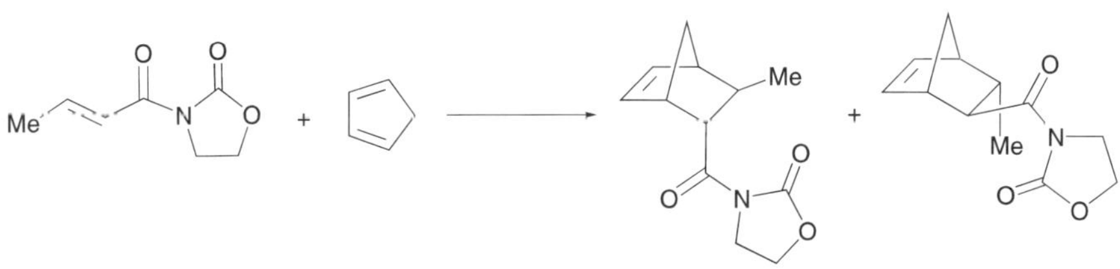

Scheme 9

4.2 Solid Phase Immobilization

4.2.1 Cationic Exchange

Metal complexes of chiral bis(oxazoline) ligands, in most cases Cu(II) complexes, have been supported by cationic exchange on inorganic, organic, and composite anionic solids.

The resulting solids were used as enantioselective catalysts in cyclopropanation, Diels–Alder and ene reations. In many cases, the recoverability and reusability of these catalysts were carefully studied and this aspect has been the driving force for new research.

The first examples of cationic exchange of bis(oxazoline)-metal complexes used clays as supports [49, 50]. Cu(II) complexes of ligands *ent*-**6a**, **6b**, and **6c** (Fig. 15) were supported on three different clays: laponite (a synthetic clay), bentonite, and montmorillonite K10. The influence of the copper salt from which the initial complexes were prepared, as well as that of the solvent used in the cationic exchange, was analyzed.

The solids prepared were characterized to demonstrate the presence of the bis(oxazoline)-copper complexes in the solid, as well as to rule out the presence of non-chiral copper centers. The nature of the complex–support interactions was thoroughly analyzed using different spectroscopic techniques, such as infrared, electron paramagnetic resonance (EPR), and X-ray absorption (EXAFS) techniques [51]. To this end the complexes of ligand **6c** with different copper salts were investigated both in solution and after cationic exchange in laponite. The interlamellar spacing variation after complex exchange was determined to be 7.6 Å by X-ray diffraction measurements. This increase in spacing is compatible with the estimated size of the **6c**-Cu(II) complex. When the starting copper salt was $CuCl_2$, it was found that the chloride anions had been completely removed from the solid after treatment with the solution containing the complex, meaning that the process was a genuine cation exchange with the copper salt anion being completely substituted by the anionic support. Furthermore, spectroscopic studies determined that the exchanged copper complexes were isolated species with an almost square-planar structure, with the clay surface acting as the counterion of the cationic complex.

The solids were used as catalysts in the benchmark cyclopropanation reaction between styrene and ethyl diazoacetate (Scheme 7). As far as the nature of the clay is concerned, laponite was found to be the best support for the catalytic complexes. The best enantioselectivity results (Table 7) were obtained with ligand **6b** (69% ee in *trans* cyclopropanes and 64% ee in *cis* cyclopropanes) but the recovered solid showed a lower activity and enantioselectivity, which was attributed to partial loss of the chiral ligand from the support. In general, the use of the three chiral ligands led to enantioselectivity results that were intermediate between those obtained in homogeneous phase with $CuCl_2$ and $Cu(OTf)_2$ as catalyst precursors. This seemed to indicate that the solid behaved as a counterion with an intermediate coordinating ability to the copper centers.

In view of this behavior, the next step involved a search for an anionic solid that was more similar to triflate. Several organic polymers with sulfonic groups were tried as supports [52]. Dowex and Deloxan were used as supports for the *ent*-**6a**-Cu(II) complex. The solid catalysts were used in the same cyclopropanation reaction (Scheme 7) and some relevant results are gathered in

Table 7 Results of cyclopropanation reactions with **6a**-$Cu(OTf)_2$ and **6b**-$Cu(OTf)_2$ in solution and exchanged on anionic supports

Ligand	Support	Yield[a] (%)	ee *trans*[a] (%)	ee *cis*[a] (%)
ent-**6a**	Homogeneous	49	60	51
ent-**6a**	Laponite	39–26	54–53	43–42
ent-**6a**	Dowex	31–33	17–2	15–2
ent-**6a**	Deloxan	33–37	38–8	31–7
ent-**6a**	Nafion[b]	52–50	59–58	45–47
ent-**6a**	Nafion-silica	38–42	57–56	46–45
6b	Homogeneous	59	94	91
6b	Laponite	36–35	69–43	64–37
6b	Nafion[b]	34–34	5–5	7–6
6b	Nafion-silica	41–34	23–14	19–14

[a] Numbers separated by dashes indicate results in successive reuses
[b] Reaction temperature 60 °C

Table 7. As can be seen, both Dowex and Deloxan led to poor enantioselectivities, which further decreased after catalyst recovery. Better results, which are comparable with those obtained in homogeneous phase, were obtained with Nafion (Table 7) [53], although it was necessary to carry out the reaction at 60 °C due to the low copper content in the solid. This low copper level is a consequence of the low surface area of this polymer ($< 0.02\ m^2\ g^{-1}$) and, for this reason, a nafion-silica nanocomposite was used as the support [53]. With this catalyst, the reaction took place at room temperature and with similar enantioselectivity (Table 7).

Both in the case of the nafion and the nafion-silica, the catalyst can be recovered and reused at least once. However, when ligand **6b** was used, the enantioselectivities were much poorer than those obtained in homogeneous phase and they decreased further after catalyst recovery. This fact was explained by the formation of non-chiral catalytic centers in the solid by loss of the chiral ligand during the exchange process (Fig. 17) as a consequence of unfavorable steric ligand-support interactions. This equilibrium may be affected by several parameters, such as type of support, leaving cation, counterion of the copper complex in solution and solvent. All of these parameters, together with structural variations in the composite supports, were investigated. However, changes in these factors only led to marginal improvements in the enantioselectivities [54]. Better enantioselectivity results were only obtained when (1*R*,2*S*,5*R*)-menthyl diazoacetate was used instead of ethyl diazoacetate. Enantioselectivities of 79% ee in *trans*- and 83% ee in *cis*-cyclopropanes were obtained and these values are almost identical to the results obtained in homogeneous phase (77% and 90% ee, respectively) [54].

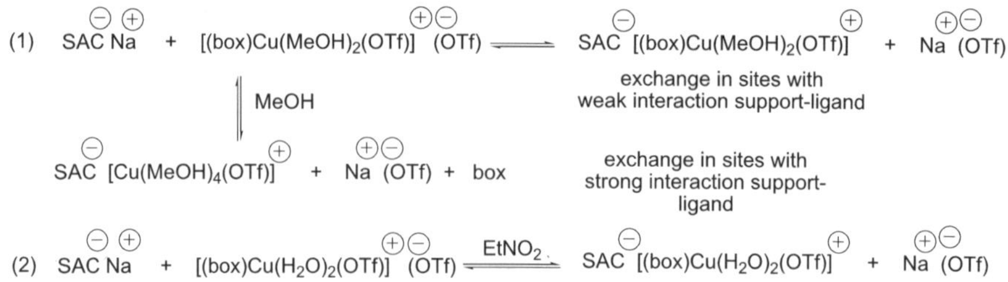

Fig. 17 The cationic exchange process

This lack of stability in the exchange process is not particular to complexes of **6b**, as shown by the results obtained with copper, zinc, and magnesium complexes of ligand **6a** supported on laponite. These solids were tested as catalysts in the Diels–Alder reaction between cyclopentadiene and *N*-crotonyloxazolidin-2-one (Scheme 9) [55]. In general, the results were poorer that those obtained in homogeneous phase, with enantioselectivities of about 10% ee (for copper and zinc complexes). Analysis of the solids demonstrated that complexes **6a**-Zn and **6a**-Mg were not efficiently exchanged and most of the chiral ligand was also lost during the exchange process.

It seems reasonable to believe that this problem could be overcome by studying more coordinating ligands with the same structural features. Very recently, it has been demonstrated [56] that the use of iminobis(oxazolines) (Fig. 18) leads to better enantioselectivities and recoverable catalysts, both with laponite and nafion-silica supports (Table 8). Theoretical calculations are consistent with the stronger coordinating ability of iminobis(oxazolines) being the origin of these results [57].

Another problem that required solving was the moderate yield obtained in the cyclopropanation reaction when only one equivalent of styrene was used. By increasing the amount of styrene up to its use as the reaction solvent, a noticeable effect on the selectivities was observed when laponite was used as the support [58]. The active role of the clay support was definitely established when the results in homogeneous and heterogeneous phases were compared (Table 9). These effects involved the reversal of the *trans* preference in solution to the *cis* preference with the laponite-supported catalyst in styrene, and also a reversal in the absolute configuration of the major *cis* enantiomer ob-

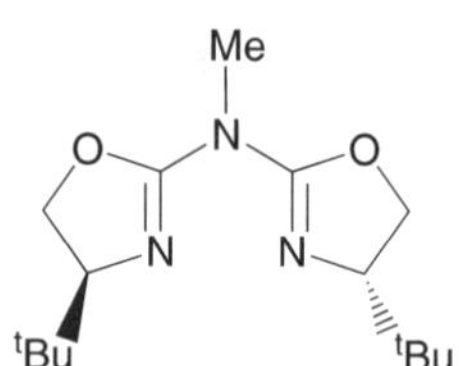

9

Fig. 18 Chiral iminobis(oxazoline) ligands used for immobilization

Table 8 Results of cyclopropanation reactions with copper complexes exchanged on anionic supports

Ligand	Support	Yield[a] (%)	ee *trans*[a] (%)	ee *cis*[a] (%)
6b	Laponite	36–35	69–43	64–37
6b	Nafion-silica	41–34	23–14	19–14
9	Laponite	46–45	83–81	76–74
9	Nafion-silica	30–21	88–84	81–77

[a] Numbers separated by dashes indicate results in successive reuses
[b] Reaction temperature 60 °C

tained. When the heterogeneously catalyzed cyclopropanation reaction was carried out in different solvents the same effects were observed with low polarity solvents (Table 9) [58].

It was demonstrated that these effects did not result from a permanent modification of the nature of the catalyst after recycling the catalyst used in styrene and reusing it in CH_2Cl_2. Indeed, under these conditions the same stereoselectivities were obtained as with the use of the fresh catalyst in CH_2Cl_2 (entry 3).

The important message from these results is that one or other major enantiomer can be obtained with the same sample of catalyst, simply by changing the reaction solvent.

Table 9 Results of cyclopropanation reactions in different solvents with **6b**-$Cu(OTf)_2$ in solution and exchanged in laponite[a]

Anion	Solvent	*Trans*/*cis*	ee *trans*[b] (%)	ee *cis*[b] (%)
TfO[c]	CH_2Cl_2	71/29	54	42
TfO[c]	Styrene	69/31	55	42
Laponite	CH_2Cl_2	61/39	49	24
Laponite	Styrene	31/69	7	−34
Laponite	Toluene	40/60	3	−21
Laponite	n-C_6H_{14}	31/69	3	−33
Laponite	n-C_7F_{16}	32/68	−1	−30

[a] Except when styrene was the solvent, the reactions were carried out with equimolecular amounts of reagents and 1% of catalyst
[b] Negative numbers indicate that the opposite enantiomer is obtained as the major product
[c] Homogeneous reactions

An explanation was suggested for these solvent and support effects and this is represented in Fig. 19. Thus, in solvents with greater dielectric permittivity, ε, the cationic complex is situated further from the clay surface and the stereoselectivities are therefore more similar to those obtained in homogeneous phase. On the other hand, in solvents with low ε, close ion pairs are formed and the surface has a larger effect on the reaction.

This surface effect opens the door to the design of chiral ligands that exploit this effect to obtain selectivities different from those obtained in solution. The initial hypothesis was that a non-C_2 symmetric ligand would interact more strongly with the surface, thus enhancing support effects. This possibility was explored in a recent study [59] in which two pyridineoxazoline ligands (Fig. 20) were synthesized, their copper complexes exchanged in laponite, and the resulting materials used as catalysts in the usual benchmark cyclopropanation reaction. Some relevant results obtained with these catalysts are gathered in Table 10.

It is concluded from these results that with this kind of non-C_2 symmetric ligand (that led necessarily to poor enantioselectivities in homogeneous phase), it is possible to exploit support effects to change the *trans/cis* selectivity and to improve the enantioselectivity. This is demonstrated for the *trans*-cyclopropanes obtained with ligand **10a** in styrene. Due to the relative disposition of the ester and phenyl groups in the transition state, support ef-

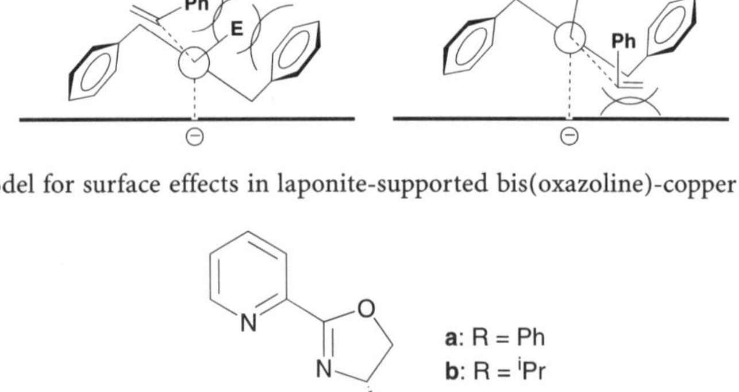

Fig. 19 Model for surface effects in laponite-supported bis(oxazoline)-copper complexes

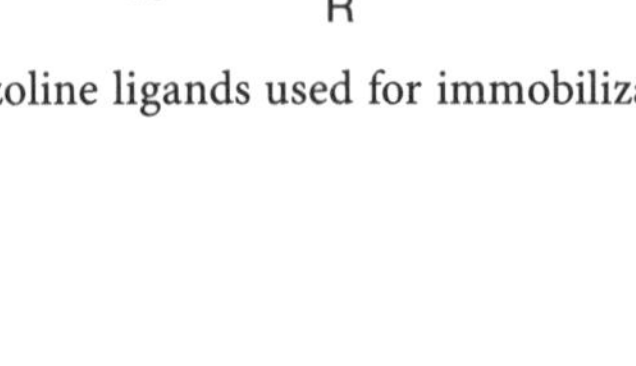

10

Fig. 20 Chiral pyridineoxazoline ligands used for immobilization

Table 10 Results of cyclopropanation reactions with **10a**-$Cu(OTf)_2$ and **10b**-$Cu(OTf)_2$ in solution and exchanged in laponite[a]

Ligand	Anion	Solvent	*Trans/cis*	ee *trans*[b] (%)	ee *cis* [b] (%)
10a	TfO[c]	CH_2Cl_2	67/33	5	17
	Laponite	CH_2Cl_2	45/55	27	27
	Laponite	Styrene	31/69	65	24
10b	TfO[c]	CH_2Cl_2	69/31	4	8
	Laponite	CH_2Cl_2	46/54	13	−1
	Laponite	Styrene	33/67	27	−12

[a] Except when styrene was the solvent, the reactions were carried out with equimolecular amounts of reagents and 1% of catalyst
[b] Negative numbers indicate that the opposite enantiomer is obtained as the major product
[c] Homogeneous reactions

fects are expected to be more important for the *trans*-cyclopropanes, as seems to happen for ligand **10a** (Fig. 21).

In parallel with the work on clays and nafion-like supports, zeolites were also tested as inorganic supports for electrostatic immobilization of bis(oxazoline) complexes. The first immobilization of a bis(oxazoline)-copper complex in a zeolite was described in 1998 [60, 61]. The ligands used in this study were **6a**, **6b**, **11a**, and **12** (Fig. 22). Two different procedures for the cationic exchange were tested. In the first approach, Cu(II) was first exchanged in zeolite HY and the chiral ligand was then added to the resulting CuHY zeolite. In the second method, the corresponding bis(oxazoline)-

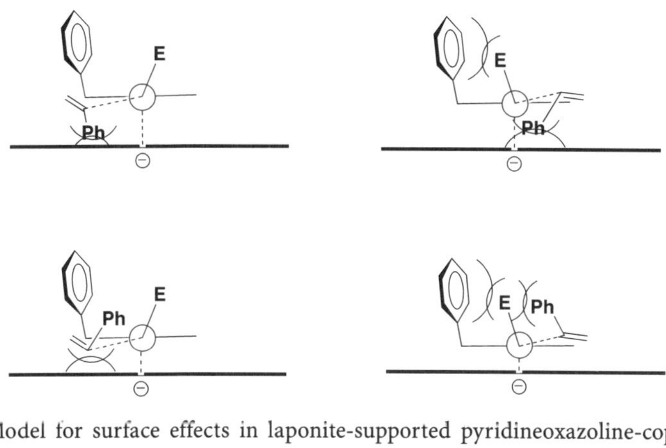

Fig. 21 Model for surface effects in laponite-supported pyridineoxazoline-copper complexes

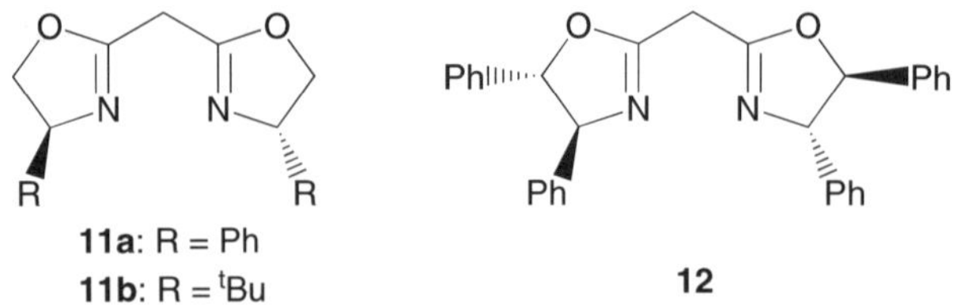

Fig. 22 Chiral bis(oxazoline) ligands used for immobilization in zeolites

copper complex was directly exchanged in the zeolite. Both procedures led to similar results in the catalytic tests.

The reaction used to test these solid catalysts was the aziridination of styrene with *N*-tosyliminophenyliodinane (PhI = NTos) (Scheme 10). In most cases, enantioselectivities were low or moderate (up to 60% ee). The loss of enantioselectivity on changing from ligand **11a** to ligand **12** was attributed to the fact that ligand **12** is too big for the copper complex to be accommodated into the zeolite supercages. Further studies carried out with ligands **11a** and **11b** [62] demonstrated that the reaction is more enantioselective with the supported catalyst (82% ee with **11a** and 77% ee with **11b**) than in solution (54% ee with **11a** and 31% ee with **11b**). This trend supports the confinement effect of the zeolite structure on the stereoselectivity of the reaction.

A number of other factors, such as the nitrene source and its influence on copper leaching, the styrene/nitrene donor ratio, solvent, and chiral auxiliary structure, were considered in further studies by the same group [62–65]. The best results were obtained with [*N*-(*p*-nitrophenylsulfonyl)imino]phenyliodinane (PhI = NNs), which led to $\geq$ 90% ee and high yields ($\geq$ 85%). A comparison between homogeneous and heterogeneous reactions with different chiral ligands (Table 11, Fig. 22) indicates that the heterogeneous catalyst gives enhanced enantioselection for a range of bis(oxazolines) in comparison to the homogeneous catalyst. As already observed in the case of the clay-supported catalyst, this indicates that a good catalyst in homogeneous phase may not be the best choice for designing a heterogeneous catalyst.

One interesting effect observed in these aziridination reactions was the increase in percentage ee with reaction time, both in homogeneous and heterogeneous phases [66]. The origin of this effect was thoroughly studied in a series of experiments that demonstrated that the aziridine products were able to react with sulfonamide byproducts and with nitrene donors, in the

Scheme 10

Table 11 Effect of bis(oxazoline) on the aziridination of styrene with zeolite-supported catalysts[a]

Ligand	PhI = NTs		PhI = NNs	
	Yield (%)	ee (%)	Yield (%)	ee (%)
6a	78(91)	76(73)	78(96)	85(81)
6b	58(78)	24(35)	68(79)	82(43)
11a	70(74)	77(28)	80(85)	82(54)
11b	80(85)	28(8)	72(42)	77(31)

[a] Numbers in parentheses correspond to homogeneous reactions

presence of the chiral catalyst, to give an enantioselective interconversion between aziridine enantiomers. When a racemic mixture of aziridines was used as the starting material, deracemization process occurred to give a final value of 38% ee. The authors of this work proposed that the interconversion occurs via the intermediacy of ring-opened species, formed by reaction of the aziridine with a nitrogen nucleophile (derived from either the nitrene donor or the sulfonamide) mediated by N-coordination of the aziridine to a Cu(II) ion (Scheme 11). Some theoretical calculations were carried out to explain this mechanism [67].

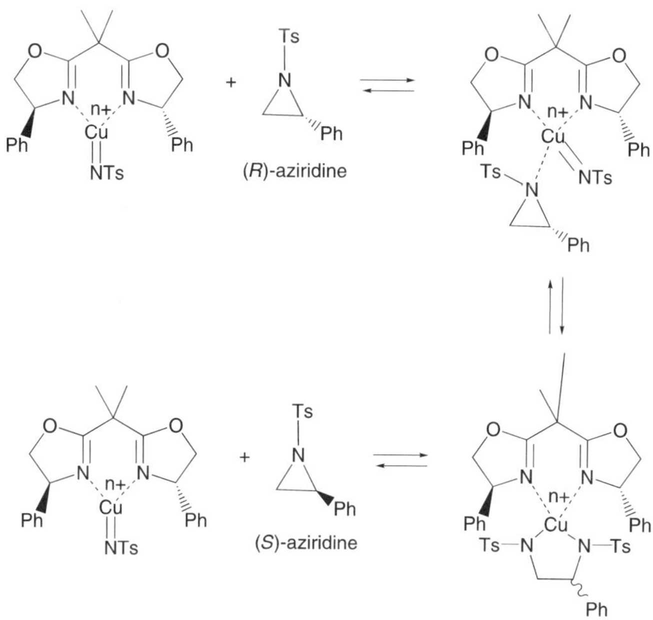

Scheme 11

More recently, bis(oxazoline)-metal complexes supported in micro- and mesoporous solids have been used as catalysts of hetero-Diels–Alder and ene reactions.

In the case of the Diels–Alder reaction [68] (Scheme 12), several solids (AlSBA-15, MCM-41, MSU-2 and zeolite HY) were tested as supports for the bis(oxazoline)-copper complexes. The best enantioselectivity results were obtained with the zeolite HY, although the yield was the poorest (16% yield, 41% ee). As happened with the aziridination reaction, the enantioselectivity changed with time. Short reaction times led to the same major enantiomer as observed in homogeneous reactions. However, at higher conversions, i.e., longer reaction times, the opposite major enantiomer was obtained.

Carbonyl- and imino-ene reactions were also catalyzed by the bis(oxazoline)-copper complexes of ligands **6a**, **6b** and **11b** supported on zeolite Y (Scheme 13) [69]. The enantioselectivities obtained with the supported catalysts were similar or better than those obtained in homogeneous phase with the same ligands. Some relevant examples are shown in Table 12.

Less than 1% copper leaching was detected under the reaction conditions and it was demonstrated that the leached copper does not contribute to

Scheme 12

Scheme 13

Table 12 Results of reactions of ethyl glyoxylate with different alkenes, catalyzed by several bis(oxazoline)-copper complexes immobilized on Y zeolite

Ligand	Alkene	Yield (%)	ee (%)
6a	α-Methylstyrene	85	77
6a	Methylene cyclohexane	65	94
6a	Methylene cyclopentane	71	93
6b	α-Methylstyrene	87	93
11b	α-Methylstyrene	91	85

the catalysis, meaning that the system is truly heterogeneous. Furthermore, the catalyst can be efficiently recovered by filtration and washing. The catalyst was reused up to three times with the reagent alternatively changed from ethyl glyoxylate to methyl piruvate without loss of activity or enantioselectivity.

4.2.2
Adsorption

The only example of immobilization of a bis(oxazoline) complex by adsorption onto silica was published very recently [70]. The complexes **6a**-$Cu(OTf)_2$ and **6b**-$Cu(OTf)_2$ were adsorbed onto a chromatographic grade of silica gel and the resulting solids used as catalysts in two Diels–Alder reactions.

The enantioselectivity obtained in the hetero-Diels–Alder reaction (Scheme 12) was low (18% ee). This is, in part, due to the important temperature effect. For example, 50% ee was obtained in reactions carried out in homogeneous phase at – 60 °C and 95% ee in reactions at – 78 °C. However, at 0 °C the enantioselectivity dropped to 28% ee, a value closer to that obtained with the immobilized catalyst at the same temperature. Recycling was investigated and the solid was used four times with the same activity maintained. The **6b**-$Cu(OTf)_2$ catalyst proved to be less effective for this reaction and less stable in terms of recycling, a situation in agreement with the results obtained with exchanged catalysts [53].

In the case of the reaction between *N*-acryloyloxazolidin-2-one and cyclopentadiene, both catalysts showed activities and enantioselectivities similar to those observed in homogeneous phase. However, a reversal of the major *endo* enantiomer obtained with the immobilized **6a**-$Cu(OTf)_2$ catalyst, with regard to the homogeneous phase reaction, was noted. Although this support effect on the enantioselectivity remains unexplained, it resembles the surface effect on enantioselectivity of cyclopropanation reaction with clay supports [58].

5
Miscellaneous Chiral Ligands

5.1
Diamines and Related Ligands

Probably the first non-covalent immobilization of a chiral complex with diazaligands was the adsorption of a rhodium-diphenylethylenediamine complex on different supports [71]. These solids were used for the hydride-transfer reduction of prochiral ketones (Scheme 2) in a continuous flow reactor. The inorganic support plays a crucial role. The chiral complex was easily leached from alumina but the catalyst was not eluted from silica under the same conditions. It was postulated that the improved retention on silica is due to the formation of hydrogen bonds between the silanol groups on the silica surface and the diamine ligand (Fig. 23).

The catalytic results depend on several factors, such as particle size, flow rate, and substrate concentration. Under optimal conditions turnover numbers of 300 in the reduction of methyl phenylpyruvate were obtained, which compare favorably with the value of 20 obtained in the homogeneous reaction. With regard to enantioselectivity, the heterogeneous (^{i}PrOH/heptane) and the homogeneous (^{i}PrOH) catalysts led to the same results; 99% ee for methyl phenylpyruvate and 65% ee for acetophenone. One of the major advantages of the immobilized catalyst was the possibility of working at low substrate concentration. Total conversion was obtained with longer reactors and lower flow rates. Under such conditions a positive effect on the enantioselectivity was observed in the case of α,α,α-trifluoroacetophenone, leading to 27% ee in comparison with 16% ee obtained in solution.

More recently, the same type of ligand was used to form chiral iridium complexes, which were used as catalysts in the hydrogenation of ketones. The inclusion of hydrophilic substituents in the aromatic rings of the diphenylethylenediamine (Fig. 23) allowed the use of the corresponding complexes in water or water/alcohol solutions [72]. This method was optimized in order to recover and reuse the aqueous solution of the catalyst after product extraction with pentane. The combination of chiral 1,2-bis(*p*-methoxyphenyl)-*N*,*N'*-dimethylethylenediamine and triethyleneglycol monomethyl ether in methanol/water was shown to be the best method, with up to six runs with total acetophenone conversion and 65–68% ee. Only in the seventh run did the yield and the enantioselectivity decrease slightly.

Encapsulated rhodium complexes were prepared from Rh-exchanged NaY zeolite by complexation with (*S*)-prolinamide or *N*-*tert*-butyl-(*S*)-prolinamide [73, 74]. Although these catalysts showed higher specific activity than their homogeneous counterparts in non-enantioselective hydrogenations, the hydrogenation of prochiral substrates, such as methyl (*Z*)-acetamidocinnamate [73] or (*E*)-2-methyl-2-pentenoic acid [74], led to low

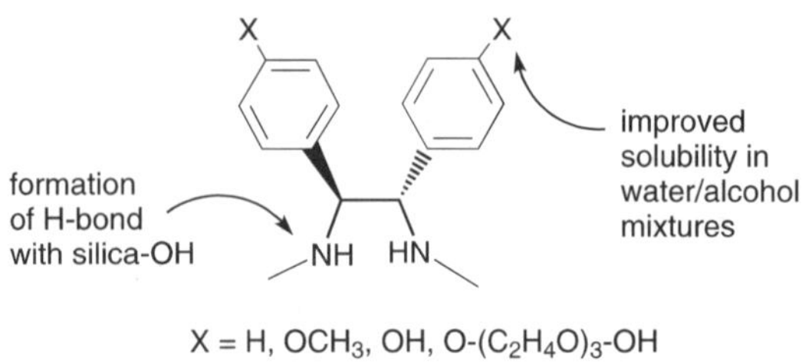

Fig. 23 Chiral diamines used for immobilization

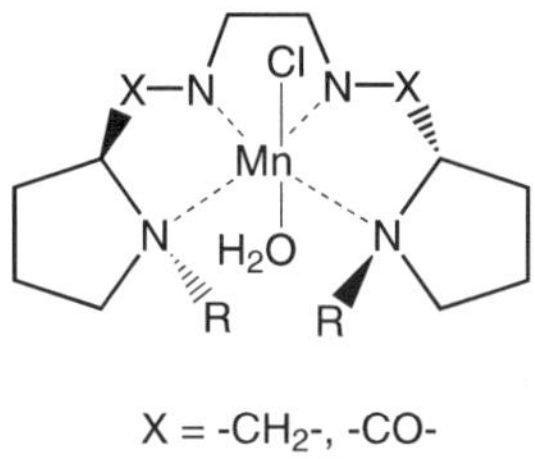

Fig. 24 Complex of Mn with a tetradentate aminated ligand

enantioselectivities (up to 20% ee). In any case, the use of the immobilized catalysts did improve on the results obtained in solution.

Encapsulation in Y zeolite was also the method chosen to immobilize Mn complexes of C_2-symmetric tetradentate ligands (Fig. 24) [75]. These materials were used as catalysts for the enantioselective oxidation of sulfides to sulfoxides with NaOCl. The lack of activity when the larger iodosylbenzene was used as an oxidant was interpreted as an indication that the reaction took place inside the zeolite microporous system. Both the chemo- and enantioselectivity were dependent on the structure of the sulfide. (2-Ethylbutyl)phenylsulfide led to better results than methylphenylsulfide, although in all cases the enantioselectivity was low (up to 21% ee).

(*R*,*R*)-1,2-Diphenylethylenediamine has been also used as a secondary chiral ligand in the preparation of immobilized Ru catalysts. In this case a chiral rigid BINAP polymer was used as the support (Fig. 25) and the diamine was added to the preformed pol-BINAP-Ru complex. This complex showed excellent performance in the hydrogenation of arylmethylketones, with enantioselectivities in the range 80–92% ee [76]. The same complex was employed in a tandem asymmetric reaction on *p*-acetylbenzaldehyde [77]. Diethylzinc was first added onto the formyl group and subsequent ketone hydrogenation led to two different stereogenic alcohols.

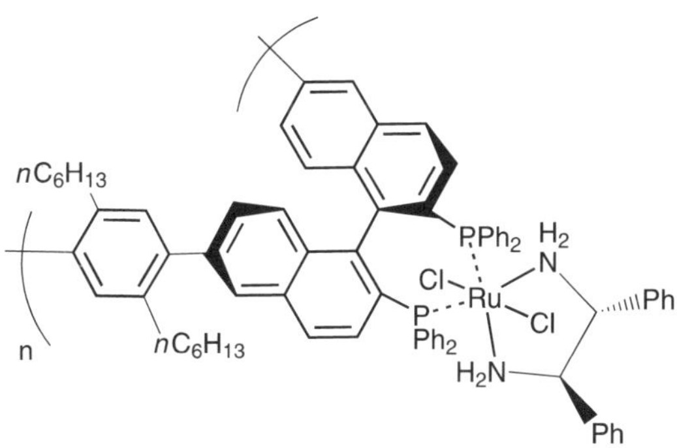

Fig. 25 Poly-BINAP-Ru-diamine complex

5.2 Porphyrins

Very few examples have been described for the non-covalent immobilization of chiral porphyrin complexes (Fig. 26). In the first case, the porphyrin-dichlororuthenium^{IV} complex was encapsulated in silica, which was prepared around the complex by a sol-gel method [78], in an attempt to prevent deactivation observed in solution in the epoxidation of different alkenes with 2,6-dichloropyridine *N*-oxide. In fact, the heterogeneous catalyst is much more active, with TON up to 10 800 in the case of styrene compared to a maximum of 2190 in solution. Enantioselectivities were about the same under both sets of conditions, with values around 70% ee.

The same type of porphyrin-Ru complex was immobilized by coordinative adsorption on aminopropylsilicas (Fig. 26) as either amorphous or crystalline supports [79]. Mesoporous crystalline MCM-48 was the best support, as shown by the improved results obtained in the epoxidation of styrene with 2,6-dichloropyridine *N*-oxide (TON > 13 000 and 74% ee). The versatility of this catalyst was demonstrated in the intramolecular cyclopropanation of *trans*-cinnamyl diazoacetate. TON was ten times higher than that obtained in solution and 85% ee was observed. The solid was recycled and reused, although partial loss of selectivity occurred.

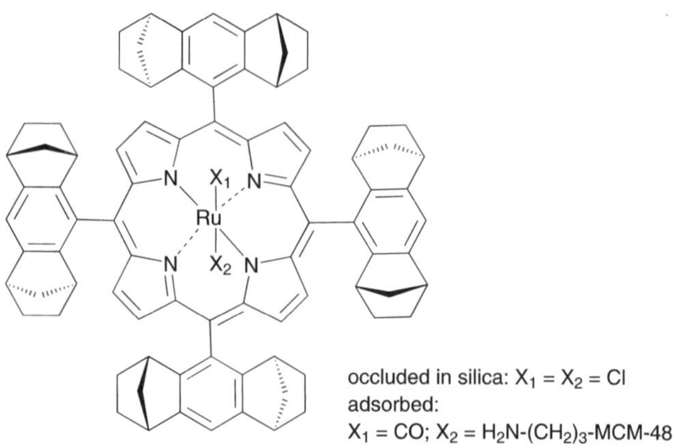

Fig. 26 Chiral porphyrin-Ru complex used in immobilization

5.3 Natural Aminated Polymers

Several examples have been described in which a chiral natural polymer, such as silk fibroin or chitosan, act as chiral ligand and support at the same time. In such cases, the chiral ligand (the monomer or monomers coordinating

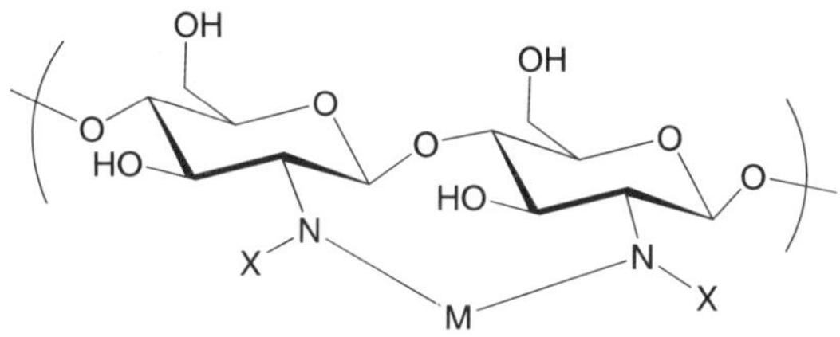

Chitosan: X = H; M = Pd
Chitin: X = $COCH_3$; M = Pt

Fig. 27 Natural aminated polymers used for immobilization

the metal) can be considered as covalently bonded to the support (the rest of the polymer). However, there are several examples where these polymer–metal complexes have been immobilized onto silica, without forming covalent bonds between them.

Chitosan (Fig. 27) was deposited on silica by precipitation. The palladium complex was shown to promote the enantioselective hydrogenation of ketones [80] with the results being highly dependent on the structure of the substrate. In the case of aromatic ketones, both yield and enantioselectivity depend on the N/Pd molar ratio. Low palladium contents favored enantioselectivity but reduced the yield. Very high conversions were obtained with aliphatic ketones, although with modest enantioselectivities. More recently, the immobilized chitosan-Co complex was described as a catalyst for the enantioselective hydration of 1-octene [81]. Under optimal conditions, namely Co content 0.5 mmol g^{-1} and 1-octene/Co molar ratio of 50, a 98% yield and 98% ee were obtained and the catalyst was reused five times without loss of activity or enantioselectivity.

Chitin (Fig. 27) was supported on silica by grinding the two solids together. The Pt complex was tested as a catalyst in the enantioselective hydrogenation of racemic 1-phenylethanol to obtain (*R*)-1-cyclohexylethanol [82]. Up to 65% yield with 100% ee was obtained and the catalyst was reused five times with almost the same results.

6 Conclusions

In conclusion, many methods are available for the immobilization of metal complexes with diazaligands. The biphasic liquid methods seem to be more versatile – especially in the case of ionic liquids, where modification of the ligands is not needed. Very good results can also be obtained with solid supports, although the surface effects, which can be positive or negative, are still unpredictable. From the data in the literature it appears that electrostatic immobilization is the most promising approach because of better complex retention on the solid and the lack of complex size limitations when mesoporous supports are used. However, the results regarding activity, enantioselectivity, and recoverability are strongly dependent on the electronic and steric requirements of the catalytic reaction, together with the coordinating ability of reagents, products, and by-products.

References

1. Blaser HU, Studer M (1999) Heterogeneous catalysis. In: Jacobsen EN, Pfaltz A, Yamamoto H (eds) Comprehensive asymmetric catalysis, vol III. Springer, Berlin Heidelberg New York, p 1353
2. De Vos DE, Vankelecom IFJ, Jacobs PA (eds) (2000) Chiral catalysts immobilization and recycling. Wiley-VCH, Weinheim
3. Oehme G (1999) Catalyst immobilization: two-phase systems. In: Jacobsen EN, Pfaltz A, Yamamoto H (eds) Comprehensive asymmetric catalysis, vol III. Springer, Berlin Heidelberg New York, p 1377
4. Blaser HU, Studer M (1999) Catalyst immobilization: solid supports. In: Jacobsen EN, Pfaltz A, Yamamoto H (eds) Comprehensive asymmetric catalysis, vol III. Springer, Berlin Heidelberg New York, p 1367
5. Pozzi G, Cinato F, Montanari F, Quici S (1998) Chem Commun 877
6. Pozzi G, Cavazzini M, Cinato F, Montanari F, Quici S (1999) Eur J Org Chem 1947
7. Cavazzini M, Manfredi A, Montanari F, Quici S, Pozzi G (2000) Chem Commun 2171
8. Cavazzini M, Manfredi A, Montanari F, Quici S, Pozzi G (2001) Eur J Org Chem 4639
9. Cavazzini M, Pozzi G, Quici S, Shepperson I (2003) J Mol Catal A 204:433
10. Maillard D, Nguefack C, Pozzi G, Quici S, Valadé B, Sinou D (2000) Tetrahedron: Asymmetry 11:2881
11. Millard D, Pozzi G, Quici S, Sinou D (2002) Tetrahedron 58:3971
12. Cavazzini M, Quici S, Pozzi G (2002) Tetrahedron 58:3943
13. Shepperson I, Cavazzini M, Pozzi G, Quici S (2004) J Fluorine Chem 125:175
14. Song CE, Roh EJ (2000) Chem Commun 837
15. Song CE, Roh CR, Roh EJ, Choo DJ (2000) Chem Commun 1743
16. Oh CR, Choo DJ, Shim WH, Lee DH, Roh EJ, Lee S, Song CE (2003) Chem Commun 1100
17. Baleizão C, Gigante B, Garcia H, Corma A (2002) Green Chem 4:272
18. Baleizão C, Gigante B, Garcia H, Corma A (2003) Tetrahedron Lett 44:6813
19. Vankelecom IFJ, Tas D, Parton RF, Van de Vyver V, Jacobs PA (1996) Angew Chem Int Ed 35:1346
20. Janssen KBM, Laquiere I, Dehaen W, Parton RF, Vankelecom IFJ, Jacobs PA (1997) Tetrahedron: Asymmetry 8:3481
21. Ogunwumi SB, Bein T (1997) Chem Commun 901
22. Sabater MJ, Corma A, Domenech A, Fornés V, García H (1997) Chem Commun 1285
23. Gbery G, Zsigmond A, Balkus Jr KJ (2001) Catal Lett 74:77
24. Kahlen W, Wagner HH, Hölderich WF (1998) Catal Lett 54:85
25. Gigante B, Corma A, García H, Sabater MJ (2000) Catal Lett 68:113
26. Doménech A, Formentin P, García H, Sabater MJ (2000) Eur J Inorg Chem 1339

27. Frunza L, Kosslich H, Landmesser H, Höft E, Fricke R (1997) J Mol Catal A 123:179
28. Zhou XG, Yu XQ, Huang JS, Li SG, Li LS, Che CM (1999) Chem Commun 1789
29. Dioos BML, Jacobs PA (2003) Tetrahedron Lett 44:8815
30. Fraile JM, García JI, Massam J, Mayoral JA (1998) J Mol Catal A 136:47
31. Heinrichs C, Hölderich WF (1999) Catal Lett 58:75
32. Schuster C, Hölderich WF (2000) Catal Today 60:193
33. Schuster C, Möllmann E, Tompos A, Hölderich WF (2001) Catal Lett 74:69
34. Piaggio P, McMorn P, Langham C, Bethell D, Bulman-Page PC, Hancock FE, Hutchings GJ (1998) New J Chem 1167
35. Piaggio P, Langham C, McMorn P, Bethell D, Bulman-Page PC, Hancock FE, Sly C, Hutchings GJ (2000) J Chem Soc Perkin Trans 2:143
36. Piaggio P, McMorn P, Murphy D, Bethell D, Bulman-Page PC, Hancock FE, Sly C, Kerton OJ, Hutchings GJ (2000) J Chem Soc Perkin Trans 2:2008
37. Kim GJ, Kim SH (1999) Catal Lett 57:139
38. Kim GJ, Shin JH (1999) Catal Lett 63:83
39. Xiang S, Zhang Y, Xin Q, Li C (2002) Chem Commun 2696
40. Kim JH, Kim GJ (2004) Catal Lett 92:123
41. Kureshy RI, Khan NH, Abdi SHR, Ahmad I, Singh S, Jasra RV (2003) Catal Lett 91:207
42. Kureshy RI, Khan NH, Abdi SHR, Ahmad I, Singh S, Jasra RV (2004) J Catal 221:234
43. Bhattacharjee S, Anderson JA (2004) Chem Commun 554
44. Annunziata R, Benaglia M, Cinquini M, Cozzi F, Pozzi G (2003) Eur J Org Chem 1191
45. Fraile JM, García JI, Herrerías CI, Mayoral JA, Carré D, Vaultier M (2001) Tetrahedron: Asymmetry 12:1891
46. Fraile JM, García JI, Herrerías CI, Mayoral JA, Gmough S, Vaultier M (2004) Green Chem 6:93
47. Davies DL, Kandola SK, Patel RK (2004) Tetrahedron: Asymmetry 15:77
48. Meracz I, Oh T (2003) Tetrahedron Lett 44:6465
49. Fraile JM, García JI, Mayoral JA, Tarnai T (1997) Tetrahedron: Asymmetry 8:2089
50. Fraile JM, García JI, Mayoral JA, Tarnai T (1998) Tetrahedron: Asymmetry 9:3997
51. Alonso PJ, Fraile JM, García J, García JI, Martínez JI, Mayoral JA, Sánchez MC (2000) Langmuir 16:5607
52. Fernández MJ, Fraile JM, García JI, Mayoral JA, Burguete MI, García-Verdugo E, Luis S, Harmer MA (2000) Topics Catal 13:303
53. Fraile JM, García JI, Mayoral JA, Tarnai T, Harmer M (1999) J Catal 186:214
54. Fraile JM, García JI, Herrerías CI, Mayoral JA, Harmer MA (2004) J Catal 221:532
55. Fraile JM, García JI, Harmer M, Herrerías CI, Mayoral JA (2001) J Mol Catal A165:211
56. Fraile JM, García JI, Harmer MA, Herrerias CI, Mayoral JA, Reiser O, Warner H (2002) J Mater Chem 12:3290
57. Fraile JM, García JI, Herrerías CI, Mayoral JA, Reiser O, Socuellamos A, Werner H (2004) Chem Eur J 10:2997
58. Fernández AI, Fraile JM, García JI, Herrerías CI, Mayoral JA, Salvatella L (2001) Catal Commun 2:165
59. Cornejo A, Fraile JM, García JI, Gil MJ, Herrerías CI, Legarreta G, Martínez-Merino V, Mayoral JA (2003) J Mol Catal A 196:101
60. Langham C, Piaggio P, Bethell D, Lee DF, McMorn P, Bulman Page P, Willock DJ, Sly C, Hancock FE, King F, Hutchings GJ (1998) Chem Commun 1601
61. Langham C, Bethell D, Lee DF, McMorn P, Bulman Page P, Willock DJ, Sly C, Hancock FE, King F, Hutchings GJ (1999) Appl Catal A 182:85
62. Taylor S, Gullick J, McMorn P, Bethell D, Bulman Page P, Hancock FE, King F, Hutchings GJ (2001) J Chem Soc Perkin Trans 2:1724

63. Taylor S, Gullick J, McMorn P, Bethell D, Bulman Page P, Hancock FE, King F, Hutchings GJ (2001) J Chem Soc Perkin Trans 2:1714
64. Gullick J, Taylor S, McMorn P, Bethell D, Bulman Page P, Hancock FE, King F, Hutchings GJ (2002) J Mol Catal A 180:85
65. Taylor S, Gullick J, McMorn P, Bethell D, Bulman Page P, Hancock FE, King F, Hutchings GJ (2003) Topics Catal 24:43
66. Gullick J, Taylor S, Ryan D, McMorn P, Coogan M, Bethell D, Bulman Page P, Hancock FE, King F, Hutchings GJ (2003) Chem Commun 2808
67. Taylor S, Gullick J, Galea N, McMorn P, Bethell D, Bulman Page P, Hancock FE, King F, Willock DJ, Hutchings GJ (2003) Topics Catal 25:81
68. Wan Y, McMorn P, Hancock FE, King F, Hutchings GJ (2003) Catal Lett 91:145
69. Caplan NA, Hancock FE, Bulman Page P, Hutchings GJ (2004) Angew Chem Int Ed 43:1685
70. O'Leary P, Krosveld NP, De Jong KP, van Koten G, Klein Gebbink RJM (2004) Tetrahedron Lett 45:3177
71. Gamez P, Fache F, Lemaire M (1994) Bull Soc Chim Fr 31:600
72. Ferrand A, Bruno M, Tommasino ML, Lemaire M (2002) Tetrahedron: Asymmetry 13:1379
73. Zsigmond A, Bogár K, Notheisz F (2002) Catal Lett 83:55
74. Zsigmond A, Bogár K, Notheisz F (2003) J Catal 213:103
75. Alcón MJ, Corma A, Iglesias M, Sánchez F (2002) J Mol Catal A 178:253
76. Yu HB, Hu QS, Pu L (2000) Tetrahedron Lett 41:1681
77. Yu HB, Hu QS, Pu L (2000) J Am Chem Soc 112:6500
78. Zhang R, Yu WY, Wong KY, Che CM (2001) J Org Chem 66:8145
79. Zhang JL, Liu YL, Che CM (2002) Chem Commun 2906
80. Yin MY, Yuan GL, Wu YQ, Huang MH, Jiang YY (1999) J Mol Catal A 147:93
81. Xue L, Zhou DJ, Tang L, Ji XF, Huang MY, Jiang YY (2004) React Funct Polym 58:117
82. Yuan GL, Yin MY, Jiang TT, Huang MY, Jiang YY (2000) J Mol Catal A 159:45

Top Organomet Chem (2005) 15: 191–229
DOI 10.1007/b136351
© Springer-Verlag Berlin Heidelberg 2005
Published online: 14 September 2005

Chiral Diaminocarbene Complexes, Synthesis and Application in Asymmetric Catalysis

Sylvain Roland · Pierre Mangeney (✉)

Laboratoire de chimie organique, Université Pierre et Marie Curie, 4 place Jussieu, UMR 7611 CNRS, 75252 Paris, France
sroland@ccr.jussieu.fr, mangeney@ccr.jussieu.fr

Abstract The synthetic routes to optically pure *N*-heterocyclic carbene (NHC) complexes and their application in asymmetric catalysis is reviewed. Chiral *N*-heterocyclic carbenes which are mainly generated from the corresponding azolium salts can be used as ligands to prepare various complexes with transition metals. However, only a few of them have been used in asymmetric catalysis and reported examples where chiral NHC complexes give good enantioselectivity are still rare even if excellent levels of enantioselectivity have been reached (up to > 99% ee). This chapter describes the general properties of this family of ligand, the synthesis of the chiral azolium salts, the preparation of NHC-metal complexes and the asymmetric transformations catalyzed by palladium, rhodium, ruthenium, iridium and copper complexes.

Keywords Azolium salts · *N*-heterocyclic carbenes · transition metal complexes · asymmetric catalysis

1
General Introduction

Among the several ligands recently developed for transition metal catalyzed enantioselective reactions, the electron rich carbenes and particularly the diaminocarbenes (or *N*-heterocyclic carbenes) have one of the most promising potential (Fig. 1). This is not only due to the great improvement in the preparations of transition metal complexes with these ligands but also to the fact that these complexes exhibit numerous properties which make them very valuable. Indeed diaminocarbenes strongly bind the metal (stronger than the electron rich phosphines) yielding complexes wich are often thermally- and air-stable. For all these reasons, this class of ligand has lead to major advances in catalysis as shown for instance in the Ru-catalyzed alkene metathesis. Last but not least, the diaminocarbenes are open to various and easy structural modifications including the introduction of stereogenic centers. Therefore, their application in asymmetric catalysis has emerged as a very promising field of research. The interest of the chemists in this class of derivatives is well illustrated by five excellent reviews published since 1997 [1–5].

Fig. 1

2
Structure of *N*-heterocyclic Carbenes

N-heterocyclic carbenes are neutral compounds with a divalent carbon atom located between the two nitrogens. The four types of stable diaminocarbenes used for the synthesis of chiral complexes are listed below (Fig. 2):

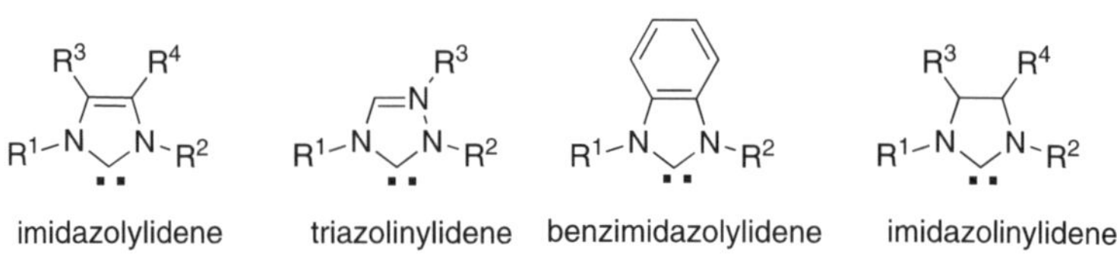

Fig. 2

Due to the presence of the two σ electron-withdrawing nitrogens, these carbenes are singlet carbenes. The nitrogens stabilize the σ nonbonding orbital increasing its s character. In the same time, they are π-electron donating substituents and the participation of the lone pairs increases the energy of the vacant p_π orbital. By this way, the σ-p_π gap is increased and the singlet state is favored (Fig. 3).

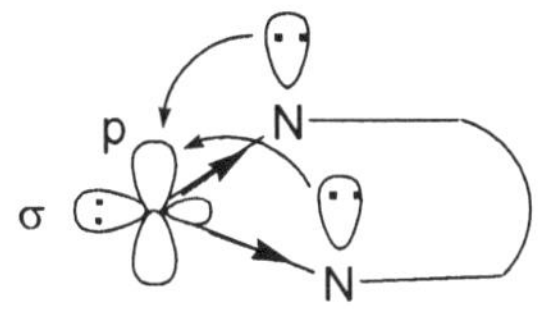

Fig. 3

In 1991, Arduengo reported the isolation of a stable and crystalline diaminocarbene **1** [6]. A 102.2° N–C–N bond angle was found by an X-ray analysis (Fig. 4).

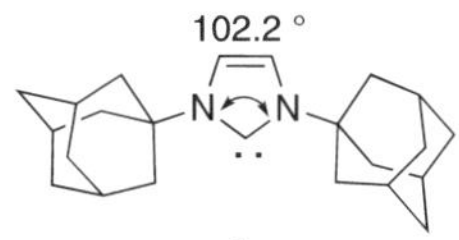

1

Fig. 4

Since that time, the solid state of several diaminocarbenes has been elucidated by the same technique and in all cases, the bond angle at the carbene center (100–110°) was in good agreement with the expected singlet nature of these carbenes. Rather short lenghts were observed for the N–C bonds (1.32–1.37 Å) giving evidence of their multiple bond character resulting of the donation of the nitrogen lone pair into the carbene vacant orbital [1–5]. This donation has a dramatic influence on the spatial structure of the diaminocarbenes, the nitrogen atoms being almost planar. This structural property, not surprising for the unsaturated diaminocarbenes which are aromatic, is also true for the imidazolinylidenes in which the nitrogens are not stereogenic centers as they are in the imidazolidines of type **2** (Fig. 5) [7].

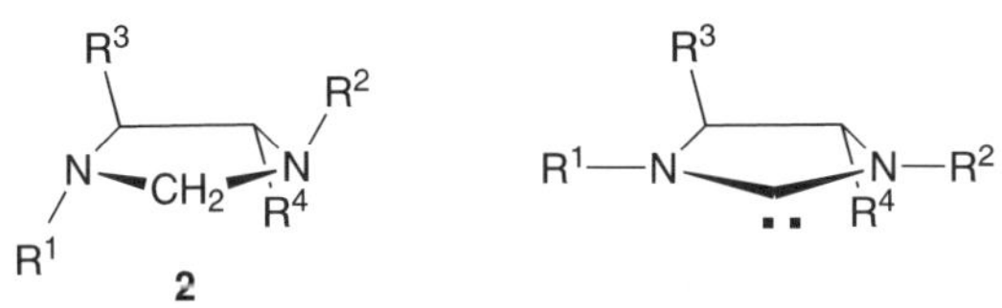

Fig. 5

3
Synthesis of Chiral N-heterocyclic Carbenes and of Their Complexes

3.1
Introduction

According to all their structural features, diaminocarbenes are good σ-donors and poor π-cceptors. They can easily displace most of the usual transition metal ligands including bridging halides, alkenes, nitriles, carbonyls, arenes and phosphines to give complexes showing often an exceptional air and thermal stability. The X-ray analysis of numerous complexes has shown single metal-carbon bonds illustrating the predominant σ-donor character of the diaminocarbenes even if non-negligible π interactions between the metal and the carbene p_π orbital has been observed [8]. As a consequence, a rotation is possible around the carbenic carbon–metal bond. Therefore, the most accurate representation of these complexes should be **A** instead of the **B** usually found in the literature (Fig. 6). For greater convenience and as proposed by Burgess [5], we will use this more conventional representation in this chapter.

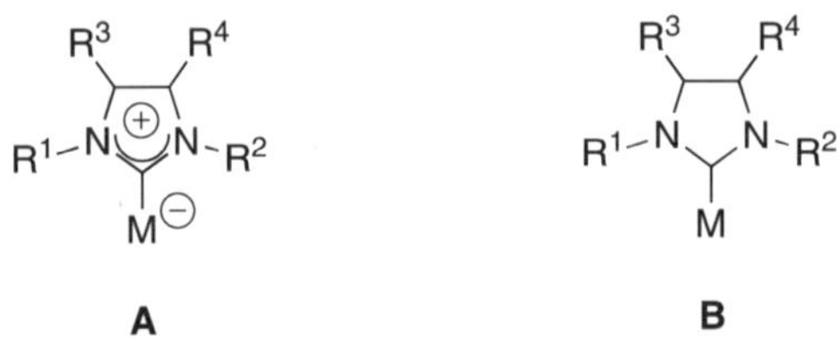

Fig. 6

Despite the planar conformation of nitrogens, several possibilities are available for the introduction of chirality. It is possible to prepare benzimidazolylidenes, triazolylidenes, imidazolylidenes or unsubstituted-backbone imidazolinylidenes with a stereogenic center on one or two N-substituents (carbenes **I** and **II**). The other possibility is to relay the imidazolinylidenes backbone stereogenecity via the N-substituents or to combine stereogenic N-substituents with a chiral backbone (carbenes **III** or **IV**). It is at least possible to prepare bis-carbenes of type **V** with one (or two) stereogenic link between the two carbenes (Fig. 7).

Lappert was the first to report, in 1983, the synthesis of chiral rhodium(I) and cobalt(I) imidazolinylidene complexes by heating an enantiopure electron rich olefin **3** in the presence of a complex precursor (Scheme 1) [9].

Except this particular example, the other possibilities such as the desulfurization of cyclic thioureas or the vacuum thermolysis of methoxy derivatives [1–5], has never been used for the preparation of chiral diaminocar-

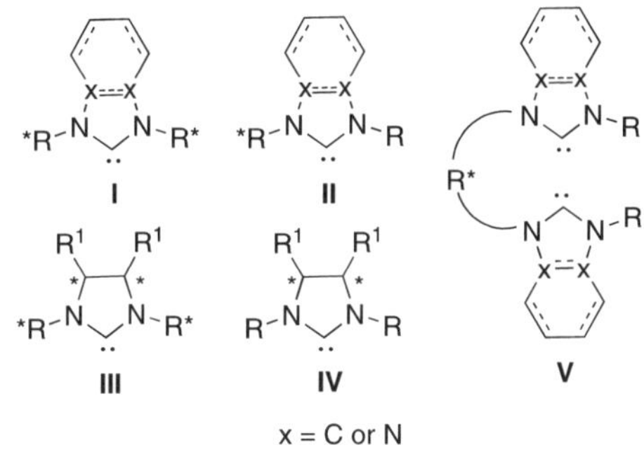

Fig. 7

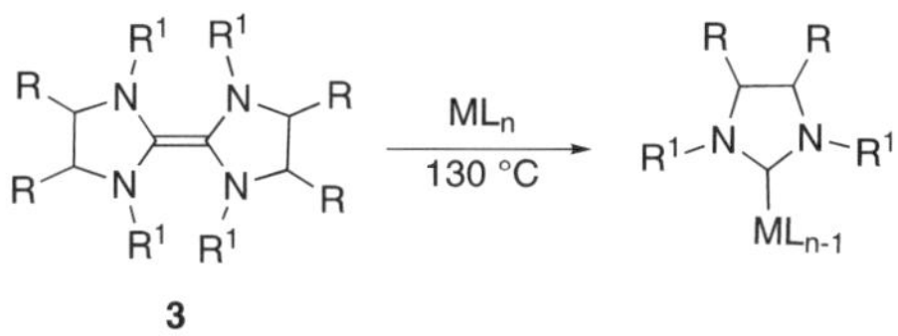

Scheme 1

benes. These compounds are always obtained by deprotonation of the corresponding azolium salts. The pK_a of *N*,*N*-dialkylimidazolium and benzimidazolium salts has been measured in THF, DMSO or in water and found to be in the range of 20–23 [10]. Therefore, the deprotonation of these salts to give the corresponding carbenes is an easy process. However the preparation of diaminocarbenes can be complicated by a possible dimerization. Such a reaction has been observed by Wanzlick during an attempt of preparation of the achiral diaminocarbene **4** by a thermal elimination of chloroform in **5** [11]. Only the electron rich olefin was obtained and it was postulated that this compound was the result of an irreversible dimerization of the carbene (Scheme 2).

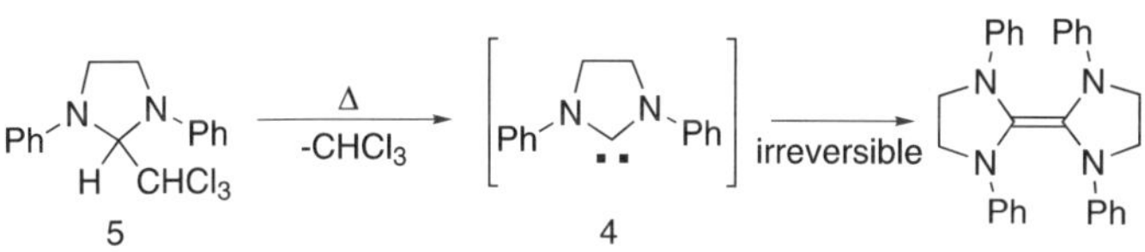

Scheme 2

Since then, it has been observed that the dimerization of imidazolylidenes and triazolinylidenes does not occur at room temperature. In contrast, the dimerization of imidazolinylidenes is a fast and irreversible process but is prevented by using bulky nitrogen substituents. The benzimidazolylidenes are in equilibrium with the dimer, this equilibrium being affected by steric factors. Therefore except the imidazolylidenes which can be isolated as free carbenes, the other diaminocarbenes are generally formed in the presence of the metallic complex precursor [1–5]. No example of a free chiral diaminocarbene is reported in the literature.

Several protocols, discussed in detail in the following sections, have been developed in which the final complex and/or the intermediates are isolated or generated in situ (Scheme 3). The most general method consists in a deprotonation of the salt to generate the carbene which is then transfered onto the metal precursor. It is possible to perform the deprotonation in a first separate step (Eq. 1) or directly in the presence of the metal precursor (Eq. 2). Alternatively chiral azolium salts can be converted into transition metal complexes by treatment with a complex containing basic ligands such as $Pd(OAc)_2$ or $[Rh(\mu\text{-O}t\text{-Bu})(nbd)]_2$. An interesting possibility is the use of silver oxide as base thus avoiding the use of strong bases and polar solvents. The intermediate silver(I) complexes can be isolated as air-stable solids. They act as efficient carbene transfer agents towards other metal complex precursors (Eq. 3). The new NHC-metal complexes are either isolated or generated in situ.

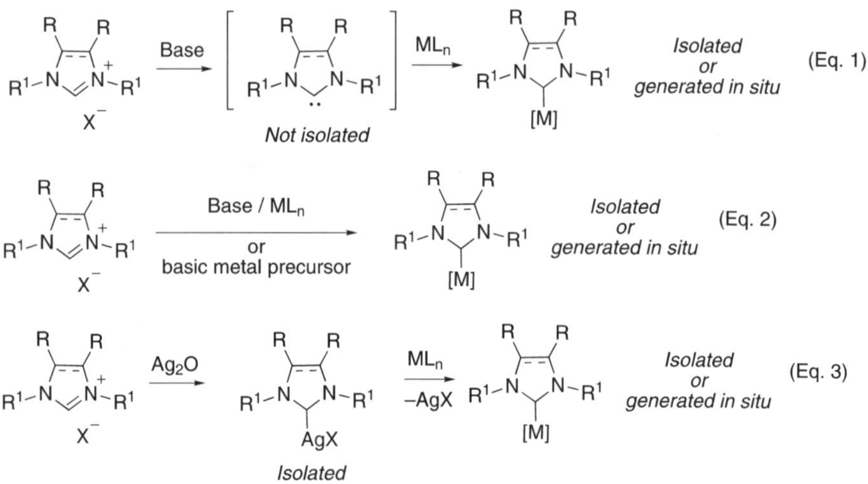

Scheme 3

Mono or bis-carbene complexes are possible depending on the carbene/metal precursor ratio and the steric bulk of the carbene. Most of the metal precursors and bases used for the synthesis of chiral complexes are presented below: *Metal precursors*:
Pd: $Pd(OAc)_2$, $PdCl_2(CH_3CN)_2$, $Pd(dba)_2$, $[Pd(\eta^3\text{–}C_3H_5)Cl]_2$
Rh: $[Rh(\mu\text{-O}t\text{-Bu})(nbd)]_2$, $[Rh(cod)Cl]_2$, $[Rh(acac)(C_2H_4)_2]$
Ir: $[Ir(cod)Cl]_2$
Cu: $Cu(OTf)_2$, $Cu(OAc)_2$, CuTC

Cr: $Cr(CO)_6$
Ni: $Ni(CO)_4$, $Ni(acac)_2$, $NiCl_2(PPh_3)_2$
W: $W(CO)_6$
Ru: $(PR_3)_2Cl_2Ru = CHPh$, $[Cp^*RuCl]_4$
Bases:
t-BuOLi, *t*-BuOK, *t*-BuONa, $(CF_3)_2(CH_3)COK$, NaH, NaH/NH_3, NEt_3, NaOAc, Ag_2CO_3, Ag_2O, *n*-BuLi

Using these procedures, many chiral diaminocarbene-transition metal complexes have been synthesized but only a few of them have been used for asymmetric catalysis. The chiral complexes which were isolated but did not receive any application in asymmetric catalysis, are presented at the end of the chapter.

The azolium salts being the main precursors of the chiral diaminocarbenes, the next section will concern their preparations.

3.2
Preparation of Imidazolium Salts

Symmetrical *N*, *N′*-disubstituted imidazolium salts are usually obtained by addition of paraformaldehyde on a bis-imine of glyoxal under acidic conditions. A one-pot procedure has been developed. Several enantiomerically pure amines were used to prepare the corresponding symmetrical salts **6** (Scheme 4) [12, 13].

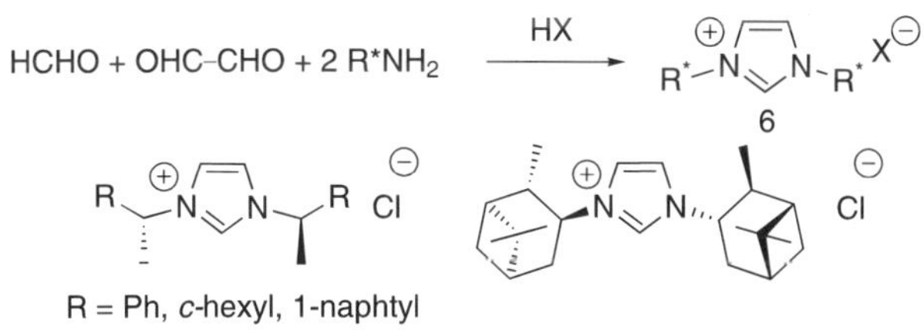

Scheme 4

With slight modifications of these conditions it is possible to prepare mono *N*-substituted imidazoles (Scheme 5), the reaction working well with aliphatic amines but not with many aromatic amines. The unsymmetrical

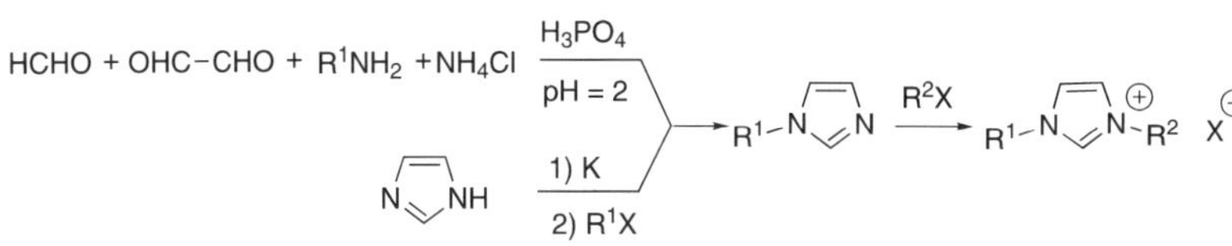

Scheme 5

salts are then obtained by *N*-alkylation. 1-Substituted imidazoles are also accessible by alkylation of the imidazolide anion (Scheme 5). In these preparations the chirality can be introduced via the amine (R^1NH_2), and/or via alkylating reagents (R^1X and/or R^2X) [4, 5].

Herrmann has prepared several unsymmetrical salts **7** from 1-alkylimidazoles (Scheme 6). The chirality was introduced, after *N*-alkylation of the imidazole by chloroacetonitrile, by addition of enantiomerically pure aminoalcohols onto the nitrile to form an oxazolidine ring [14].

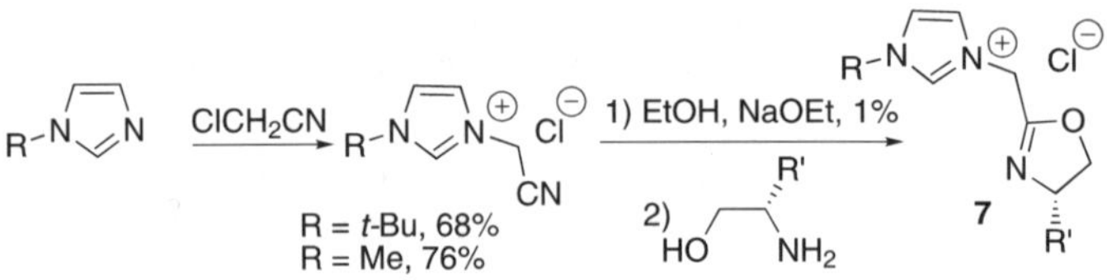

R	R'	yield (%)
Me	Benzyl	72
Me	*i*-Pr	68
t-Bu	Benzyl	64
t-Bu	*i*-Pr	68

Scheme 6

Burgess followed a similar strategy for the preparation of the salts **8** (Scheme 7). On that occasion several routes to mono-*N*-substituted imidazoles were explored yielding the desired compounds in variable yields depending on the nature of the amines. The chirality was introduced via alkylating reagents **9** bearing chiral oxazolines [15].

Gade and Bellemin-Laponnaz have reported the synthesis, in good yields, of chiral oxazoline-imidazoliums salts **10a** (Scheme 8) obtained by reaction of 2-bromo-4(*S*)-*t*-butyl oxazoline with several mono-*N*-substituted imidazoles [16]. Similaly an imidazolium salt **10b** bearing a paracyclophane substituent was prepared by Bolm [17].

The synthesis of the unsymmetrical imidazolium salt **11** bearing a planar-chiral ferrocene was described by Bolm starting from (R_p)-[2-(trimethysilyl)-ferrocenyl] methanol **12** which afforded the salt in good yield after reaction with *N*,*N*-carbonyl diimidazole and methylation (Scheme 9) [18].

A tripodal imidazolium salt **13** was obtained by Howarth from enantiopure imidazoles **14** (Scheme 10) by reaction with 1,3,5-tris(bromomethyl)-2,4,6-trimethyl benzene [19].

RajanBabu reported the first preparation of a bis-imidazolium salt **15** bearing a chiral linker (Scheme 11). The starting material was the enantiomerically pure (*S*)-1,1′-bi-2-naphtol bis(trifluoromethanesulfonate) which was transformed in two steps into the dibromomethyl derivative **16** and then into the bis-imidazole. Quaternarization of this compound afforded **15** [20].

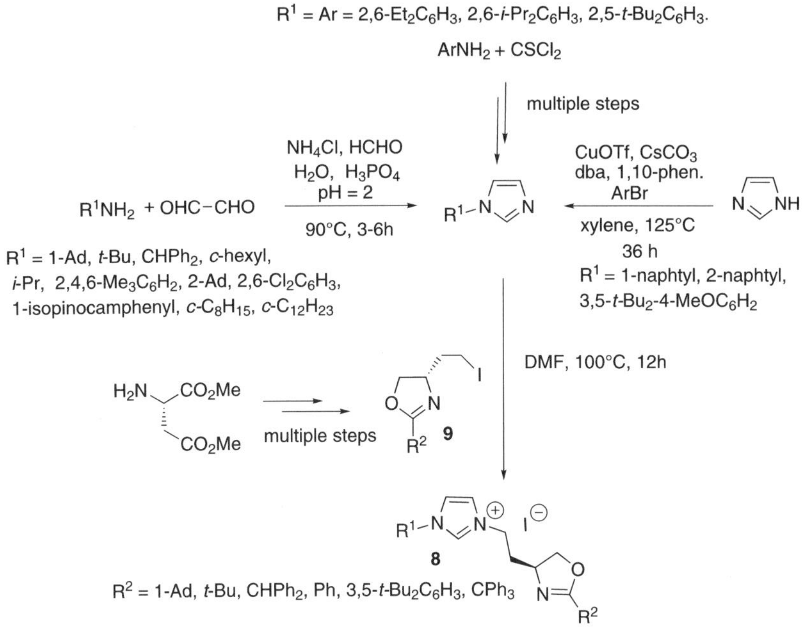

Scheme 7

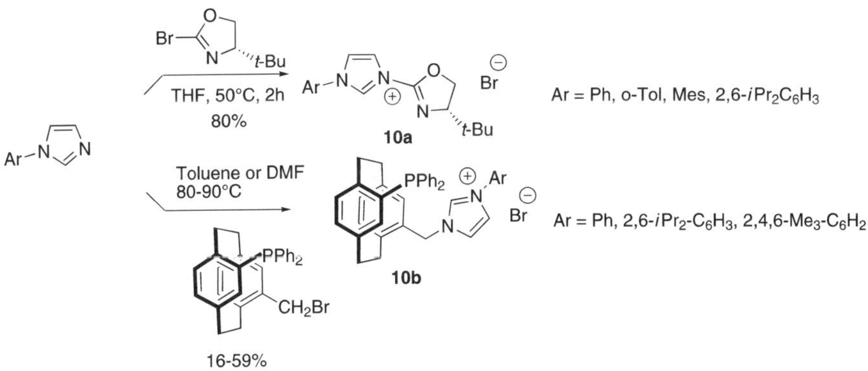

Scheme 8

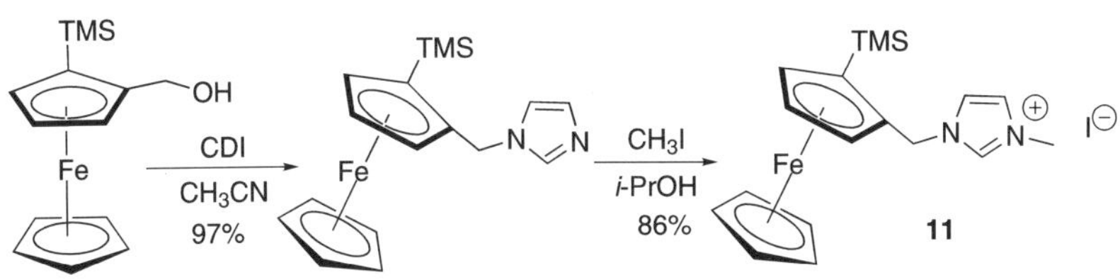

12

Scheme 9

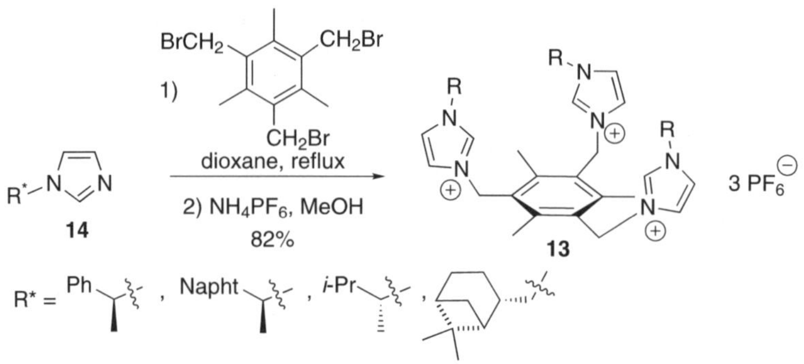

Scheme 10

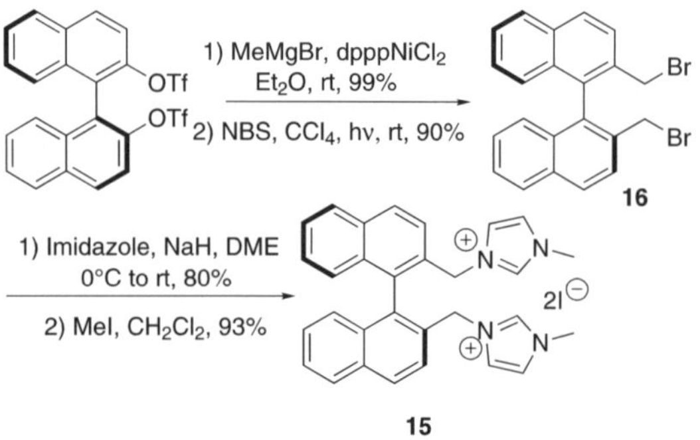

Scheme 11

1,1′-Binaphtalenyl-2,2′-diamine was the starting material used by Shi to prepare the bis-benzimidazolium salt **17** (Scheme 12). This salt was obtained in good overall yield in a four step procedure. The first one is a palladium catalyzed amination of **18** with 2-bromo-nitrobenzene. Reduction of the ob-

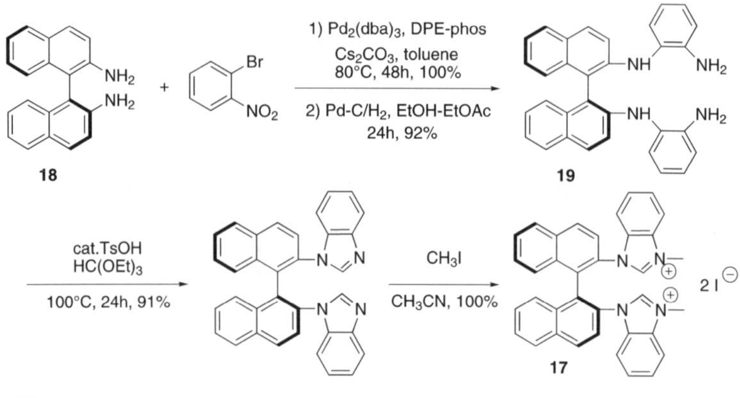

Scheme 12

tained derivative afforded the tetraamine **19** which was treated first with triethyl orthoformate and then with methyl iodide [21].

A bis-imidazolium salt was also prepared by Burgess, starting from the dichloride **20** derived from optically pure *N*,*N′*-dimethyl-*trans*-1,2-diaminocyclohexane (Scheme 13). The salt **21** was obtained by addition of this compound to several 1-alkylimidazoles [22].

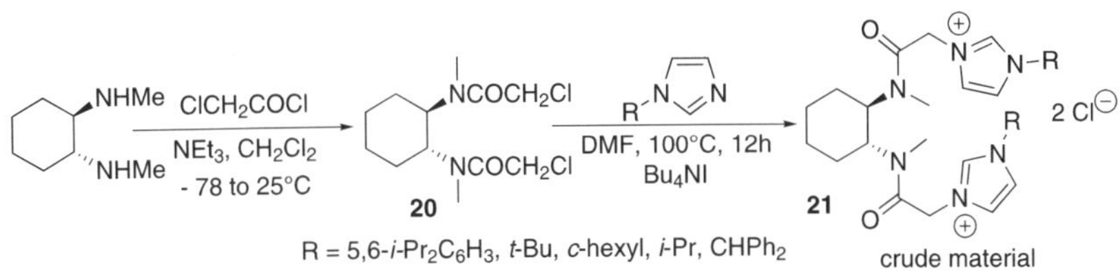

Scheme 13

An other type of bis-imidazoliums salts **22**, resulting of a base-induced 1,3-cycloaddition of tosyl-methylisocyanide (TosMIC) to the enantiopure diimine **23**, was obtained by Douthwaite according to Scheme 14 [23].

Starting from the benzylidene diimine **24** (Scheme 15) and only one equivalent of TosMIC, the same reaction afforded the imidazole-imine derivative and then, after alkylation, the corresponding salt **25**. After hydrolysis, an

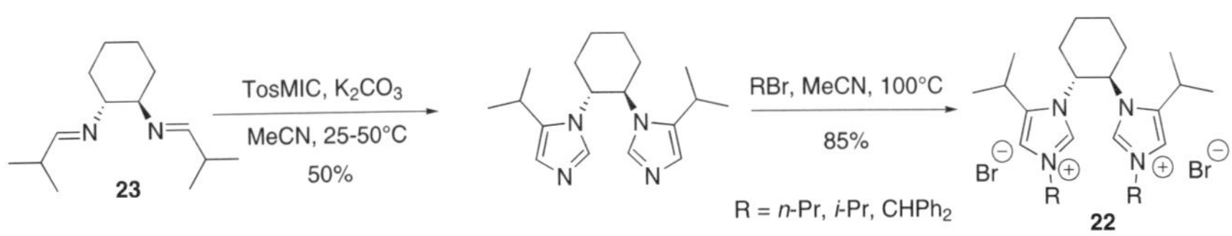

Scheme 14

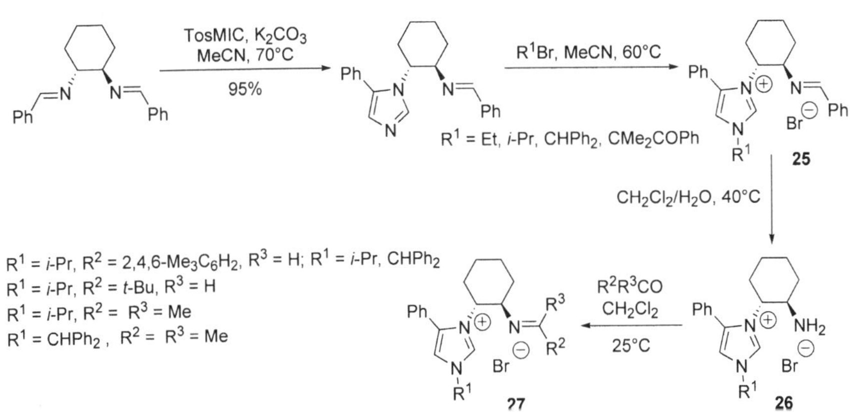

Scheme 15

imidazolium-amine salt **26** was obtained and transformed, by reaction with several ketones, into the imidazolium-ketimines salts **27** [24].

3.3 Preparation of Triazolium Salts

1,4-Disubstituted triazolium salts are prepared from alkyl or aryl hydrazines *via* an oxadiazolium salt **28** (Scheme 16). Addition of a chiral amine on this salt resulted in a "ring opening – ring closure" reaction affording the triazolium salts **29**.

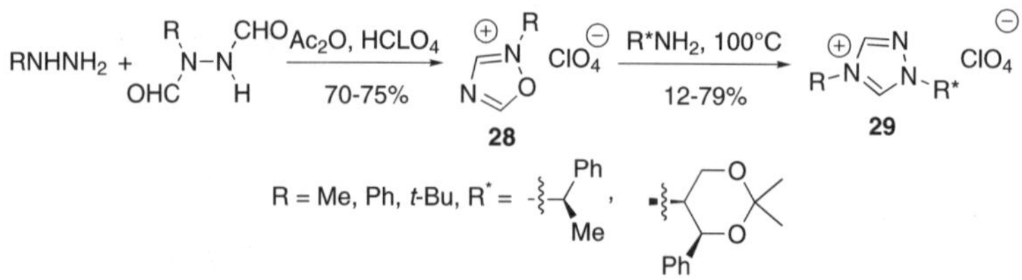

Scheme 16

1-Alkyl-3,4-disubstituted-4*H*-1,2,4-triazol-1-ium salts **30** were prepared by reaction of *N*-alkyl-*N*-formyl hydrazines with an imidoyl chloride, prepared in situ, and then cyclization with acetic anhydride (Scheme 17). By this way, several chiral salts, including a polymer supported one, were prepared [3].

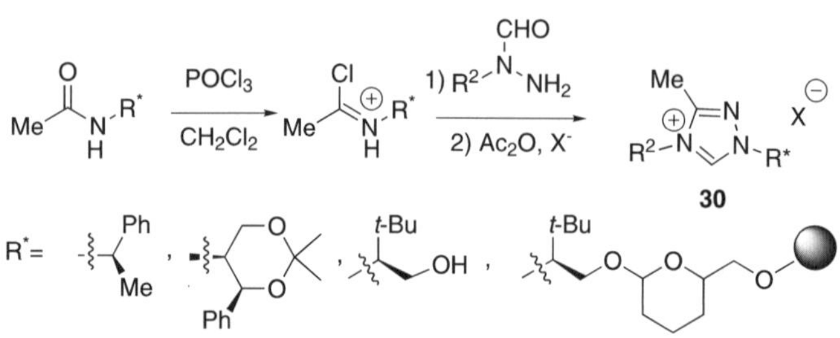

Scheme 17

An enantiopure bicyclic triazolium salt was prepared by a three step conversion of the oxazolidinone **31** (Scheme 18). The first step was a methylation affording an iminoether which was transformed into the phenylhydrazone **32**. A final cyclization gave the salt **33** as a cristalline solid [25].

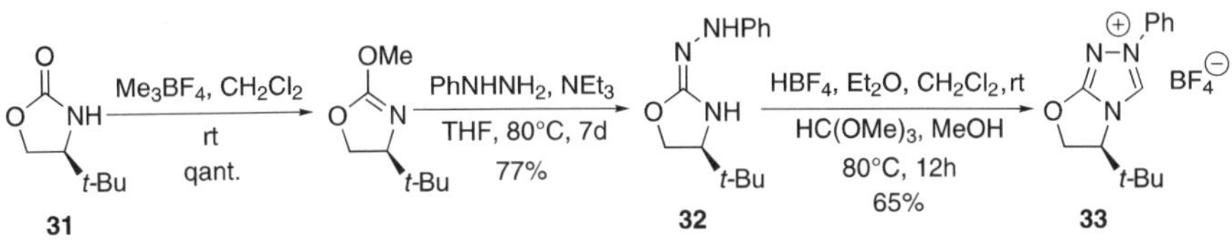

Scheme 18

3.4 Preparation of Imidazolinium Salts

These salts are obtained from 1,2 (or 1,3)-diamines. The cyclization step is generally the condensation of the diamines on ethyl orthoformate in the presence of ammonium tetrafluoroborate (Scheme 19). By using enantiopure diamines, chiral salts **34** bearing stereogenic centers on the backbone, on the nitrogen substituents, or on both were prepared [1, 2, 4, 5, 26]. Several dicyclophane imidazolinium salts have been prepared by one-pot three-step procedure [27].

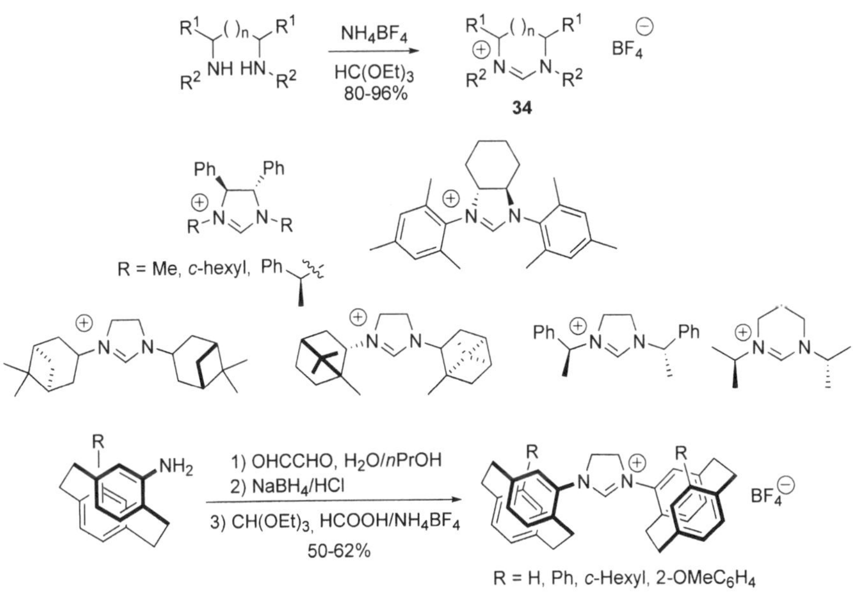

Scheme 19

Grubbs reported the synthesis of several N,N'-aryl substituted imidazolinium salts **35** from chiral N,N'-aryl diamines obtained by palladium-catalyzed amination of the appropriate aryl bromide with (1*R*,2*R*)-diaminocyclohexane

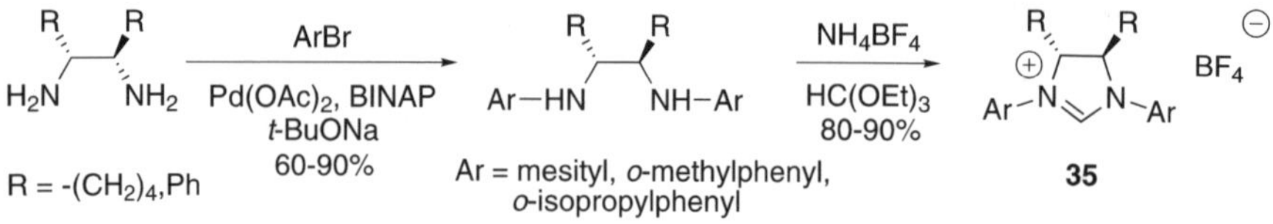

Scheme 20

or (1*R*,2*R*)-diphenylethylenediamine (Scheme 20). The formation of the salt was then obtained as usual with orthoformiate [28, 29].

Some chiral imidazolidinium salts bearing tert-butyl groups on the backbone were prepared from the enantiopure imidazololidine **36** (Scheme 21). The key step of these preparations involves oxidation of the imidazolidine ring, in the presence of palladium hydroxyde and ammonium formate. The imidazolidine **37**, resulting from an hydrogenolysis of the chiral *N*-substituents and an oxidation of the heterocycle, is obtained. The salts **38** are then obtained by *N*-alkylation. Alternatively, starting from **36** it is also possible to prepare the salt **39** in the presence of Pd/C in acidic conditions or in the presence of iodine [30–32].

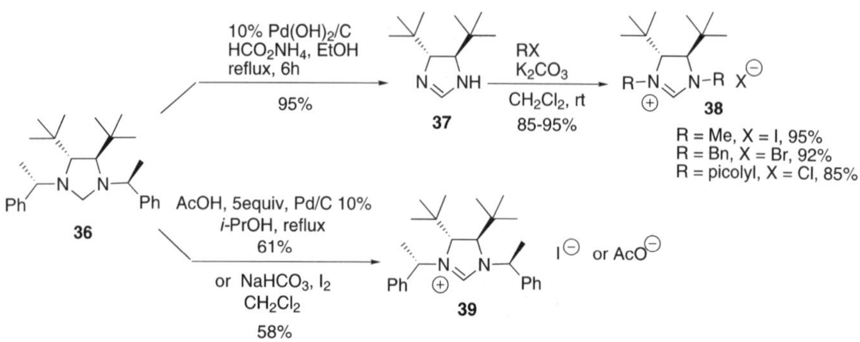

Scheme 21

A non symmetric salt **40** was prepared, in 83% overall yield, by Hoveyda starting from an optically pure 2′-amino-(1, 1′)-binaphtanelyl-2-ol (Scheme 22). This compound was transformed into the aminoalcohol **41** by

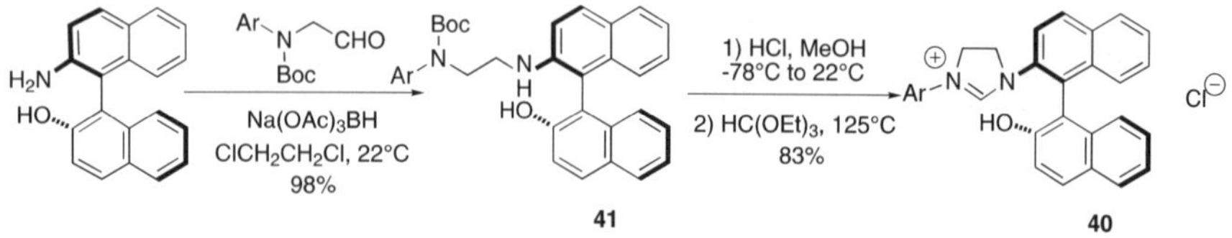

Scheme 22

a reductive alkylation. The construction of the imidazolinium cycle was then realized in two steps [33].

3.5
Preparation of Benzimidazolium Salts

The key step for the synthetic approach of these salts involves a two steps Pd-catalyzed amination of *o*-dibromobenzene. By this procedure, it is possible to prepare symmetrical or unsymmetrical 1,2-benzenediamines. The monobromoanilines **42** were obtained trough a controlled monoamination reaction. The diphosphine BINAP was found to be the most versatile ligand for this reaction. The benzymidazolium salt **43** was obtained by a second amination followed by a ring closure reaction (Scheme 23). When an optically pure amine was used, carefull attention to reaction conditions proved to be critical to suppress a possible epimerization. The use of a strong acid (HCl or $HClO_4$) was necessary for the cyclization step. As usual, a counterion exchange afforded easily isolable salts [34, 35].

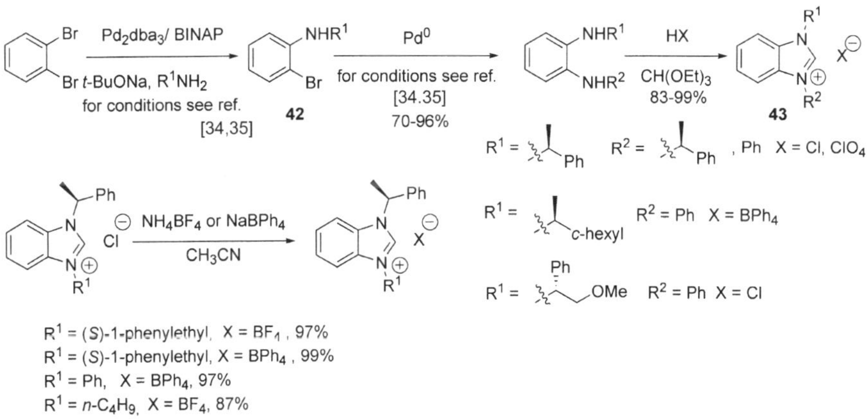

Scheme 23

4
Application in Asymmetric Catalysis

4.1
Palladium *N*-heterocyclic Carbene Complexes

N-heterocyclic carbene complexes of Pd(II) or Pd(0) were extensively used in various reactions and several groups have reported syntheses of chiral complexes [5]. However, only a few examples of asymmetric catalysis are reported in the literature. Applications in the Heck reaction, α-arylation of amides, allylic alkylation and oxidative kinetic resolution of secondary alcohols were developed. The synthesis of the first chiral *N*-heterocyclic palladium(II) complexes was described by Enders and coworkers in 1996 [36]. Bis-diaminocarbene Pd(II) complexes **44a–b** were obtained by deprotonation of the corresponding perchlorate azolium salts using potassium *tert*-butoxide in the presence of $Pd(OAc)_2$ and an excess of NaI (Scheme 24). These NHC-Pd complexes were tested for the first time in asymmetric catalysis. However, it was only stated that low asymmetric inductions (ee < 8%) could be achieved in Heck-type reactions. The substrates tested and the conditions were not mentioned. This is the only attempt reported to date of an asymmetric version of the Heck reaction using NHC – Pd complexes. The Pd(II) complexes are air- and water-stable solids and were isolated as a mixture of *trans* and *cis* isomers. Being stable on silicagel the *trans* and *cis* complexes could be separated by chromatography. Crystallographic evidence of their square-planar geometry was obtained by X-ray analysis of the trans-syn triazolinylidene palladium(II) complex **44a**. The possibility to recycle these stable non-supported chiral complexes is considered in this work.

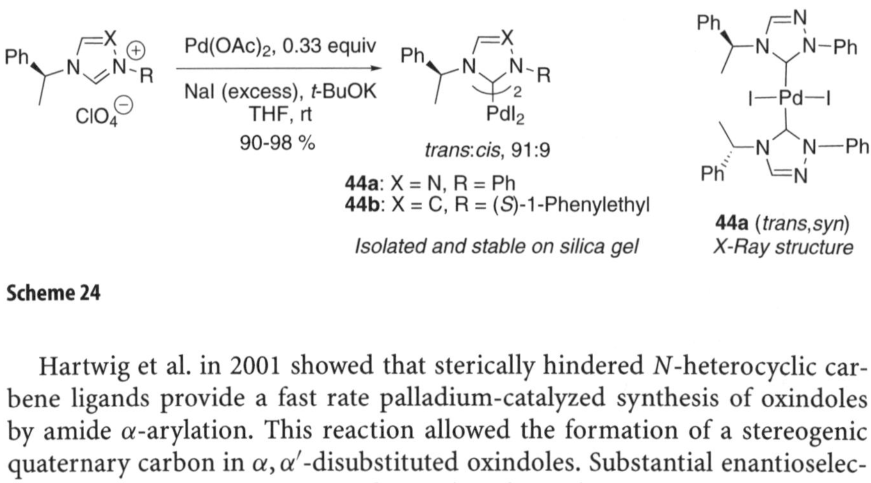

Scheme 24

Hartwig et al. in 2001 showed that sterically hindered *N*-heterocyclic carbene ligands provide a fast rate palladium-catalyzed synthesis of oxindoles by amide α-arylation. This reaction allowed the formation of a stereogenic quaternary carbon in α,α'-disubstituted oxindoles. Substantial enantioselectivities, reaching 76% ee, were obtained in the cyclization of the α-naphthyl α-methyl amide **45** (Scheme 25) using the azolium salt **46** derived from bornylamine as the carbene ligand precursor [37]. The reaction occurred with high yield and under mild conditions in DME with $Pd(dba)_2$ as palladium source and sodium *tert*-butoxide as base. The active palladium diaminocarbene complex was not isolated but generated in situ by deprotonation of the imidazolinium salt by sodium *tert*-butoxide (in excess) to generate the carbene which is transfered onto the Pd(0). A 1 : 1 ratio of $Pd(dba)_2$ and carbene ligand was used, suggesting that a monocarbene palladium complex is formed. The authors showed that the base, the counterion and the solvent were all important in the control of the enantioselectivity whereas the ligand-to-metal ratio was not. In the same reaction, the salt **47** derived from isopinocampheylamine gave 67% ee at 25 °C. By comparison, the ligand obtained from the salt **48** (derived from cyclohexanediamine) gave lower enantioselectivities.

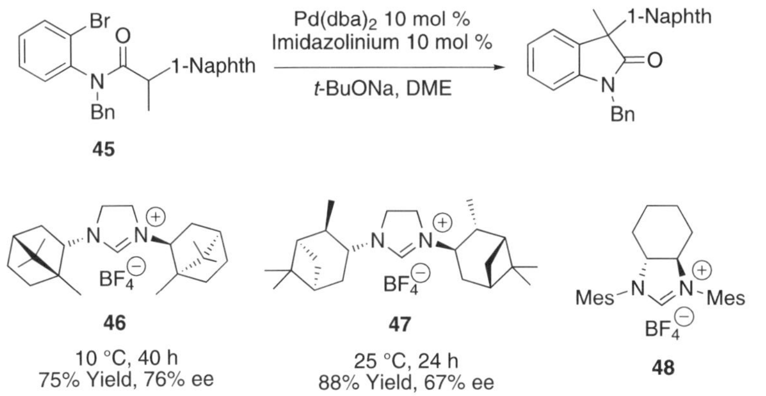

Scheme 25

In 2003, Douthwaite et al. reported up to 92% ee in the allylic alkylation reaction catalyzed by palladium complexes [24] bearing chelating NHC-imine ligands derived from trans-1,2-diaminocyclohexane (Scheme 26). The palladium(II) catalyst precursors used in the reaction were not isolated but generated in solution prior use by transfer of the carbene ligands from silver complexes **49a–h** onto 0.5 equiv of palladium allyl chloride dimer. The precursor diaminocarbene silver(I) complexes were easily obtained by treatment of the corresponding imidazolium salts with Ag_2O [38]. This new class of chiral ligands was tested in the allylic alkylation of (*E*)-1,3-diphenylprop-3-en-1-yl acetate with dimethylmalonate. This is the first report of an asymmetric version of this reaction with chiral NHC – Pd complexes. An achiral variant was previously reported by Mori in 2003 [39]. The dependence of enantioselectivity on structure modification was studied. An increase of the bulkiness of the R^1 group and a decrease of the one of the imine (R^2 and R^3 groups) appears to favor the selectivity for the *S* enantiomer. The best enantioselectivities were reached with the diaminocarbene obtained from the silver complex **49a** bearing an imino group derived from acetone and an isopropyl group on the nitrogen of the heterocycle. The best ligand-to-metal ratio is 1 : 1 and the ee decreases on addition of more than 1 equiv of ligand suggesting that monocarbene-imino Pd(II) allyl chloride complexes are first formed. These catalyst precursors can be reduced into active Pd(0) complexes by reaction with nucleophiles.

To probe the coordination mode and conformations of these ligands, a palladium(II) dichloride complex **50** was prepared by addition of the silver

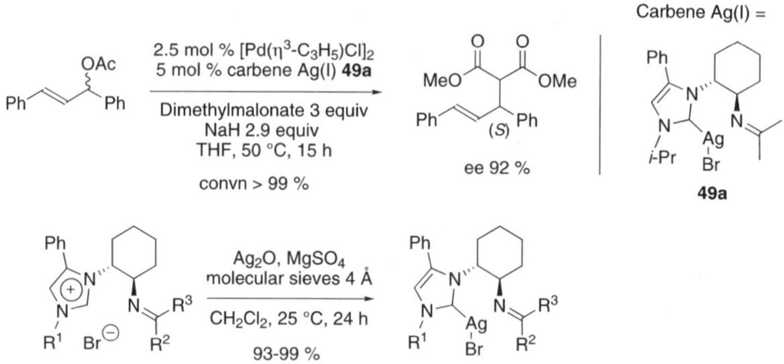

49a: R^1 = *i*-Pr, R^2 = Me, R^3 = Me
49b: R^1 = $CHPh_2$, R^2 = Me, R^3 = Me
49c: R^1 = *i*-Pr, R^2 = Ph, R^3 = H
49d: R^1 = *i*-Pr, R^2 = Me, R^3 = H
49e: R^1 = *i*-Pr, R^2 = *t*-Bu, R^3 = H
49f: R^1 = Et, R^2 = Ph, R^3 = H
49g: R^1 = $CHPh_2$, R^2 = Ph, R^3 = H
49h: R^1 = $CMe_2C(O)Ph$, R^2 = Ph, R^3 = H

Scheme 26

complex **49c** onto $PdCl_2(CH_3CN)_2$ (Scheme 27). This air-stable complex was studied by single-crystal X-ray diffraction. This molecular structure confirmed that such ligands are chelating and are able to form a six-atom metallocycle in a boatlike conformation. Currently the enantioselectivity and kinetics are relatively low compared to many chiral phosphine ligands.

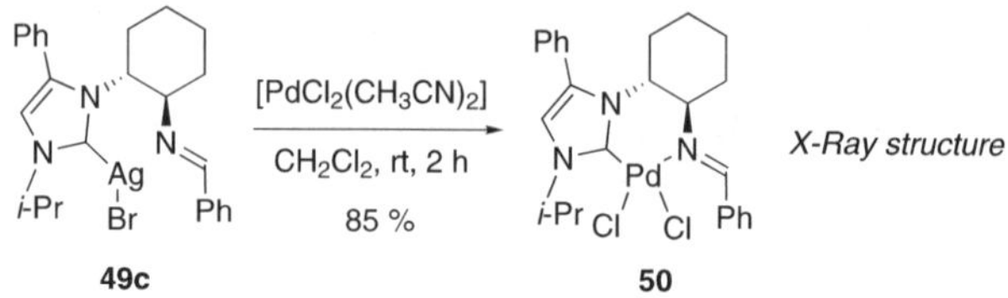

Scheme 27

In 2003, Sigman et al. reported the use of a chiral carbene ligand in conjunction with the chiral base (–)-sparteine in the palladium(II) catalyzed oxidative kinetic resolution of secondary alcohols [26]. The dimeric palladium complexes **51a–b** used in this reaction were obtained in two steps from N,N′-diaryl chiral imidazolinium salts derived from (*S*, *S*) or (*R*, *R*) diphenylethane diamine (Scheme 28). The carbenes were generated by deprotonation of the salts with *t*-BuOK in THF and reacted in situ with dimeric palladium allyl chloride. The intermediate NHC – Pd(allyl)Cl complexes **52** are air-stable and were isolated in 92–95% yield after silica gel chromatography. Two diastereomers in a ratio of approximately 2 : 1 are present in solution ($CDCl_3$).

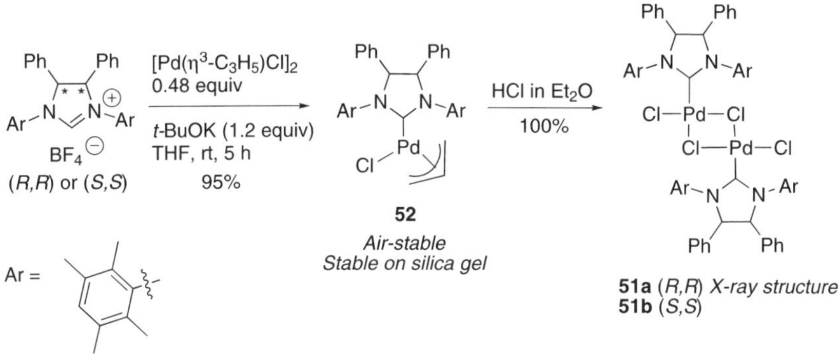

Scheme 28

Protonolysis of the allyl group with HCl in ether proceeds smoothly to liberate propene and deliver the $PdCl_2$-carbene dimers **51a–b** in quantitative yield and excellent purity. The (R,R) **51a** complex was analyzed by X-ray crystallography.

The authors demonstrated that both achiral and chiral *N*-heterocyclic carbene ligands in conjuction with (–)-sparteine, are effective for the Pd(II)-catalyzed oxidative kinetic resolution. In the case of a chiral ligand, due to the "matched" or "mismatched" diastereomeric interactions with (–)-sparteine, antipodes of the ligand afforded different k_{rel} values (Scheme 29). A significantly higher k_{rel} value of 11.8 (42% ee for 34.6% convn) was observed for catalyst (S,S) **51b** as compared with the (R,R) **51a**. This observation of a matched interaction demonstrates that a chiral carbene ligand and chiral base can act in concert to enhance the kinetic resolution. To further highlight the contribution of the ligand in the matched oxidative kinetic resolution, the chiral complex **51b** (S,S configuration) was evaluated using silver acetate instead of sparteine as base. In these conditions, the complex preferentially oxidized the same enantiomer of alcohol as oxidation using (–)-sparteine but with a lower k_{rel} of 1.6.

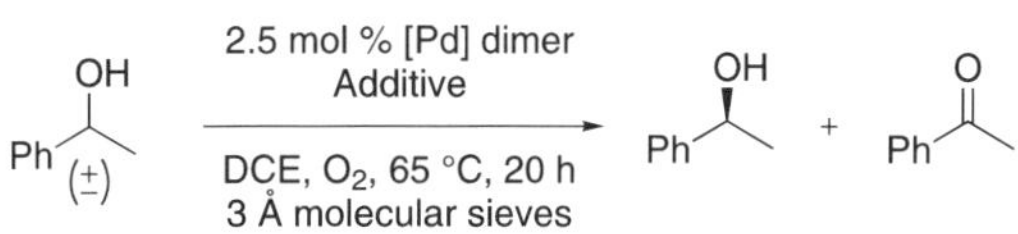

Dimer	Additive	% convn (% ee)	k_{rel}
51a	(–)-sparteine (20 mol %)	39.7 (36.4)	4.5
51b	(–)-sparteine (20 mol %)	34.6 (42.0)	11.8
51b	AgOAc (10.5 mol %)	34.5 (10.2)	1.6

Scheme 29

4.2 Rhodium *N*-heterocyclic Carbene Complexes

Chiral diaminocarbene complexes of rhodium have been merely used in asymmetric hydrosilylations of prochiral ketones but also in asymmetric addition of aryl boron reagents to enones.

Herrmann et al. reported for the first time in 1996 the use of chiral NHC complexes in asymmetric hydrosilylation [12]. An achiral version of this reaction with diaminocarbene rhodium complexes was previously reported by Lappert et al. in 1984 [40]. The Rh(I) complexes **53a–b** were obtained in 71–79% yield by reaction of the free chiral carbene with 0.5 equiv of $[Rh(cod)Cl]_2$ in THF (Scheme 30). The carbene was not isolated but generated in solution by deprotonation of the corresponding imidazolium salt by sodium hydride in liquid ammonia and THF at – 33 °C. The rhodium complexes **53** are stable in air both as a solid and in solution, and their thermal stability is also remarkable. The hydrosilylation of acetophenone in the presence of 1% mol of catalyst **53b** gave almost quantitative conversions and optical inductions up to 32%. These complexes are active in hydrosilylation without an induction period even at low temperatures (– 34 °C). The optical induction is clearly temperature-dependent: it decreases at higher temperatures. No significant solvent dependence could be observed. In spite of moderate ee values, this first report on asymmetric hydrosilylation demonstrated the advantage of such rhodium carbene complexes in terms of stability. No dissociation of the ligand was observed in the course of the reaction.

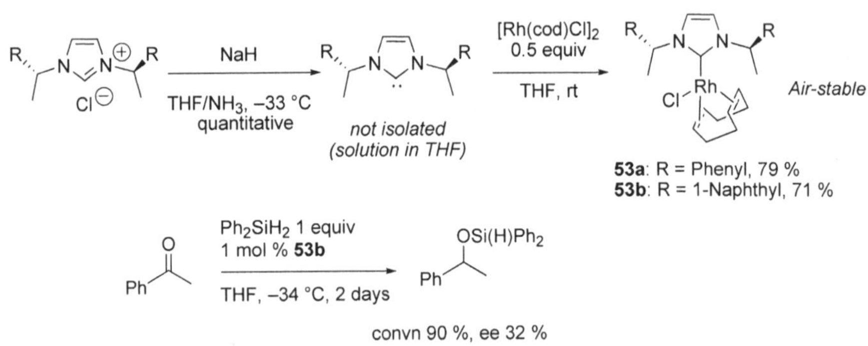

Scheme 30

In 1998, Enders et al. reported the use of the rhodium(cod) complexes **54a–f** containing chiral triazolinylidenes in the same reaction [41]. Complexes **54** were prepared in THF in 65–95% yield, by reaction of the triazolium salts with 0.45 equiv of $[Rh(cod)Cl]_2$ in the presence of NEt_3 (Scheme 31). The carbene ligand in such complexes is nonchelating with possible hindered rotation around the carbene carbon–rhodium bond. Due to

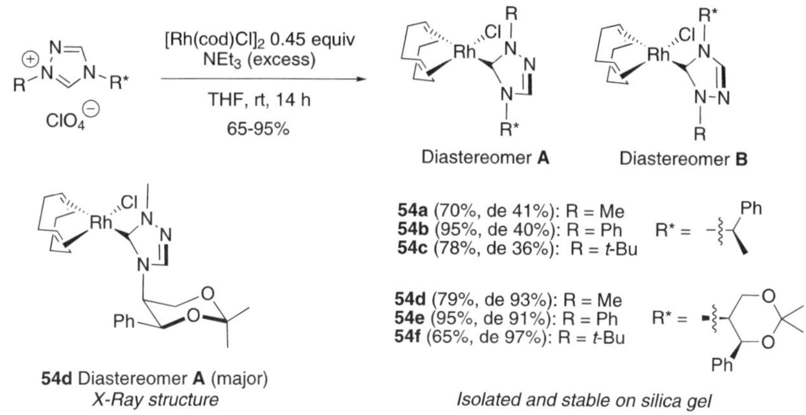

Scheme 31

the axis of chirality two diastereomers **A** and **B** can be formed with a perpendicular orientation of the ligand core with the square-plane of the metal complex. The diastereomeric excess of the complexes was determined by ^{1}H-NMR spectroscopy. The better excesses (91–97%) were obtained using the 2,2-dimethyl-4-phenyl-1,3-dioxanyl residue as chiral subsituent (complexes **54d–f**). An optimum diastereoselectivity (97%) was reached in this case with the bulky *t*-butyl group (complex **54f**). The major diastereomer **A** of complex **54d** (R = Me) could be crystallized and its solid state structure was determined by X-ray analysis. This structure shows the rhodium atom with a square-planar arrangement of the ligands.

In typical hydrosilylations, methyl ketones were reduced to the corresponding alcohols in 40–90% yield with diphenylsilane and 1% mol of the catalyst in THF. The best enantioselectivities (up to 44%) were achieved with complexes **54d–f** with similar results for aromatic and aliphatic ketones (Scheme 32). The configuration of the resulting alcohol depends upon the achiral group R of the triazolinylidene ligand. For example, the enantiomeric excess changes from 44% (*S*) for R = Me (complex **54d**) to 43% (*R*) for R = *t*-Bu (complex **54f**) with acetylcyclohexane as the starting material. The optimal reaction temperature in order to achieve the highest enantioselectivities varies from + 42 °C to – 10 °C. An nonlinear temperature effect was

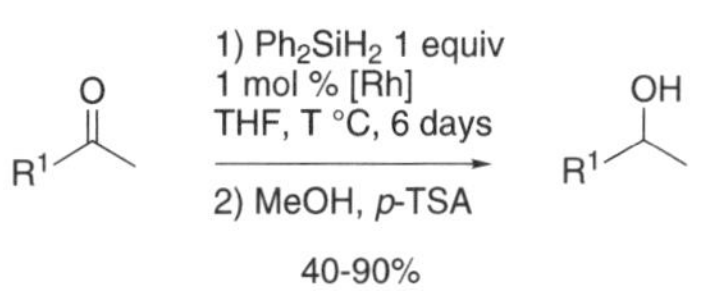

[Rh]	R¹	T (°C)	Time	ee (%)
54d	Ph	22	4 h	20 (*S*)
54f	Ph	11	6 d	40 (*S*)
54d	*c*-Hex	−10	6 d	44 (*S*)
54e	*c*-Hex	2	4 d	43 (*S*)
54f	*c*-Hex	22	3 d	43 (*R*)

Scheme 32

observed with a decrease in enantioselectivity with increasing or decreasing temperature with respect to the optimal conditions.

In 2003, Shi et al. reported the synthesis and use of axially chiral Rh – NHC complexes **55** and **56** derived from axially dissymmetric 1,1′-binaphthalenyl-2,2′-diamine giving excellent chiral induction in the enantioselective hydrosilylation of methyl ketones [21]. A Rh(III) **55** and a Rh(I) complex **56** were obtained by treatment of the precursor dibenzimidazolium salt with $[Rh(cod)Cl]_2$ in acetonitrile in the presence of NaOAc and KI (Scheme 33). These complexes were isolated respectively in 25 and 16% yield after separation by silica gel column chromatography. They are stable at ambient atmosphere and their structure was determined by X-ray diffraction. For the Rh(III) complex **55**, the two carbenes coordinate the same Rh atom leading to a bidentate bis-diaminocarbene complex with a square bipyramidal coordination geometry. For the chiral Rh(I) complex **56**, each *N*-heterocyclic carbene binds one atom of rhodium. The X-ray structure showed that the Rh atom has a square planar arrangement. An initial investigation of these catalysts (1% mol) in the reduction of acetophenone with diphenylsilane revealed that the use of the chiral Rh(III) complex **55** resulted in significantly higher enantioselectivity (98% ee compared to 23% with **56**).

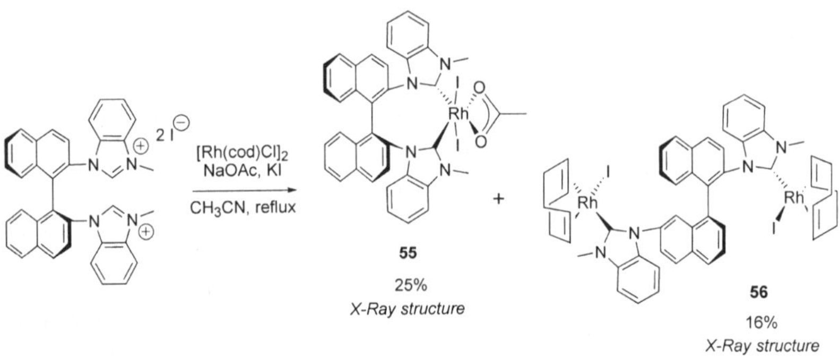

Scheme 33

Various aryl-alkyl ketones and dialkyl ketones could be reduced using the Rh(III) – NHC catalyst **55** in high yields (82–96%) and with good to excellent enantioselectivities (67–98% ee) (Scheme 34).

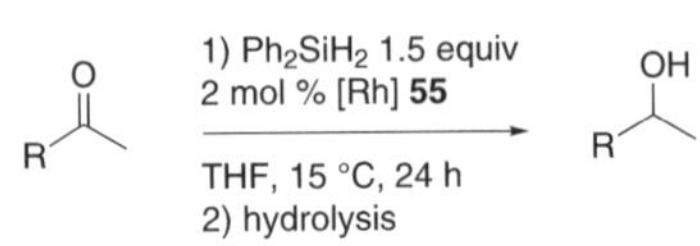

R	Yields (%)	ee (%)
Ph	87	98 (*R*)
2-Naphth	91	96 (*R*)
p-bromo-Ph	88	95 (*R*)
1-Ad	96	96 (*R*)
n-C_7H_{15}	86	67 (*R*)

Scheme 34

In 2004, the group of Gade and Bellemin-Laponnaz reported the use of rhodium complexes with chelating NHC–oxazoline ligands, still in asymmetric hydrosilylation of aryl-alkyl or dialkyl subsituted ketones [16]. The rhodium complexes **57a–d** were generated in 80–89% yield by reaction of the imidazolium salts with 0.5 equiv of [Rh(μ-O-t-Bu)(nbd)]$_2$ in THF (Scheme 35). Crystals suitable for X-ray structure analysis were obtained for complex **57c** in which Ar = Mes. This complex has a slightly distorted square-pyramidal geometry with the bromo ligand in the apical position.

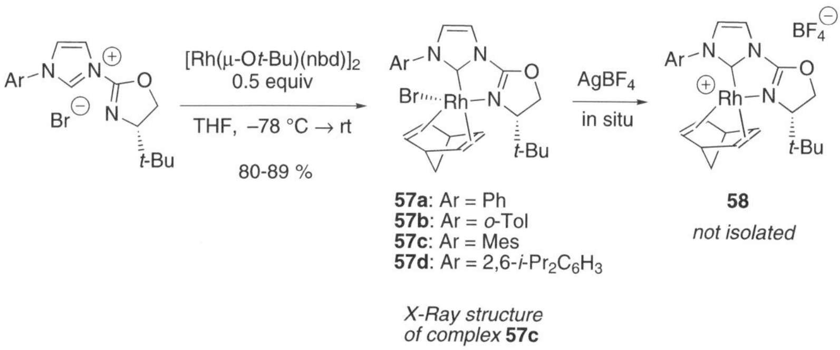

Scheme 35

The hydrosilylation of acetophenone by diphenylsilane in CH_2Cl_2 at rt was used as a test reaction to compare the selectivity obtained with the carbene ligands (Scheme 36). The reactions were performed in the presence of a slight excess of $AgBF_4$ (1.2% mol). In these conditions, the N-mesityl-substituted catalyst **57c** (1% mol) gave the highest selectivity (65% ee). The in situ formation of square-planar cationic rhodium species **58** as active catalysts appears to be crucial since the same reaction performed without silver salt gave both poor yield (53%) and enantioselectivity (13%).

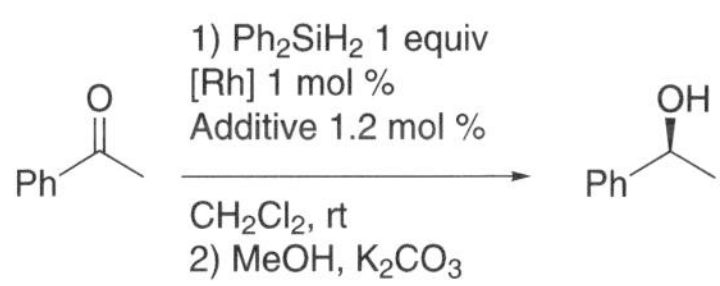

[Rh]	Additive	Yields (%)	ee (%)
57a	$AgBF_4$	90	0
57b	$AgBF_4$	82	31
57c	$AgBF_4$	90	65
57c	none	53	13
57d	$AgBF_4$	89	20

Scheme 36

Lowering the reaction temperature led to a significant increase in stereoselectivity. The catalytic runs performed at – 60 °C gave the best results with acetophenone being hydrosilylated with 90% ee and 92% yield in the presence of **57c**. Similar enantioselectivities (88–91%) were obtained in the reduction of a variety of aryl-alkyl ketones. Moreover, the rhodium-catalyzed hydrosilylation of unsymmetrical dialkyl ketones, which have proved to be difficult substrates in this reaction, gave also high enantioselectivities (77–95% ee) in the same conditions (Scheme 37).

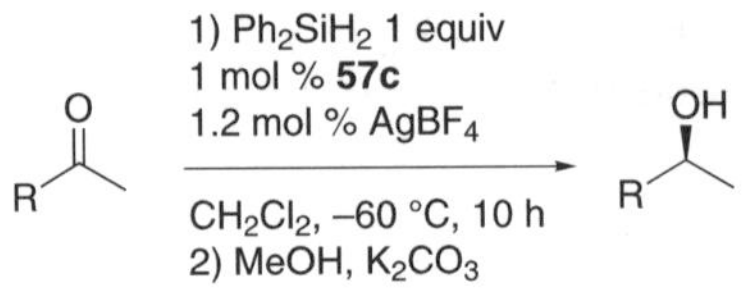

R	Yields (%)	ee (%)
Ph	92	90 (*S*)
2-Naphth	99	91 (*S*)
p-fluoro-Ph	90	91 (*S*)
n-Hex	95	79 (*S*)
i-Pr	53	88 (*S*)
t-Bu	96	95 (*S*)

Scheme 37

In 2003, Andrus et al. reached enantioselectivities up to 98% in asymmetric conjugate addition of aryl borane reagents to enones using catalytic chiral dicyclophane imidazoliniums (3 mol%) and $[Rh(acac)(C_2H_4)_2]$ (2 mol%) [27]. In the absence of base, the anion of acetylacetone may deprotonate the imidazolinium salt to generate the Rh – NHC complex. Four C_2-symmetric dicyclophane imidazoliniums **59a–d** were screened in the reaction of phenyl boronic acid (1.5 equiv) and 2-cyclohexenone under standard conditions in THF/water (Scheme 38). High yields were obtained at 60 °C for all ligands investigated. The dianisyl imidazolinium **59d** gave the higher enantioselectivity (98% ee) and 96% isolated yield after 3 h of reaction.

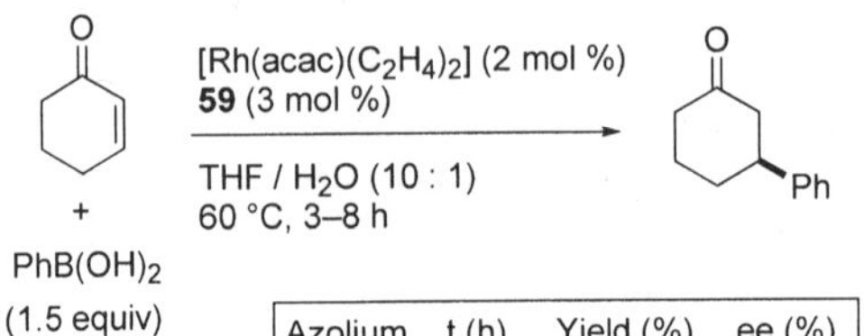

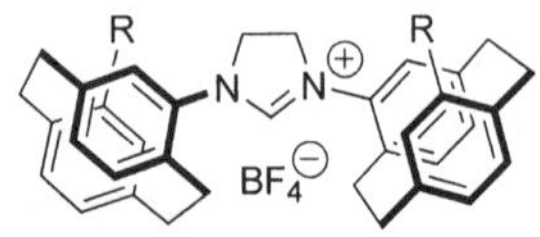

59a : R = H
59b : R = Ph
59c : R = *c*-Hex
59d : R = *o*-MeO-Ph

Azolium	t (h)	Yield (%)	ee (%)
59a	4	69	61
59b	8	86	91
59c	6.5	89	97
59d	3	96	98

Scheme 38

The optimal reaction conditions were applied with **59d** in the addition of various aryl boronic acids and potassium trifluoroborates to several cyclic and acyclic enones (Fig. 8). Arylboronic acids added to cyclic enones in high yields (89–97%) and with good to excellent selectivities (85–98% ee). Under these conditions, the potassium trifluoroborate reagents reacted at faster rates, but with slightly lower selectivities (83–96% ee). The reactions of acyclic enones with aryl boron reagents gave also excellent yields (83–96%).

Enones :

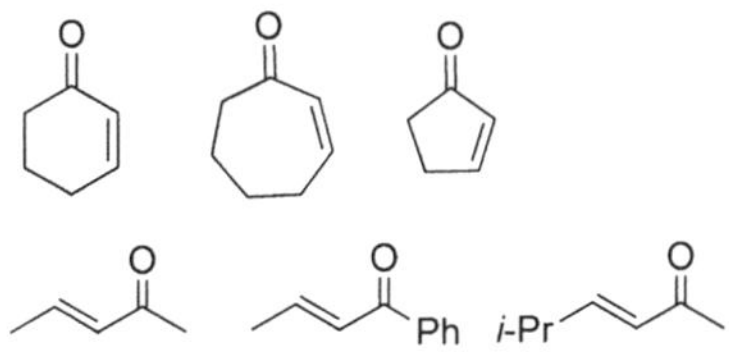

Aryl boron reagents :

$ArB(OH)_2$ or $ArBF_3K$

Ar = Ph, *p*-MeO-C_6H_4, *o*-Me-C_6H_4, *p*-MeCO-C_6H_4, *p*-CF_3-C_6H_4

Conditions : $[Rh(acac)(C_2H_4)_2]$ (2 mol %), **59d** (3 mol %), THF / H_2O (10 : 1), 60 °C

Fig. 8

Enantiomeric excesses of 81–91% were obtained with the boronic acids and of 73–81% with potassium trifluoroborates.

4.3 Ruthenium *N*-heterocyclic Carbene Complexes

Chiral diaminocarbene complexes of ruthenium were used in asymmetric olefin metathesis. Studies on achiral NHC-ruthenium systems revealed that the carbene ligand does not dissociate from ruthenium during the reaction and proved the advantage of using such complexes in terms of air and thermal stability. The development of chiral NHC-ruthenium metathesis catalysts was then investigated and was expected to expand dramatically the scope and utility of these transformations compared to the previously reported molybdenum catalysts that exhibit high enantioselectivities but require rigorous exclusion of air and moisture. Both asymmetric ring-closing and ring-opening variants of the metathesis process have been devised.

In 2001, Grubbs reported up to 90% ee in the desymmetrization of an achiral triene by ring-closing metathesis [29]. Various ruthenium complexes **60a–b** and **61a–d** were tested with chiral carbene ligands derived from 1,2-diphenyl-1,2-diaminoethane or 1,2-diaminocyclohexane, having aryl groups on the nitrogen atoms (Fig. 9). The initial mesityl groups of the achiral carbene were preserved (complexes **60a–b**) or replaced by mono-*o*-substituted phenyl groups in order to relay the backbone chirality to the *N*-substituents (complexes **61a–d**).

The ruthenium complexes were prepared in 50–80% yield by treatment of the imidazolium salts with potassium hexafluoro-*t*-butoxide, and then by $(PCy_3)_2Cl_2Ru=CHPh$. A single phosphine is displaced by the carbene affording the desired complexes as air-stable solids that were purified by silica gel

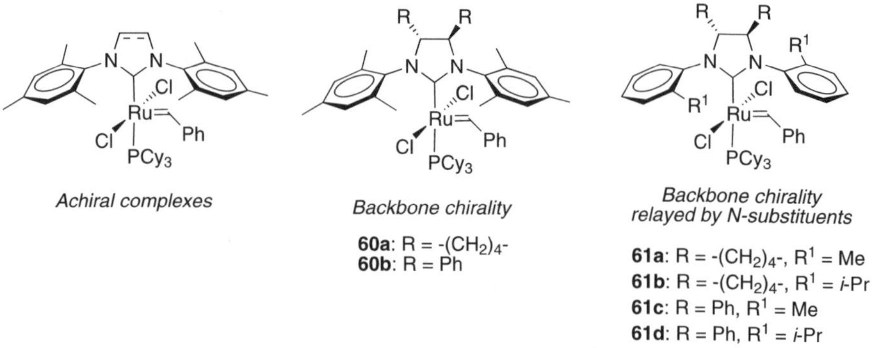

Fig. 9

column chromatography (Scheme 39). Most of the complexes were isolated as a mixture of atropoisomers. Crystallographic evidence of the conformation of the chiral NHC ligands has been obtained by X-ray analysis of the bis(pyridine) derivative **62** of the complex **61c** (R = Ph, R^1 = Me). In this structure, the *o*-methyl groups are oriented *anti* to the phenyl subsituents of the imidazole ring. The phenyl group of the benzylidene is also oriented *anti* to the *o*-methyl substituent of the proximal aryl ring. This *anti-anti* arrangement suggests that the stereochemistry of the phenyl substituents on the imidazole is effectively transferred to the metal center.

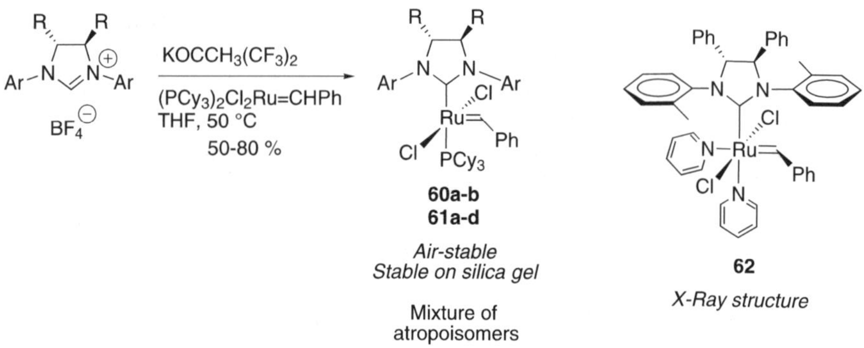

Scheme 39

These different catalysts were first tested in the desymmetrization of the achiral triene **63** (Scheme 40). The best enantioselectivities (up to 39%) were obtained with complexes **60b** and **61c–d** bearing carbene ligands derived from 1,2-diphenyl-1,2-diaminoethane (R = Ph). Ligands derived from 1,2-diaminocyclohexane gave poor enantioselectivities (< 9% ee). Replacement of the mesityl group in complex **60b** by *o*-methyl- or *o*-isopropylphenyl groups (complexes **61c–d**) slightly increases the enantioselectivity (from 13 to 23% ee). Changing the halide ligands of the catalysts **61c** and **61d** from Cl^- to I^- by adding an excess of NaI in the reaction further improves the enantioselectivity (from 23 to 38–39% ee) although the conversion decreases dramatically in these conditions (Scheme 40).

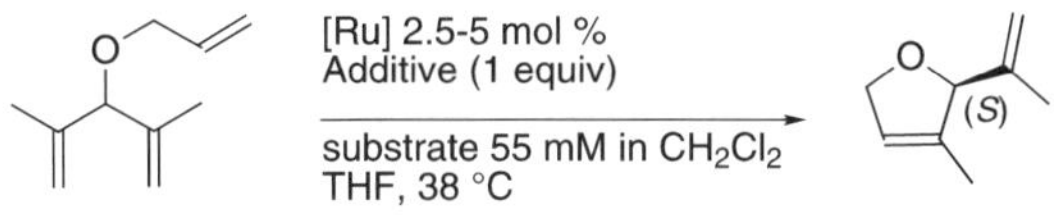

63

[Ru]	Additive	Convn (%)	ee (%)
60b	none	57	13
61c	none	95	23
61d	none	96	23
61c	NaI	18	38
61d	NaI	20	39

Scheme 40

However, high conversions (82–91%) and high enantioselectivities (up to 90% ee) could be obtained in the cyclization of the (*E*)-trisubstitued olefin **64** catalyzed by complex **61d** (Scheme 41). In this reaction neither solvent nor temperature has a significant effect on the enantioselectivity. In the case of the corresponding (*Z*)-trisubstituted olefins, conversions are high, but enantioselectivities are lower (ee < 36%).

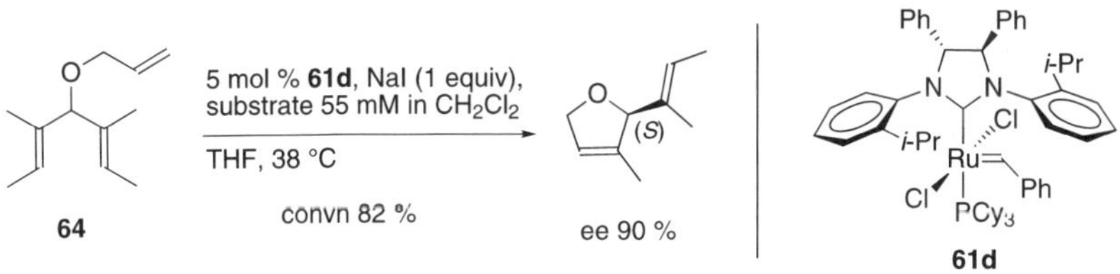

64 **61d**

Scheme 41

A model consistent with the stereochemical outcome of the reaction and with the role of the halide has been proposed (Fig. 10). This model suggests side-on olefin binding and reorganization of the halide ligands. In such geometry, a steric interaction between the unbound olefin and apical halide may justify the dramatic increase in enantioselectivity observed upon changing the halide from Cl^- to I^-.

In 2002, Hoveyda et al. reported the synthesis, structure and reactivity of a chiral bidentate Ru-based catalyst **65**, bearing a binaphthyl moiety, for olefin metathesis [33]. Preference for a bidentate chiral imidazolinylidene was based on the hypothesis that such a ligand would induce chirality more efficiently. This catalyst was designed by analogy with similar achiral complexes **66** that

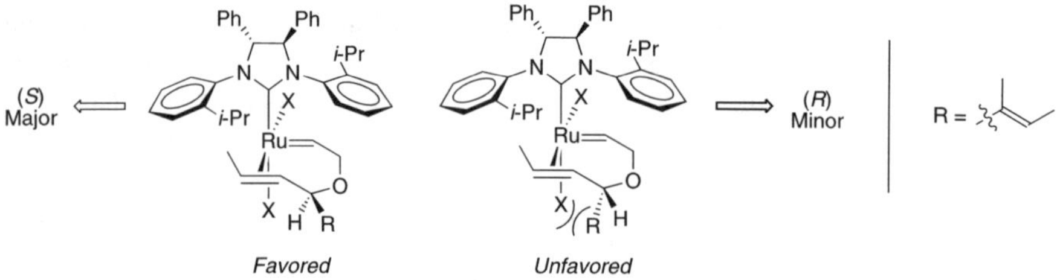

Fig. 10

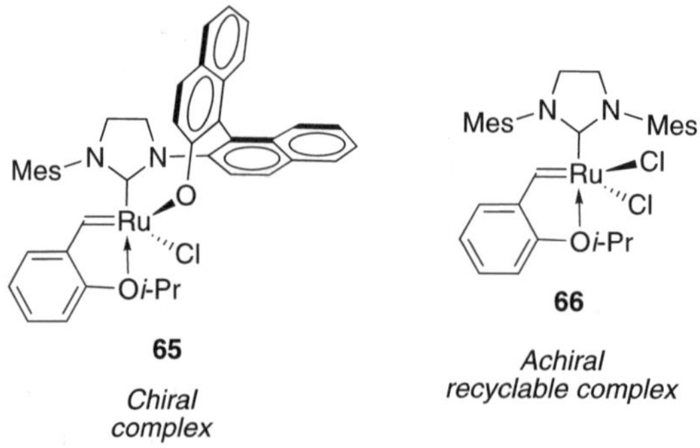

Fig. 11

were recyclable and exhibited reactivity and selectivity profiles not observed with other Ru systems (Fig. 11).

The complex **65** was synthesized by reaction of the imidazolinium salt with the precursor ruthenium complex **67** (catalytically inactive) in the presence of silver carbonate (Scheme 42). The complex being air-stable and stable on silicagel was isolated in 52% yield after chromatography. The diastereomeric and enantiomeric purity of **65** was determined by HPLC analysis and found to be above 98% (de and ee). The molecular structure was determined by X-ray analysis and showed the unusual twist geometry of this complex.

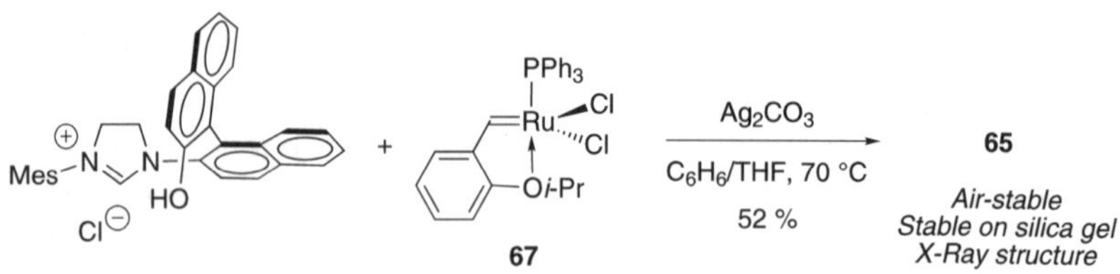

Scheme 42

Initial studies indicated that this ruthenium complex is an effective chiral catalyst for enantioselective metathesis. For example, desymmetrization of the anhydride **68** (Scheme 43) in the presence of 10 mol% of **65** and 10

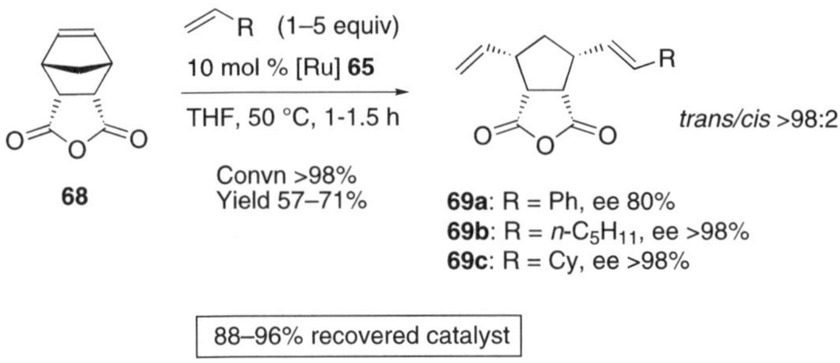

Scheme 43

equiv of styrene (R = Ph) in THF (50 °C) leads to the formation of the chiral diene **69a** in 80% ee and 71% isolated yield. However, the chiral complex **65** is less active than the achiral parent complex **66** since longer reaction times and higher temperatures are required for complete conversion. The selectivity is even better with aliphatic olefins instead of styrene. With 1-heptene ($R = n\text{-}C_5H_{11}$) and vinylcyclohexane (R = *c*-Hex), the corresponding dienes **69b–c** were obtained in > 98% ee and 57–60% yield. In addition, the chiral catalyst could be recovered after chromatography in 88–96% yield and reused without significant loss of enantioselectivity or reactivity. Byproducts from homodimerization of the terminal olefins or additional cross metathesis were not observed. Due to polymerization of these substrates, none of these transformations could be effected with chiral Mo-based catalysts.

The significant potential of the ruthenium complex **65** was further underlined in the catalytic asymmetric ring-opening/cross metathesis of the cyclic alkene **70** (Scheme 44). This transformation is catalyzed by 5% mol of **65** at room temperature, in air, and with undistilled and nondegassed THF to deliver the corresponding diene **71** in 96% ee and 66% isolated yield. In standard conditions (distilled and degassed THF), the alkene **70** reacts in 75 min to give the diene in 95% ee and 76% yield, with only 0.5 mol % of catalyst.

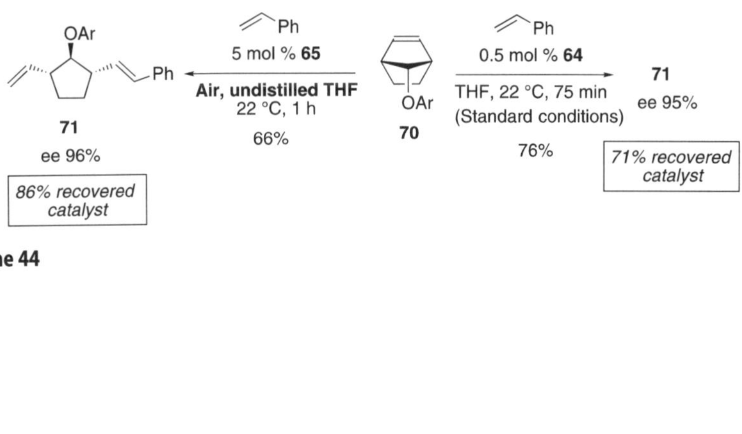

Scheme 44

4.4
Iridium *N*-heterocyclic Carbene Complexes

Chiral diaminocarbene complexes of iridium were used in asymmetric hydrogenation of olefins. Burgess et al. reported in 2001 the synthesis of bidentate cationic iridium complexes with chelating carbene–oxazoline ligands that gave enantioselectivities of up to 98% in hydrogenation of (*E*)-aryl alkenes [15, 42]. The complexes **72a–g** were prepared by reaction of the corresponding imidazolium salts with 1.5 equiv of lithium *t*-butoxide and 0.5 equiv of $[Ir(cod)Cl]_2$ in THF at 70 °C followed by anion exchange with NaBARF (sodium tetrakis(3,5-bis(trifluoromethyl)phenyl)borate) (Scheme 45). Various combinations of R^1 and R^2 groups were tested. In some cases, especially with bulky substituents the desired complex could not be isolated. Most of the complexes **72** are air-stable and were easily purified by flash chromatography.

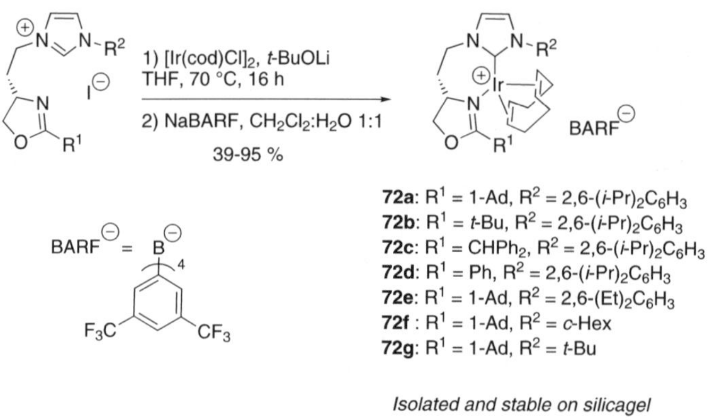

Scheme 45

The three iridium complexes **72d, 72f** and **72g** were analyzed by X-ray diffraction. Unfortunately the iridium complex **72a**, the most efficient in many reactions, failed to give suitable crystals for analysis but the corresponding crystalline rhodium complex **73** could be analyzed. According to the results obtained, the coordination sphere of the Ir atom and of the Rh atom can be described as *pseudo*-square planar (Fig. 12).

Hydrogenation of (*E*)-1,2-diphenylpropene was chosen as a model reaction to test the potential of these catalysts [Scheme 46]. The best results (99% yield and 98% ee) were obtained with the complex **72a** in which R^1 = 1-Ad and R^2 = $2,6\text{-}(i\text{-}Pr)_2C_6H_3$. The corresponding *t*-butyl-, diphenylmethyl- and phenyl-substituted oxazoline complexes (**72b, 72c** and **72d**) were found to be less effective. The complex **72f** obtained with R^1 = 1-Ad and by changing the R^2 group into a cyclohexyl group was inactive. In the case of R^1 = 1-Ad and just

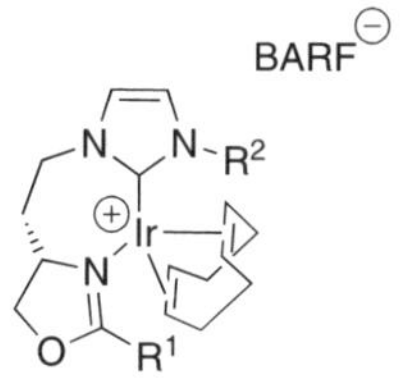

X-Ray structure for:

72d (R^1 = Ph and R^2 = 2,6-(*i*-Pr)$_2C_6H_3$)
72f (R^1 = 1-Ad and R^2 = *c*-Hex)
72g (R^1 = 1-Ad and R^2 = *t*-Bu)

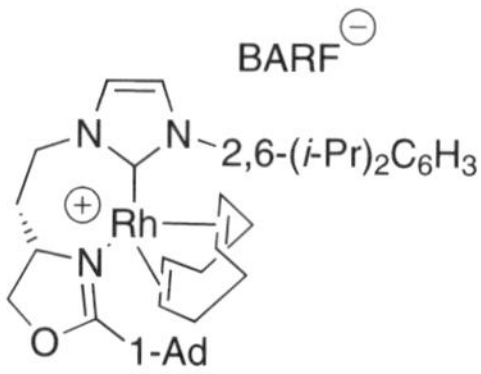

73

X-Ray structure

Fig. 12

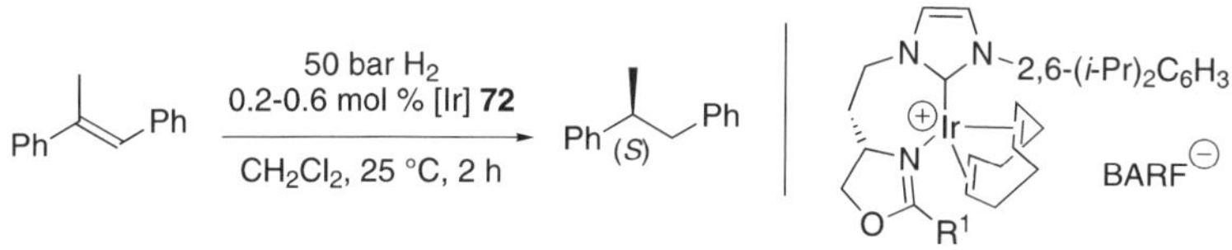

72a-d

Complex	R^1	yield (%)	ee (%)
72a	1-Ad	99	98
72b	*t*-Bu	81	50
72c	$CHPh_2$	12	25
72d	Ph	25	13

Scheme 46

by replacing the initial 2,6-diisopropylphenyl group by a 2,6-diethylphenyl group, less stereoselective reactions were observed (complex **72e**).

To determine the potentiality of catalyst **72a**, various alkenes with methyl, ethyl and *iso*-propyl substituents in *cis* or *trans* orientations relative to the aryl substituent were hydrogenated (Scheme 47). In general, (*E*)-alkenes produced better results than their *cis* isomers.

The variation of enantioselectivities with temperature and pressure was investigated. The effects of these two factors are very substrate dependent and difficult to generalize even in a single substrate serie. However, it seems that enantioselectivities are slighly better at 25–40 °C than at lower temperatures (0 °C or less). The stereoselectivity can be inverted for specific alkenes (formation of the *S* or *R* enantiomer preferentially). For several substrates, the reactions tend to proceed to completion with optimal ee's when performed at lower hydrogen pressure (2 bar) instead of 50 bar (Fig. 13). Pronounced variation of enantioselectivities with hydrogen concentration in solution may indicate the presence of two (or even more) different mechanisms which happen to give opposite enantiomers for some substrates.

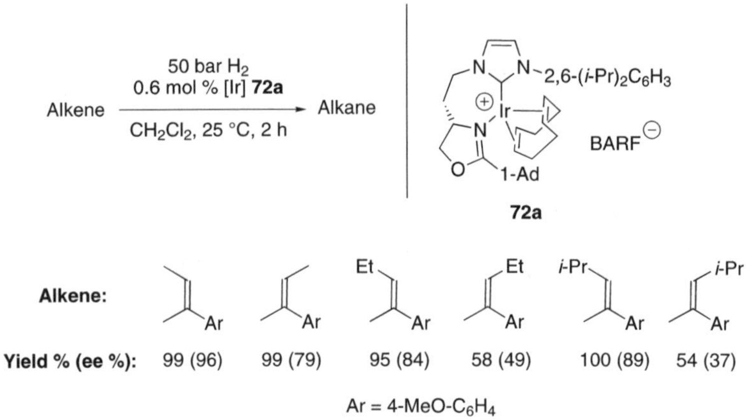

Scheme 47

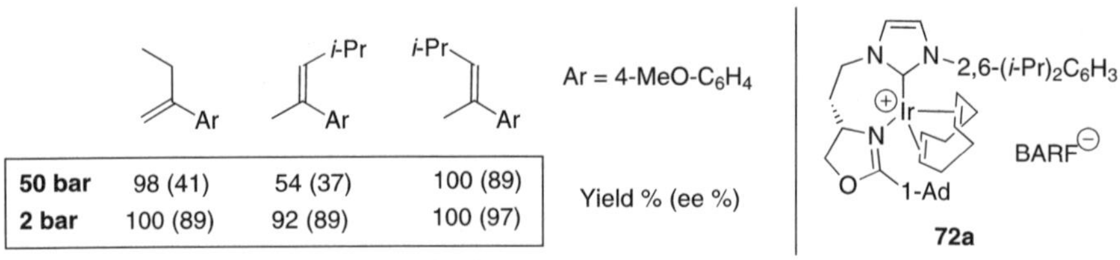

Fig. 13

In 2004, Bolm et al. reported the use of chiral iridium complexes with chelating phosphinyl-imidazolylidene ligands in asymmetric hydrogenation of functionalized and simple alkenes with up to 89% ee [17]. These complexes were synthesized from the planar chiral [2.2]paracyclophane-based imidazolium salts **74a–c** with an imidazolylidenyl and a diphenylphosphino substituent in pseudo ortho positions of the [2.2]paracyclophane (Scheme 48). Treatment of **74a–c** with *t*-BuOLi or *t*-BuOK in THF and subsequent reaction of the in situ formed carbenes with $[Ir(cod)Cl]_2$ followed by anion exchange with NaBARF afforded complexes (*Rp*)-**75a–c** in 54–91% yield. The chela-

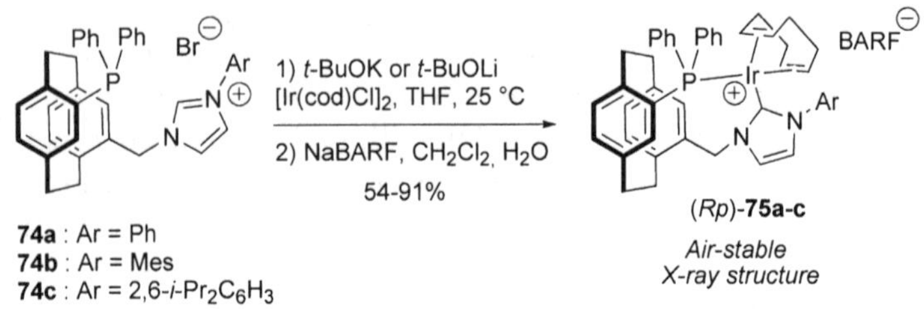

Scheme 48

tion of the phosphino group was confirmed by NMR spectroscopy and by X-ray structure analysis of an analog of **75c** in wich the $BARF^-$ counter-ion is replaced by PF_6^-.

Iridium complexes **75** (1 mol %) were first tested in the asymmetric hydrogenation of (*E*)-1,2-diphenyl-propene under 50 bar H_2 and exhibited low to good catalytic activity (Scheme 49). Complex (*Rp*)-**75a** gave the best results leading to the reducted product with 89% conversion and 82% ee. Complexes **75b–c** bearing sterically more demanding substituents on the carbene fragment afforded products with lower conversion and ee values. The catalytic activity of complex **75a** is temperature dependent. The same reaction performed at 50 °C led to a better conversion (99%) but to a lower enantioselectivity (73% ee). The hydrogenation of various functionalized alkenes with these complexes was studied. The enantioselectivity and stereochemistry of the reaction (formation of the *S* or *R* enantiomer preferentially and ee) were found to be highly pressure dependent but also dependent on the starting alkene stereochemistry. Some of the results obtained at 25 °C are presented in Scheme 49 with the reaction conditions. The best enantioselectivity (89% ee) was obtained in the hydrogenation of dimethyl itaconate using complex **75c** and under 50 bar of hydrogen.

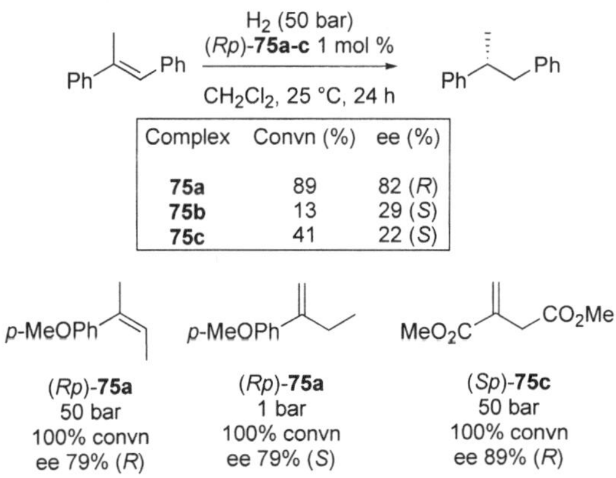

Complex	Convn (%)	ee (%)
75a	89	82 (*R*)
75b	13	29 (*S*)
75c	41	22 (*S*)

Scheme 49

4.5 Copper *N*-heterocyclic Carbene Complexes

Chiral diaminocarbene complexes of copper were used in asymmetric conjugate addition of diethylzinc to Michael acceptors. Achiral copper carbene complexes derived from imidazolium salts were synthesized and characterized for the first time by Arduengo in 1993 [43]. In 2001, Woodward reported the use of such Arduengo-type carbene in copper-catalyzed conjugate addition and showed their strong accelerating effect [44]. The same year, Alexakis [45] and Roland [46] reported asymmetric versions of the reaction using chiral carbene ligands. In their first study, Alexakis et al. reached 51% ee in the addition of Et_2Zn to enones, with a ligand **76** having a chiral backbone and chiral substituents on the nitrogen atoms (Scheme 50). Different copper(I) or copper(II) salts were used as precursors of the active copper(I) catalyst. The best results were obtained with $Cu(OTf)_2$. An increased reactivity was also observed in Et_2O in which reactions could be performed at – 78 °C to give enhanced stereoselectivities.

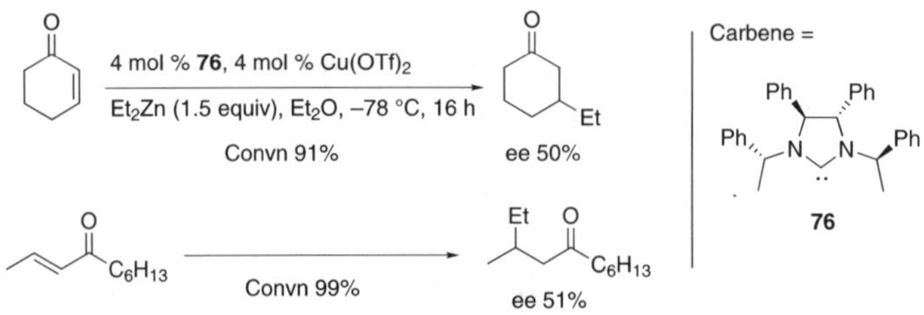

Scheme 50

The carbene-copper complex was not isolated but generated in situ by deprotonation of the imidazolinium salt **77** by *n*-BuLi in ether, in the presence of copper(II) triflate (Scheme 51). The use of two equivalents of ligand caused a dramatic decrease in reactivity.

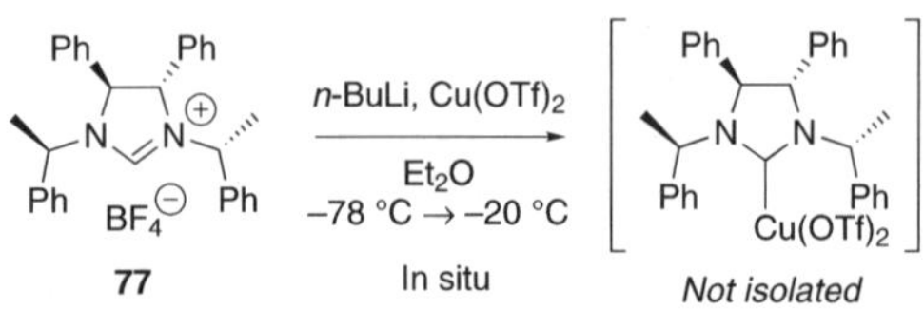

Scheme 51

Roland et al. obtained 23% ee in the addition of Et_2Zn to cyclohexenone using the silver(I) complex **78** having a chiral backbone and methyl groups on the nitrogen atoms. This complex acts as an efficient carbene transfer agent towards $Cu(OTf)_2$. The conjugate addition proceeds rapidly in toluene at 0 °C (Scheme 52).

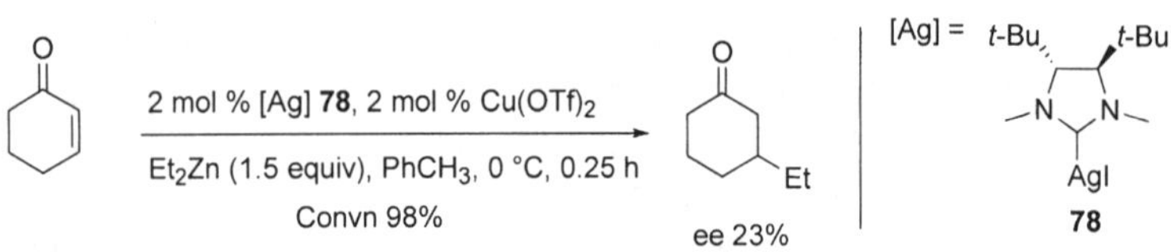

Scheme 52

This group showed that isolable silver(I) diaminocarbene complexes can be use in situ instead of free carbenes, to generate the copper carbene complex. The silver salts that precipitates during the formation of the copper complex have not any negative effect on the conversion. This method is advantageous since most of the silver complexes are isolable, air-stable and easily obtained by treatment of the corresponding imidazolinium salt by 0.5 equiv of silver oxide (Scheme 53). The solid structure of **78** was analyzed by X-ray diffraction.

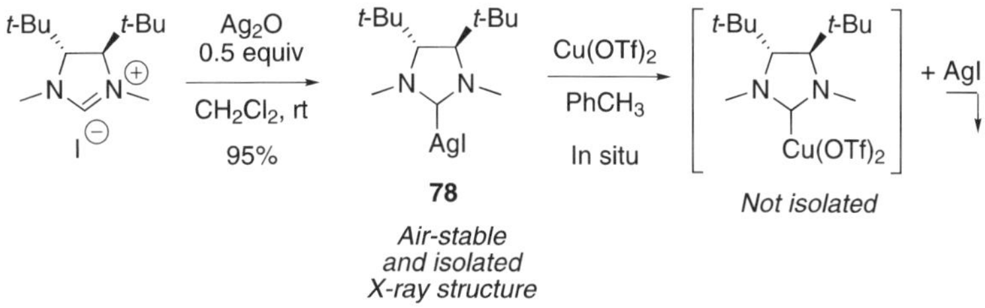

Scheme 53

In a further study made in collaboration and reported in 2003 [31], the two groups investigated the effect of the method of the preparation of the copper carbene complex on conversion and enantioselectivity. With the same ligand, enantioselectivities and conversions could be increased significantly using the silver complex precursor **79** instead of the tetrafluoroborate imidazolium salt **80** (Fig. 14). This result is probably due to a better conversion in the in situ formation of the copper complex starting from the silver(I) complex, the tetrafluoroborate azolium salts being relatively hygroscopic. The best selectivities were generally observed with $Cu(OAc)_2$ or CuTC (copper thiophene carboxylate) as the copper source, compared to the initially used copper(II) triflate.

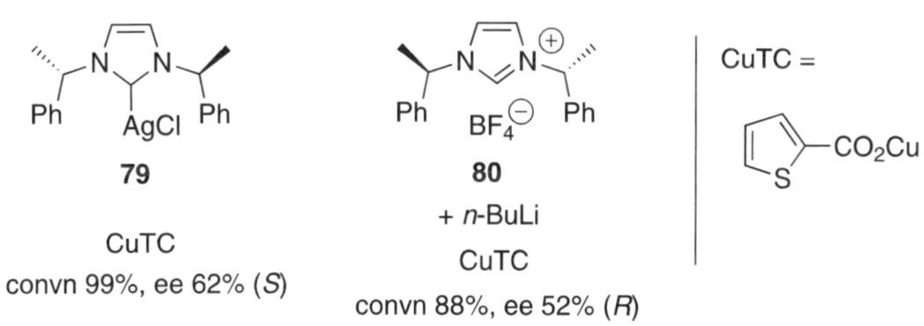

Fig. 14

Two families of complexes having either exocyclic chirality borne by the nitrogen atoms (**79, 81, 82**) or endocyclic chirality (*t*-butyl groups) relayed by benzylic groups on the nitrogen atoms (**83, 84**) were tested in the optimized conditions (Fig. 15).

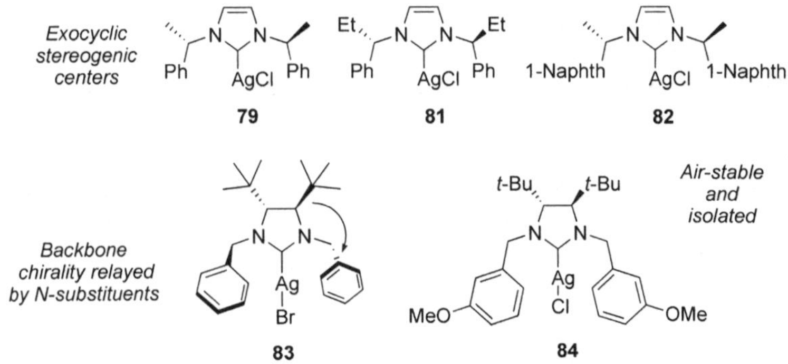

Fig. 15

Several cyclic and acyclic enones were screened, as well as nitroalkene (Fig. 16). Cyclohept-2-enone is by far the most favourable substrate leading to conversions of 95–100% and 76–93% ee. The best ligand with this substrate is the one derived from 1-naphthylethylamine (complex **82**, 93% ee). An enantioselectivity of 88% could also be obtained with the easily accessible ligand derived from phenylethylamine (complex **79**) and with the ligand derived from di-*t*-butylethane diamine bearing benzyl groups on the nitrogen atoms, monosubstituted by methoxy groups at the meta position (complex **84**). The latter ligand gave the best enantioselectivity (69% ee) with cyclohex-2-enone. By comparison, the acyclic enones tested gave lower yields (44–60%) and enantioselectivities (42–49%). Nitrostyrene, which is an excellent Michael acceptor, gave up to 75% ee.

Substrate	Cu salt	Silver(I) complex	Convn (%)	Ee (%)
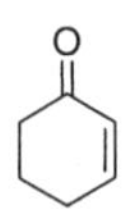	CuTC	**79**	99	62 (*S*)
	$Cu(OAc)_2$	**81**	92	55 (*S*)
	$Cu(OAc)_2$	**82**	87	59 (*S*)
	CuTC	**83**	100	58 (*S*)
	CuTC	**84**	99	69 (*S*)
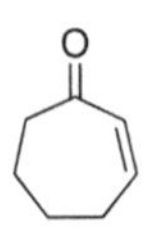	$Cu(OAc)_2$	**79**	99	88 (*S*)
	$Cu(OAc)_2$	**81**	97	76 (*S*)
	$Cu(OAc)_2$	**82**	95	93 (*S*)
	CuTC	**83**	99	76 (*S*)
	CuTC	**84**	100	88 (*S*)
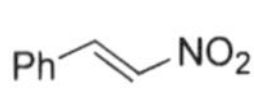	$Cu(OAc)_2$	**79**	100	75 (*R*)
	CuTC	**82**	98	69 (*S*)

Conditions: Copper salt 4 mol %, silver carbene 4 mol %
Et_2Zn 1.5 equiv, Et_2O, –78 °C, 16 h

Fig. 16

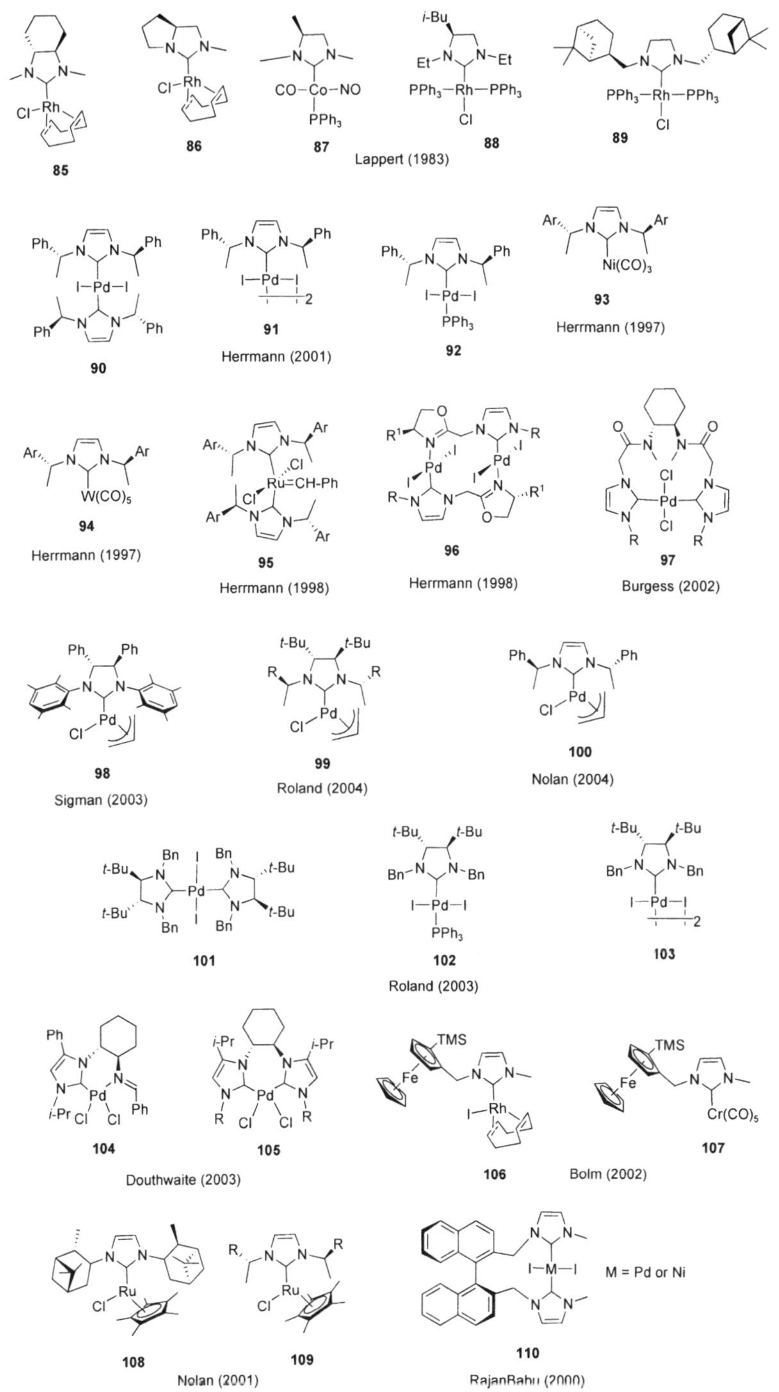
85
86
87
88
Lappert (1983)
89
90
91
Herrmann (2001)
92
93
Herrmann (1997)
94
Herrmann (1997)
95
Herrmann (1998)
96
Herrmann (1998)
97
Burgess (2002)
98
Sigman (2003)
99
Roland (2004)
100
Nolan (2004)
101
102
103
Roland (2003)
104
105
Douthwaite (2003)
106
107
Bolm (2002)
108
109
Nolan (2001)
110
M = Pd or Ni
RajanBabu (2000)

Fig. 17

5 Isolated Complexes Without Application in Asymmetric Catalysis

Many other chiral NHC-complexes with various transition metals have been isolated but not used in asymmetric catalysis. These complexes are presented in Fig. 17: **85–89** [9], **90–92** [47], **93–94** [48], **95** [49], **96** [14], **97** [22], **98** [26], **99** [32], **100** [50], **101–103** [51], **104** [24], **105** [23], **106–107** [18], **108–109** [13], **110** [20].

6 Conclusion

The use of chiral diaminocarbenes as transition metal ligands for catalyzed asymmetric synthesis is certainly an emerging field of research. They are relatively easy to prepare and they allow numerous structural modifications. Their transition metal complexes shows very usefull properties such as the thermal and air stability. Even if there is only a few reports of effective asymmetric transformations promoted by these class of catalyst, all these pioneering works open the route to the discovery of efficient new catalysts.

References

1. Hermann WA, Köcher C (1997) Angew Chem Int Ed 36:2163
2. Bourrissou D, Guerret O, Gabbaï FP, Bertrand G (2000) Chem Rev 100:39
3. Enders D, Gielen H (2001) J Organomet Chem 617–618:70
4. Hermann WA (2002) Angew Chem Int Ed 41:1290
5. Perry MC, Burgess K (2003) Tetrahedron: Asymmetry 14:951
6. Arduengo III AJ, Harlow MK (1991) J Am Chem Soc 113:361
7. Alexakis A, Lensen N, Tranchier JP, Mangeney P, Feneau-Dupont, Declercq JP (1995) Synthesis 1038
8. Hu X, Castro-Rodriguez I, Olsen K, Meyer K (2004) Organometallics 23:755
9. Coleman AW, Hitchkcock PB, Lappert MF, Maskell RK, Müller JH (1983) J Organomet Chem C9–14:250
10. a) Alder RW, Allen PR, Williams SJ, (1995) Chem Commun 1267. Kim YJ, Streiwieser AJ (2002) J Am Chem Soc 124:5757. b) Amyes TL, Diver ST, Richard JP, Rivas FM, Toth K (2004) J Am Chem Soc 126:4366
11. Wanzlick HW, Shikora E (1960) Angew Chem 72:494
12. Herrmann WA, Goossen L, Köcher C, Artus GRJ (1996) Angew Chem Int Ed Engl 35:2805
13. Huang J, Jafarpour L, Hillier AC, Stevens ED, Nolan SP (2001) Organometallics 20:2878
14. Herrmann WA, Goossen LJ, Spiegler M (1998) Organometallics 17:2162
15. Perry MC, Cui X, Powell MT, Hou DR, Reibenspies JH, Burgess K (2003) J Am Chem Soc 125:113

16. Gade LH, César V, Bellemin-Laponnaz S (2004) Angew Chem Int Ed 43:1014
17. Focken T, Raabe G, Bolm C (2004) Tetrahedron: Asymmetry 15:1693
18. Bolm C, Kesselgruber M, Raabe G (2002) Organometallics 21:707
19. Howarth J, Al-Hashimy NA (2001) Tetrahedron Lett 42:5777
20. Clyne DS, Jin J, Genest E, Gallucci JC, RajanBabu TV (2000) Org Lett 2:1125
21. Duan WL, Shi M, Rong GB (2003) Chem Commun 2916
22. Perry MC, Cui X, Burgess K (2002) Tetrahedron: Asymmetry 13:1969
23. Bonnet LG, Douthwaite RE, Hodgson R (2003) Organometallics 22:4384
24. Bonnet LG, Douthwaite RE (2003) Organometallics 22:4187
25. Enders D, Kallfass U (2002) Angew Chem Int Ed 41:1743
26. Jensen DR, Sigman MS (2003) Org Lett 5:63
27. Ma Y, Song C, Ma C, Sun Z, Chai Q, Andrus MB (2003) Angew Chem Int Ed 42:5871
28. Scholl M, Ding S, Lee CW, Grubbs RH (1999) Org Lett 1:953
29. Seiders TJ, Ward DW, Grubbs RH (2001) Org Lett 3:3225
30. Pytkowicz J, Roland S, Mangeney P (2001) J Organomet Chem 631:157
31. Alexakis A, Winn CL, Guillen F, Pytkowicz J, Roland S, Mangeney P (2003) Adv Synth Catal 345:345
32. Roland S, Audouin M, Mangeney P (2004) Organometallics 23:3075
33. Van Veldhuizen JJ, Garber SB, Kingsbury JS, Hoveyda AH (2002) J Am Chem Soc 124:4954
34. Rivas FM, Riaz U, Giessert A, Smulik JA, Diver ST (2001) Org Lett 3:2673
35. Rivas FM, Giessert A, Diver ST (2002) J Org Chem 67:1708
36. Enders D, Gielen H, Raabe G, Runsink J, Teles JH (1996) Chem Ber 129:1483
37. Lee S, Hartwig JF (2001) J Org Chem 66:3402
38. Wang HMJ, Lin IJB (1998) Organometallics 17:972
39. Sato Y, Yoshino T, Mori M (2003) Org Lett 5:31
40. Lappert MF, Maskell RK (1984) J Organomet Chem 264:217
41. Enders D, Gielen H, Runsink J, Breuer K, Brode S, Boehn K (1998) Eur J Inorg Chem 913
42. Powell MT, Hou DR, Perry MC, Cui X, Burgess K (2001) J Am Chem Soc 123:8878
43. Arduengo III AJ, Dias HVR, Calabrese JC, Davidson F (1993) Organometallics 12:3405
44. Fraser PK, Woodward S (2001) Tetrahedron Lett 42:2747
45. Guillen F, Winn CL, Alexakis A (2001) Tetrahedron: Asymmetry 12:2083
46. Pytkowicz J, Roland S, Mangeney P (2001) Tetrahedron: Asymmetry 12:2087
47. Herrmann WA, Böhm VPW, Gstöttmayr CWK, Grosche M, Reisinger CP, Weskamp T (2001) J Organomet Chem 617–618:616
48. Herrmann WA, Goossen LJ, Artus GRJ, Köcher C (1997) Organometallics 16:2472
49. Weskamp T, Schattenmann WC, Spiegler M, Herrmann WA (1998) Angew Chem Int Ed Engl 37:2490
50. Viciu MS, Navarro O, Germaneau RF, Kelly III RA, Sommer W, Marion N, Stevens ED, Cavallo L, Nolan SP (2004) Organometallics 23:1629
51. Pytkowicz J, Roland S, Mangeney P, Meyer G, Jutand A (2003) J Organomet Chem 678:166

Top Organomet Chem (2005) 15: 231–270
DOI 10.1007/b136352
© Springer-Verlag Berlin Heidelberg 2005
Published online: 10 September 2005

Chiral Ureas and Thiroureas in Asymmetric Catalysis

Jérémy A. J. Breuzard · M. Lorraine Christ-Tommasino · Marc Lemaire (✉)

Laboratoire de Catalyse et Synthèse Organique, CNRS: UMR 5181, UCB Lyon 1, CPE, Bât. 308, 43 bd. du 11 novembre 1918, 69622 Villeurbanne, France
Marc.Lemaire@univ-lyon1.fr

Abstract After an overview of chiral urea and thiourea synthetic methods, this review describes the main applications of urea and thiourea complexes in asymmetric catalysis. Some recent examples of thioureas as catalysts are also presented. Coordination chemistry of ureas and thioureas is briefly discussed.

Keywords Asymmetric catalysis · Coordination modes of ureas and thioureas · Thioureas · Transition metal complexes · Ureas

1 Introduction

Chiral diaza-molecules are relatively new ligands for asymmetric catalysis, but they are now widely used [1]. In several cases, they have efficiently replaced air-sensitive chiral phosphorus ligands. They have often led to high enantiomeric excesses in metal catalysed reactions, but they still suffer from limitations because of their relatively low stability to the reaction or work-up conditions. Even when immobilised, work up of the reaction has to be carefully optimised in order to preserve the nature of the ligand against chemical modifications or metal leaching. Some amine derivatisations are known to lead to functional groups with increased stability. Among these amine derivatives, ureas and thioureas possess N and O or S coordinating centres. Moreover, these molecules generally possess N – H functional groups, which could act as Brönsted acids, leading to deprotonation for metal complexation or selective formation of hydrogen bonding [2] for molecular recognition. Indeed, ureas and thioureas are often used for their molecular recognition properties [3–5]. Several supramolecular systems using the urea or thiourea groups as building blocks have been described during the last few years [6]. They exhibit original and fascinating properties, such as the autoassembly of multilayer structures [7]. Moreau has described macrochiral fibres of silica, which are formed as a result of the interactions between chiral ureas. These materials have been used to breed asymmetric induction during addition of zinc derivatives onto aldehydes. After generation of the first slight enantiomeric excess, the asymmetric autocatalytic reaction goes on and leads to a neat amplification of the reaction enantioselectivity [8]. Thioureas have also been used as complexing reagents to discriminate inorganic species [9] or as specific receptors for the detection of the presence of particular anions [10, 11]. Although these potential applications of ureas and thioureas are of broad interest both for economical and ecological reasons, we chose to focus on the use of these molecules only as ligands for transition metal catalysts or as organic catalysts (organocatalysts) for asymmetric synthesis. Binding sites with different characteristics are available on urea and thiourea functional groups. In terms of the HSAB model: the sulfur atom of thioureas should be considered as a soft Lewis base although the $N-C=O$ function is of medium hardness and the oxygen atom itself could offer a hard basic site. Thioureas and ureas are potentially able to interact with most of the organometallic precursors and, in addition, are both strong hydrogen bond donors and hydrogen bond acceptors. All these interactions could be used to organise the catalytic site and control the selectivity of the catalysed reaction. Examples of chiral ureas and thioureas used in asymmetric transition-metal catalysed reactions are described in Sect. 4. One of the advantages of ureas and thioureas lies in their easy access, due to the commercial availability of many isocyanates and isothiocyanates. The possibility of using phosgene and thiophosgene and amine substrate in order to prepare either new isocyanate or isothiocyanate, as well as ureas or thioureas, is also a major advantage for the development of such derivatives. Moreover, the addition of nucleophilic amines onto isocyanate or isothiocyanates is carried out in short reaction periods, affording very high yields and selectivities. Such reactivity allows the use of parallel and combinatorial technologies, thus leading to some of the rare successes in high throughput screening catalyst discovery [12]. Finally, ureas and thioureas are valuable building blocks for the synthesis of five- and six-membered heterocycles [13, 14] which are useful as biomimetic models [15]. Indeed, thiourea- and urea-containing compounds have biological activity [16–19]. They are useful as fungicides, herbicides and rodenticides [20, 21] or against bacteria and microbial infection [22–25]. Some molecules bearing one of these particular functional groups show enzymatic inhibition [26–28]. Such applications are out of the scope of this article.

2
Synthesis of Chiral Ureas and Thioures

Chiral amines and diamines are readily available substrates for the synthesis of ligands for transition metal-catalysed reactions since they can easily be transformed into chiral ureas and thioureas. Therefore, several groups have prepared chiral symmetrical ureas and thioureas, dissymmetrical ureas and thioureas, amino-urea and thiourea derivatives. Finally polyureas and non-soluble polythioureas were also prepared and tested as ligands for asymmetric catalysis.

2.1
Symmetrical Ureas and Thioureas

Treatment of a chiral amine with phosgene is the cheapest way to prepare symmetrical ureas [29]. Nevertheless, due to the toxicity and reactivity of that reagent, it can advantageously be replaced by triphosgene [30] or 1,1′-carbonyldiimidazole [31–34] or other derivatives such as 1,1′-carbonyldi-2(1H)-pyridinone [35]. This procedure can be extended to thiophosgene (Scheme 1) and its thio-analogues, such as 1,1′-thiocarbonyldi-2(1H)-pyridinone to produce thioureas [36]; chiral diamines can thus be transformed into the corresponding monoureas or monothioureas.

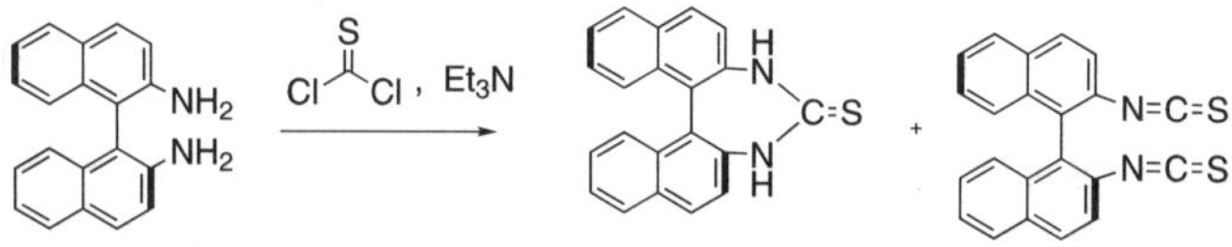

Scheme 1

2.2
Symmetrical Diureas and Dithioureas

Chiral diamines can also react with isocyanates or isothiocyanates to form chiral diureas and dithioureas. For example, a series of chiral C_2-symmetric diureas [37] or dithioureas [38] were easily prepared from the reaction of an optically active diamine with two equivalents of the desired isocyanate or isothiocyanate. The reaction takes place at room temperature in very good yields. Most of these reagents are commercially available or can be obtained by reaction of phosgene, thiophosgene or substitutes with primary amines. This procedure is also useful for the preparation of symmetrical diisocyanates or thiocyanates. Diureas and dithioureas with atropoisomeric chirality can be prepared from binaphtylamine (Scheme 1). By using chiral isocyanate [37] or thiocyanate [39] and chiral amines, urea or thiourea derivatives with four different stereogenic centres, Scheme 2 can be formed. In this way, match/mismatch effects could be observed in hydrogen transfer reactions (see Sect. 4.1.1).

2.3
Dissymetrical Ureas and Thioueas

A recent paper from Katritsky summarises all the preparations of achiral dissymmetrical thioureas and proposes a new one, based on 1-(alkyl/arylthiocarbamoyl)benzotriazoles, which act as masked isothiocyanates. As described in the previous section, other *N*-heterocyclic derivatives can be used instead

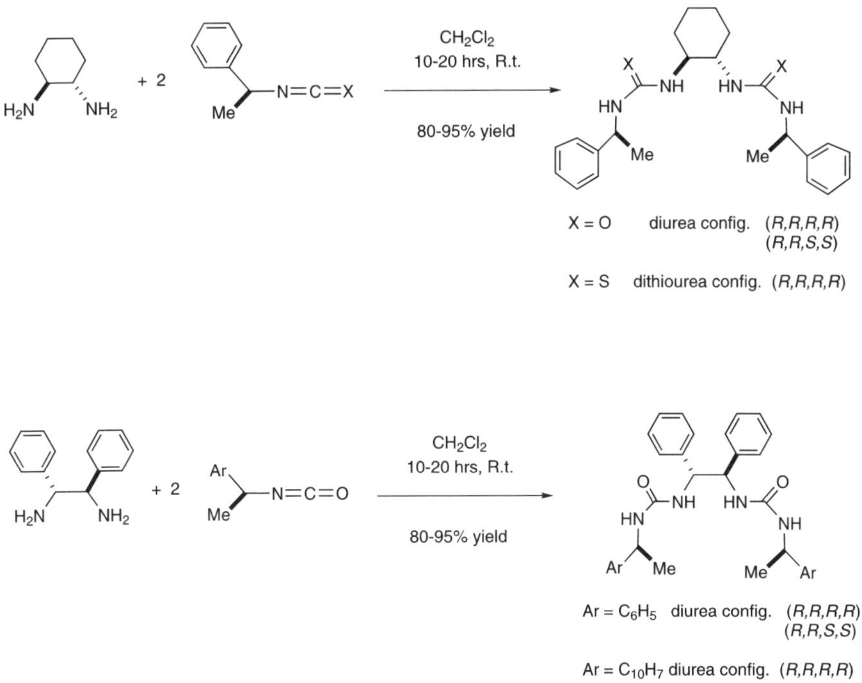

Scheme 2

of benzotriazole. The thiocarbamoylbenzotriazole derivatives being stable compounds, substituted thioureas are prepared in high yields in a one-pot reaction using this procedure (Scheme 3) [40]. However, no chiral thiourea has been prepared using this method so far.

Since chiral amines are much more available than chiral isocyanates, ureas have often been obtained from reaction of an amine used as the chiral source with a stoichiometric amount of a non-chiral isocyanate [41]. Similarly, thioureas are obtained by reacting isothiocyanates with amines. The corresponding ureas and thioureas (examples in the following sections) are

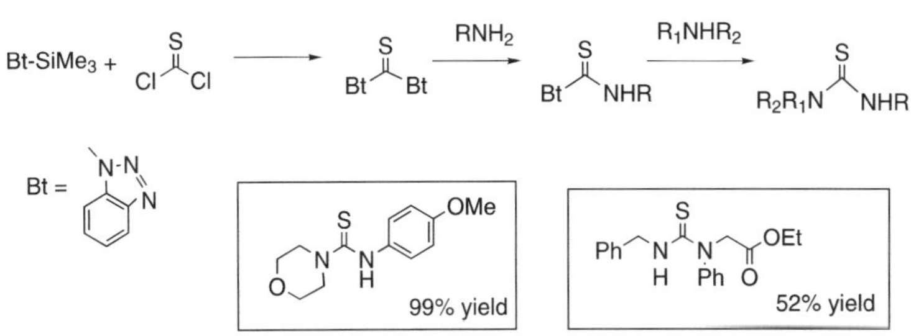

Scheme 3

obtained in excellent yields (> 70% yield for pure products). Chiral ureas can also be synthesised by reaction of chiral isocyanates with non-chiral amines. Chiral isothiocyanates have to be synthesised by known methods (see above) because only few are commercially available. However, the preparation of these derivatives is not necessary since heterocyclic urea and thiourea derivatives instead can be used for synthetic purposes: for example, the benzotriazolylthiourea intermediate is depicted in Scheme 3, but pyridyl and imidazolyl derivatives with related structures can also be used. With these compounds, the reactive function is masked during the purification step and can afterwards be substituted by an amine.

2.4 Amino-ureas and Amino-thioureas

Diamines can also react with only one equivalent of isothiocyanate to form bifunctionnal amine-thiourea ligands: 59–68% yields obtained for several alkyl isothiocyanates. However, reaction of phenylisocyanate with 1,2-diamines could also lead to the formation of the guanidine derivative by cyclisation and elimination of H_2S (Scheme 4) [42, 43].

Moreau and co-workers have also prepared (1*R*, 2*R*)-1,2-diaminocyclohexane amino-urea and thiourea derivatives [43]. Diphenylethylenediamine-substituted monothioureas are more stable than the cyclohexyldiamine counterpart, but they can also rearrange to guanidine derivatives, especially at high temperature or in the presence of metal [43]. Under the same conditions, thioureas also rearrange to guanidines in the presence of amines. Selective formation of substituted guanidines from thiourea derivatives of diaminocyclohexane or diphenylethylenediamine were also reported in a recent paper from Ishikawa [44].

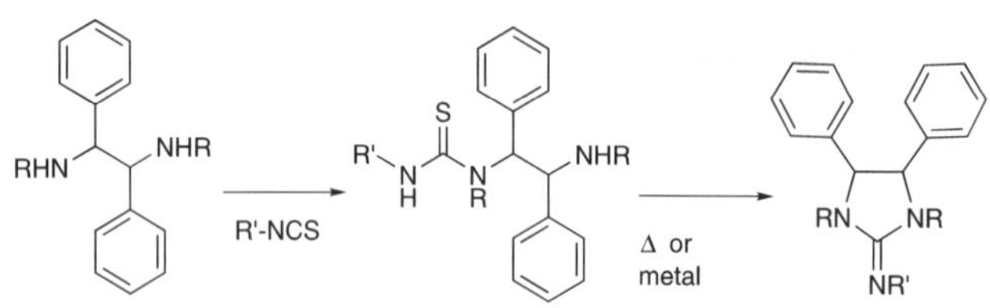

Scheme 4

2.5 Polyureas and Polythioureas

One of the limitations of the use of asymmetric catalysis comes from the difficulties of separating the chiral catalyst from the reaction medium and recycling it. Such systems are generally formed with chiral phosphane and/or

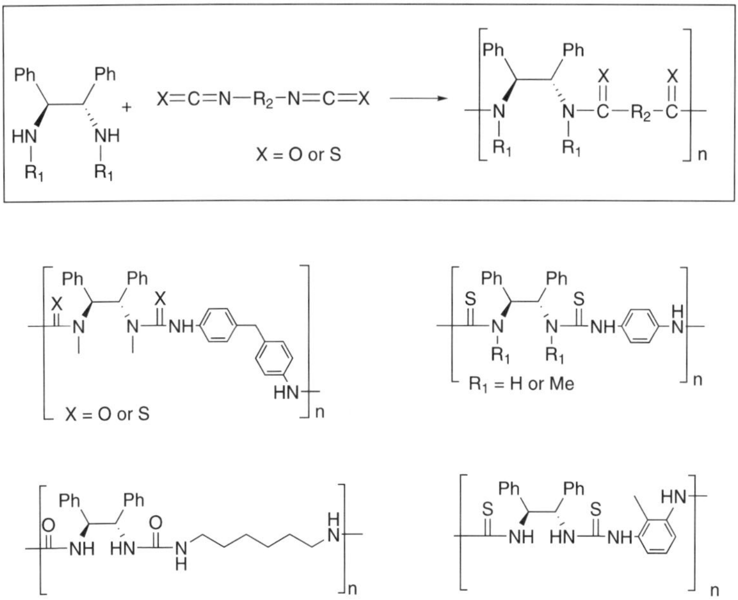

Scheme 5

diamine. Besides their relatively low stability, ligands of this type could only be transformed into insoluble catalytic materials via structural modifications, which are neither easy nor cheap [45]. Conversely, polyureas and polythioureas are readily available from diisocyanates or diisothiocyanates and chosen chiral diamines. The general preparation of polymer systems by this methodology is depicted in Scheme 5, as well as some polymer structures [46].

Linear polymers can be obtained by using diisocyanates or diisothiocyanates with diamines. In this case, the molecular weights are controlled by the solubility of the polymer in the chosen solvent. Due to the polarity and to the number of possible hydrogen bonds, such polymers are generally insoluble in classical solvents (MeOH, CH_2Cl_2) but are usually soluble in DMSO. A molecular weight of at least 2000 g/mol was estimated by NMR measurements [46]. A crosslinked insoluble material was prepared by using a mixture of di- and tri-isocyanate. In this case, the molecular imprinting effect could be performed (see Sect. 4.1.1). Insoluble materials are available by grafting ureas or thioureas onto insoluble polymers or inorganic materials. Urea and thiourea function can be used as linker for ligand immobilisation [47]. Moreau and co-workers achieved the immobilisation of polyurea ligands by sol-gel hydrolysis condensation giving hybrid materials: a left or right handed helix was autogenerated according to the chiral (*R*, *R*)-or (*S*, *S*)-diureidocyclohexane structure (Scheme 6) [48].

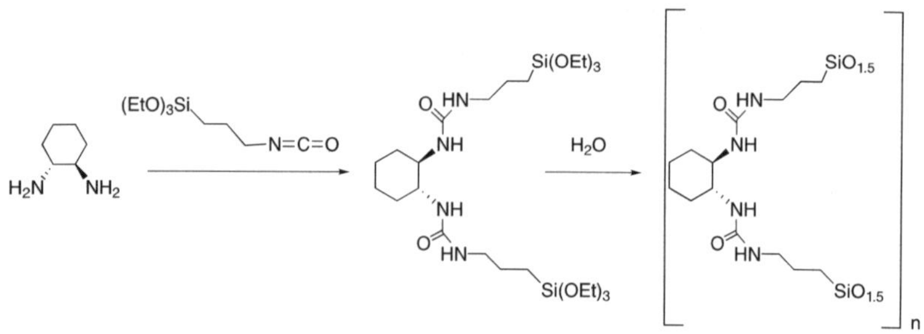

Scheme 6

Substituted thioureas have been used as ligands for transition-metal catalysed reactions and as organocatalysts for organic synthesis. These points will be discussed in Sects. 4 and 5. We first present some aspects of the coordination modes of ureas and thioureas.

3
Coordination Chemistry

Several metal complexes have been described with urea, thiourea or dimethyl derivatives [49, 50]. We will focus in this section on the coordination chemistry of substituted ureas and thioureas used as neutral ligands as well as many ureato and thioureato anions complexed to metal centres.

3.1
Urea Coordination Modes to Transition Metals

In transition metal complexes, two main coordination modes have been described for urea ligands: *N*-monohapto, or *N*,*O*-chelates. In the case of molecules containing more than one urea function, the molecules act as *N*,*N*-chelates, so that one of the urea functions always behaves as monohapto-ligand. For example, complexes of Co(III) with *N*-(2-pyridylmethyl)urea and ethylenediamine have been characterised by X-ray crystallographic analysis (Scheme 7). The urea group is coordinated through only one of its N

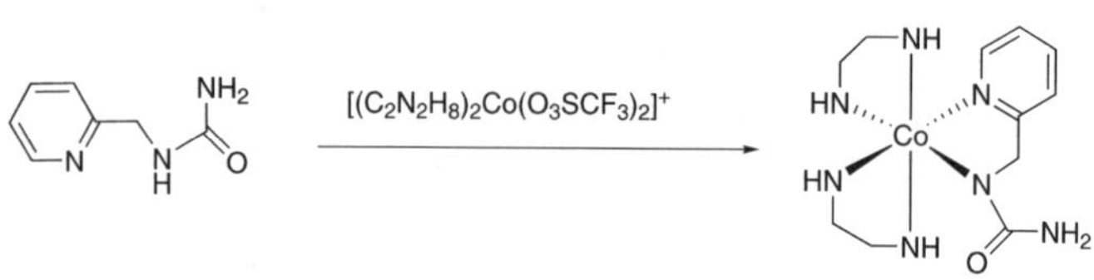

Scheme 7

atoms, while the pyridine moiety brings the second bond: thus the *N*-(2-pyridylmethyl)urea behaves as a *N,N*-bidentate ligand [51].

Iron(II) complex of tris(*N′*-*tert*-butylurea-ylato)-*N*-ethylene]aminato activates dioxygen at room temperature to afford an iron(III) complex containing a single terminal oxo ligand. X-ray structures show that the three urea molecules act as a tridentate *N,N,N*-ligand [52]. The tripodal ligand was also used to synthesise complexes of cobalt, iron or zinc with terminal hydroxo ligands (Scheme 8) [53].

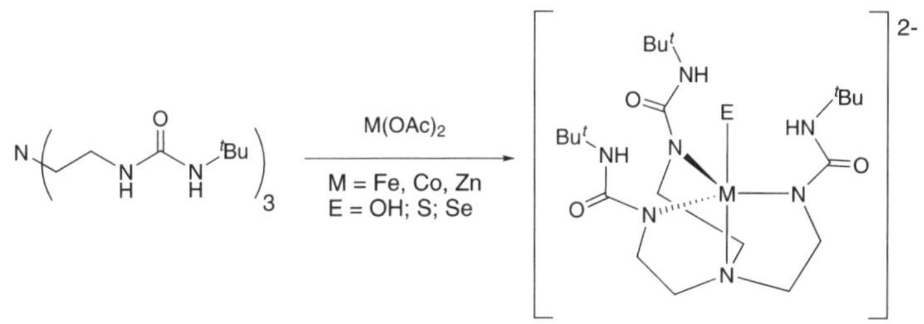

Scheme 8

3.2 Thiourea Coordination Modes to Transition Metals

Platinum-thiourea complexes have been extensively studied because of their biological activity [54], but few have been used in catalysis. Neutral thioureas are able to coordinate to metal centres through their sulfur atom (Scheme 9) [55, 56]: monomeric (**I**) and oligomeric (**II**) species are known for Rh [57], and an X-ray structure has also been determined for the chiral complex **III** [58]. In many complexes hydrogen bonding has been observed

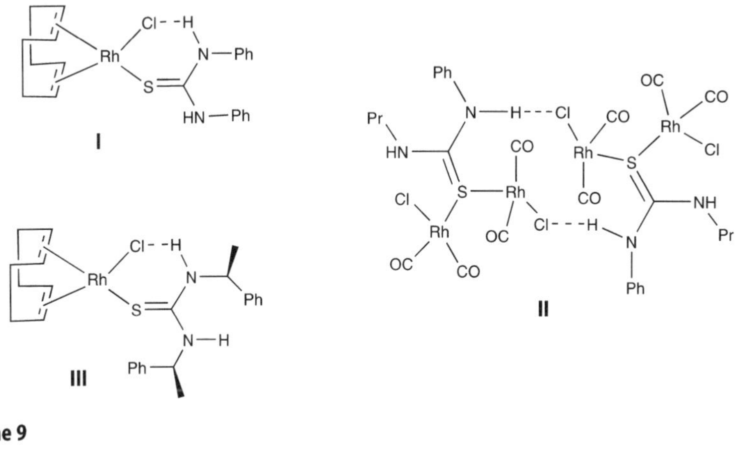

Scheme 9

between halide ligands inside the coordination sphere and one of the hydrogen atoms of the NH thiourea moiety.

Due to their stability and their easy formation, many examples of transition metal complexes containing benzoyl-substituted thiourea ligands have been described [59–62]. Most of them concern Pt species in which the thiourea ligands behave as monoanions and are bounded to the metal centre through the S and O atoms, forming a six-member ring system (Scheme 10).

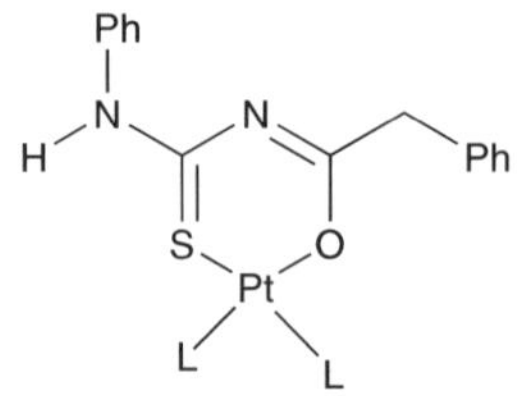

Scheme 10

Non-ionic thiourea derivatives have been used as ligands for metal complexes [63, 64] as well as anionic thioureas and, in both cases, coordination in metal clusters has also been described [65, 66]. Examples of mononuclear complexes of simple alkyl- or aryl-substituted thiourea monoanions, containing *N*,*S*-chelating ligands (Scheme 11), have been reported for rhodium(III) [67, 68], iridium and many other transition metals, such as chromium(III), technetium(III), rhenium(V), aluminium, ruthenium, osmium, platinum [69] and palladium [70]. Many complexes with *N*,*S*-chelating monothioureas were prepared with two triphenylphosphines as substituents. ^{31}P NMR spectroscopy of these phosphorus complexes has been used for structure elucidation. Depending on the bulkiness of the thiourea substituents, a single isomer (bulky) or two isomers (small size, example in Scheme 11 with two possibilities for the relative position of the cyano group)

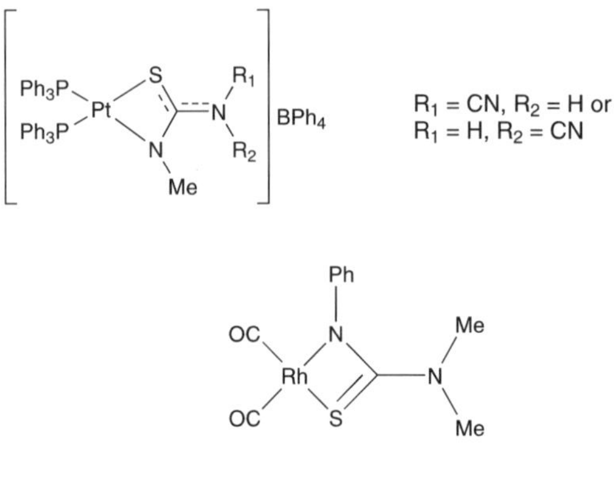

Scheme 11

have been observed with Pt. A similar structure has been proposed for the related Rh complex (Scheme 11).

Platinum and ruthenium complexes have been characterised by positive-ion electrospray ionisation mass spectroscopy (ESMS) and in some cases X-ray diffraction spectroscopy. The S – C and C – N bond lengths in these species suggest a delocalisation and a double bond character in both the C – S and C – N bonds. For example, the Ru complex depicted in Scheme 12 contains two thiourea monoanions bonded in an *N*,*S*-chelating mode [71]. Bis(monoanion)-Pt complexes with thioureato ligands have also been prepared. A Pt complex with two substituted thioureas as ligand crystallises with two independent molecules, and the bond lengths and angles are similar for both. The single-crystal X-ray structure shows S-bonded thiourea anions *trans* to each other.

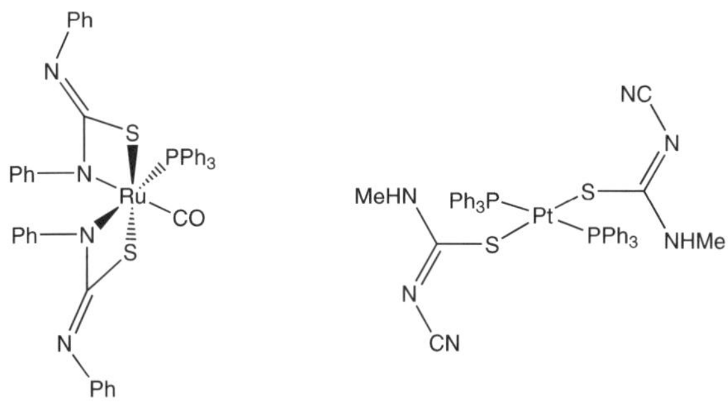

Scheme 12

Thiourea ligands can be bounded to the metal centre through one nitrogen atom, the sulfur atom, or the C = S double bond. These coordination modes were studied by density functional theory calculations for Rh-thiourea complexes (Scheme 13). No stable structure was attained by optimisation of the nitrogen coordination mode **I** but optimised geometries as trigonal-bipyramidal complexes were obtained for modes **II** and **III**. An η^2 coordination is determined for the latter complex through both S and C atoms. As this

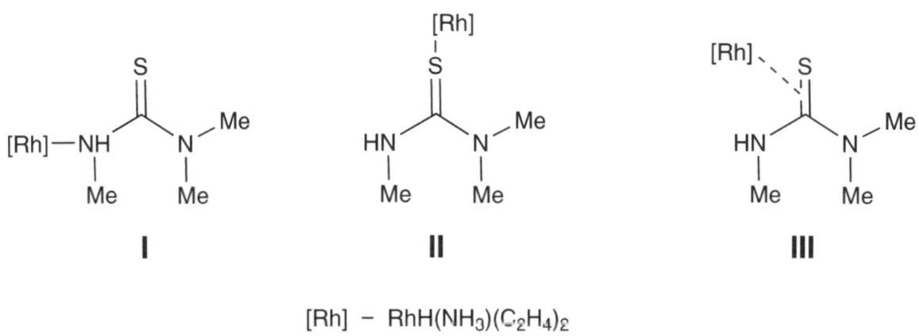

Scheme 13

structure is 9 kcal/mol higher than **III**, the dimethyl-methyl-thiourea prefers an η^1 coordination mode which can be rationalised to other thioureas since it involves the S lone pair. Thioureas and ureas have therefore very distinct coordination modes.

4 Catalytic Activity of Urea- and Thiourea-Containing Complexes

4.1 Asymmetric Reduction

4.1.1 Hydrogen Transfer Reduction of C = O Bonds

The hydrogen transfer reaction (HTR), a chemical redox process in which a substrate is reduced by an hydrogen donor, is generally catalysed by an organometallic complex [72]. Isopropanol is often used for this purpose since it can also act as the reaction solvent. Moreover the oxidation product, acetone, is easily removed from the reaction media (Scheme 14). The use of chiral ligands in the catalyst complex affords enantioselective ketone reductions [73, 74].

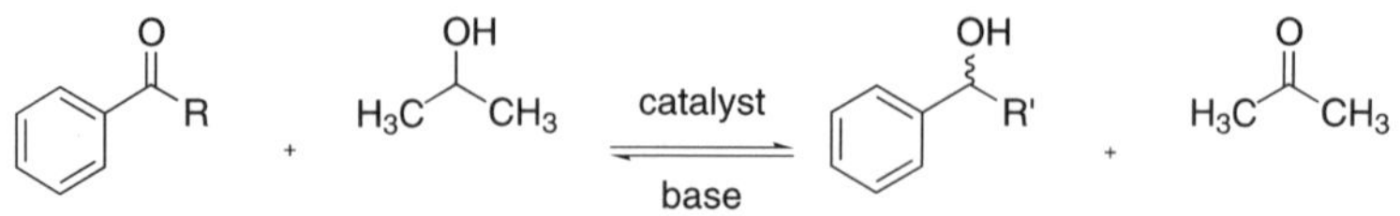

Scheme 14

Polyureas were first used in HTR as an heterogeneisation method for Rh-diamine catalysts by incorporation of the chiral ligand into polymer backbones [75]. Various polyamides and polyureas were thus prepared and the influence of the structure of the polymers on catalytic activity was also investigated. When using the polyamide or the polyurea prepared with a flexible chain copolymer (e.g. hexamethylenediisocyanate), only low ee were observed. Moreover, with such material, no recycling could be performed. A rigid polymer was obtained using a commercially available mixture of methylenediphenyldiisocyanate and its trimer. Crosslinking was instantaneous, leading to a hard and insoluble material that gave better results in the polyurea-based catalyst for acetophenone reduction (Scheme 15) [46]. When the reaction was complete, the catalyst was recovered by filtration, washed and reused twice without loss of selectivity or activity (elemental analysis confirmed Rh quantitative complexation in the polymer). It is of great importance to prepare a rigid and cross-linked material to obtain highly selective

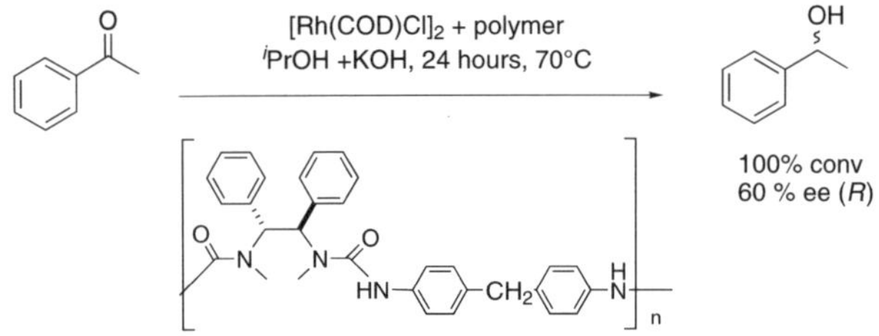

Scheme 15

and reusable catalysts. The enantioselectivity of the catalyst was improved by the application of molecular imprinting technology, in which the imprinted material allowed 70% ee after careful optimisation of the cross-linking and synthesis parameters [76, 77].

Polyurea was used as both chiral ligand and support in an heterogeneous system but it should be used at higher temperatures than the diamine complex (60 °C instead of room temperature). Nevertheless, due to the higher stability of the polymer catalyst, increased turnover numbers are observed as well as similar ee values. A series of optically pure soluble diureas were easily obtained from chiral diamines and chiral diisocyanates, thus affording two types of stereogenic centres [78]. A match/mismatch double induction effect was noticed with cyclohexyl, as well as with diphenyl-derived ligands (Scheme 16). Even if encouraging enantioselectivities were observed, long reaction times were required: 7 days for the best result, obtained for propiophenone reduction in 80% ee (*R*).

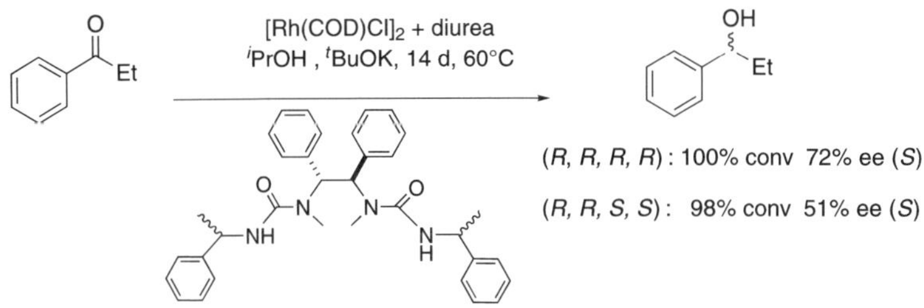

Scheme 16

Chiral dithioureas, which are sulfur analogues of diureas, were thus evaluated as ligands for the asymmetric HTR of arylketones catalysed by Ru, Ir, Rh or Co species and compared to the corresponding urea ligands [38]. Ruthenium complexes afforded the best enantioselection for the reduction of arylketones catalysed with in-situ formed species from $[Ru(C_6H_6)Cl_2]_2$ and chiral *N*,*N*′-diphenyldithiourea (Scheme 17). Dithiourea complexes were more active and enantioselective [79] than the corresponding monothioureas or guanidine derivatives [38].

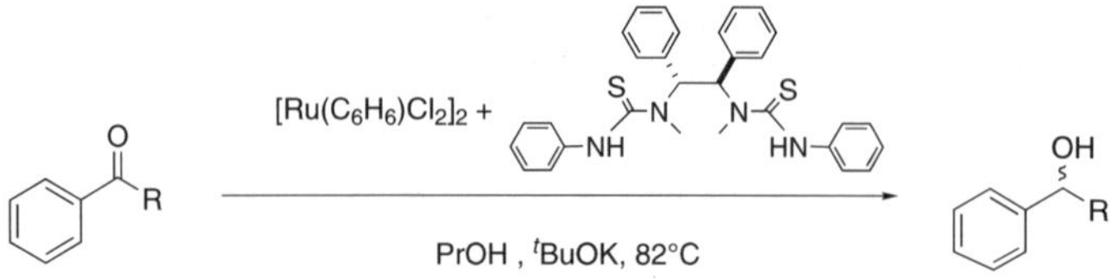

R	time (h)	% conv	% ee
Me	9	98	87 (*S*)
Et	17	96	91 (*S*)
^{i}Pr	17	92	94 (*S*)
^{t}Bu	48	93	85 (*R*)

Scheme 17

Effects on the nitrogen substitution was examined in the case of dithiourea ligands by replacing Me by H on each of the N atoms. The presence of methyl groups improves both the reaction rate and the asymmetric induction from 24% ee, 15% conv. in 3 days to 89% ee, 94% conv. in 1 day for acetophenone reduction [80]. A theoretical and experimental approach was undertaken for dithiourea-Rh catalysts to determine the influence of electron-withdrawing or electron-donating substituents as well as the influence of bulky substituents on the aromatic rings (Scheme 18). It appears that steric hindrance does not affect catalytic activity but that CF_3, CN, NO_2 causes a dramatic decrease in ee while OMe substituents slightly improve chiral induction.

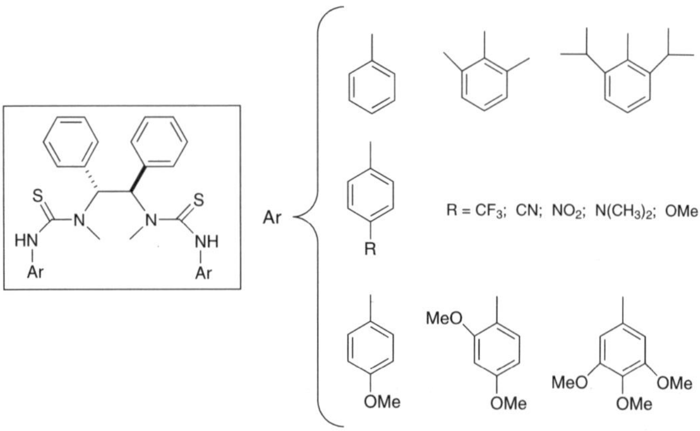

Scheme 18

Heterogeneisation of such dithiourea catalysts was achieved by the synthesis of a series of chiral polythioureas from the corresponding chiral diamine and diisothiocyanate [81]. Results of the catalytic tests have shown that it is important to preserve the C_2-symmetry inside the polymeric material [82] in order to obtain enantioselectivities close to those observed with the homogeneous catalyst. As for polyurea analogues, efficient polythiourea-based catalysts need a rigid spacer between the two thiourea units. Due to the low solubility of such polymers, the catalysts are easy to recover by filtration [83]. Using polythiourea as ligand and support associated with the Ru precursor, 77% ee was obtained for acetophenone reduction by hydride transfer. The catalyst could be recycled five times with only slight loss in reactivity and selectivity (Scheme 19) [82].

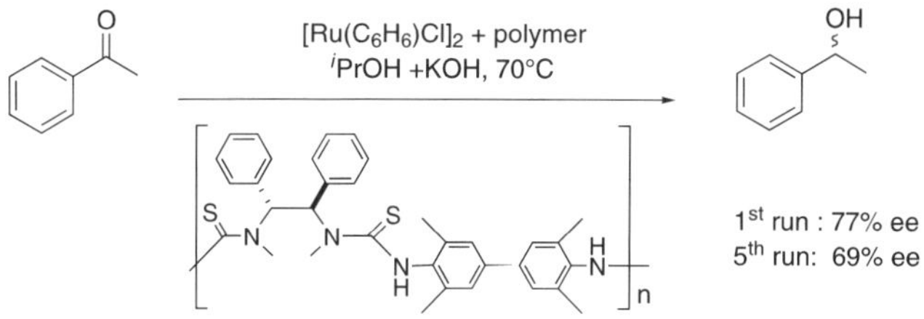

Scheme 19

Moreau and co-workers prepared (1*R*,2*R*)-1,2-diaminocyclohexane amino-urea and thiourea derivatives and combined them to Ru precatalysts for the HTR of arylketones [43]. By comparing mono- and diurea ligand effects, it was found that monourea-containing catalysts were more efficient, giving (1*R*)-phenylethanol with ee values up to 83% (Scheme 20). The monourea catalyst was also more active and enantioselective than the corresponding monothiourea species, which surprisingly led to the opposite alcohol enantiomer (*S*) with moderate enantioselectivities (Scheme 20). Immobilisation of these polyurea ligands was achieved by sol-gel hydrolysis condensation giving hybrid materials: a left- or right-handed helix was autogenerated according to the chiral (*R*, *R*)- or (*S*, *S*)-cyclohexane structure [48].

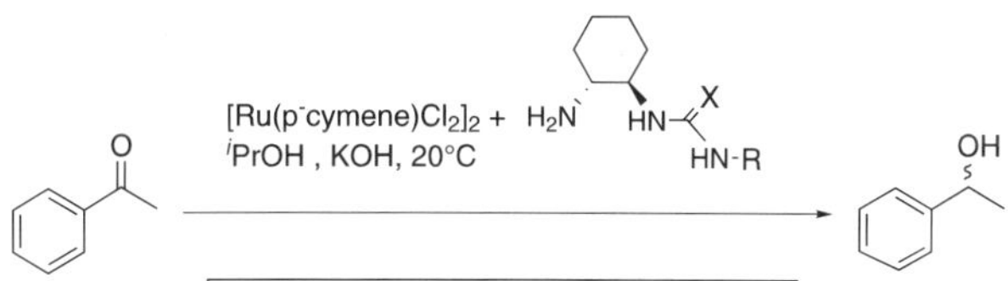

R	X	time (h)	% conv	% ee
Cy	O	4	73	83 (*R*)
Cy	S	4	20	22 (*S*)
^{t}Bu	O	18	54	66 (*R*)
^{t}Bu	S	22	72	36 (*S*)

Scheme 20

4.1.2 Hydrogenation of C = O and C = C Bonds

Enantioselective hydrogenation of C = O or C = C bonds has been widely studied over the last decade since it is a useful pathway to a wide variety of enantiomerically pure molecules. Asymmetric hydrogenation in the presence of transition metal catalysts was one of the two topics of the 2001 Nobel Prize for chemistry [84]. Many books [85–87] and reviews [88] give an overview of the catalysts employed, which often contain chiral phosphines. Among them BINAP is probably the most studied [89]. One of the most successful uses of such an ubiquitous ligand is the asymmetric hydrogenation of aromatic ketones. In such cases, the synergetic effect between diphosphine and diamine allows the reduction of aryl ketones with excellent enantioselectivities and impressive turnover rates. Nevertheless, other non-phosphorinated compounds have been developed for this purpose, in particular nitrogen-based ligands [1] such as amino alcohols, oxazolines, pyridines, phenantrolines, amines, ureas and thioureas. Various C_2-symmetric diamines, diureas and dithioureas were tested as chiral ligands combined with various Ru, Rh or Ir precatalysts for asymmetric hydrogenation of aryl ketones or enamides (Scheme 21). In both cases, thiourea complexes were much more efficient

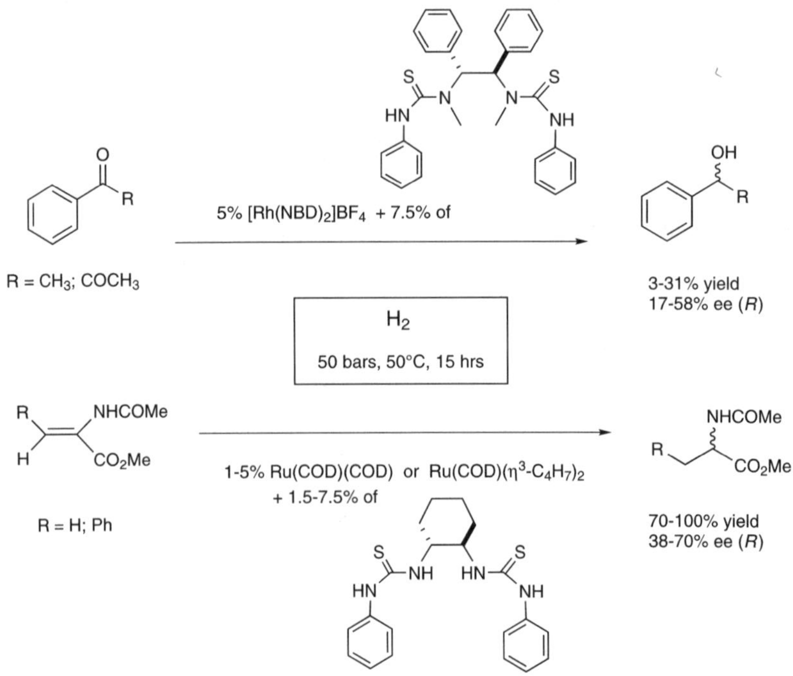

Scheme 21

than urea species: up to 58% ee for phenylglyoxylate methyl ester hydrogenation with cationic Rh/*N*,*N*′-diphenyldithiourea complex [90]. Better chiral induction was observed for the hydrogention of α-acetamidocinnamic methyl ester in 70% ee with diaminocyclohexanedithiourea/Ru catalyst [91]. Although relatively modest ee values were obtained, compared to the diphosphine counterpart, such results show that thioureas were able to act as alternative ligands for Rh- or Ru-catalysed hydrogenation.

4.1.3
Hydrosilylation

Enantioselective hydrosilylation of ketones is a versatile synthetic route to chiral secondary alcohols and it can be efficiently catalysed by several types of complexes, including Rh complexes, some Ti species and a few Fe species [92–94]. In contrast to other catalytic ketone reduction methods, asymmetric induction for hydrosilylation is often better with N ligands than with P ligands: oxazolines, pyridylimines or alkylamines. Brunner, Nishiyama and co-workers reported excellent ee values achieved with Rh Pybox [95] and Pythia [96, 97] complexes. Recently, our group developed a series of Ir catalysts for acetophenone hydrosilylation containing chiral C_2-symmetric diamines, dithioureas or monothioureas (Scheme 22). Better enantioselectivity is observed with dithiourea catalysts than with analogue diamine Ir species: up to 74% ee could be reached with a tenfold excess of ligand versus Ir precursor [39].

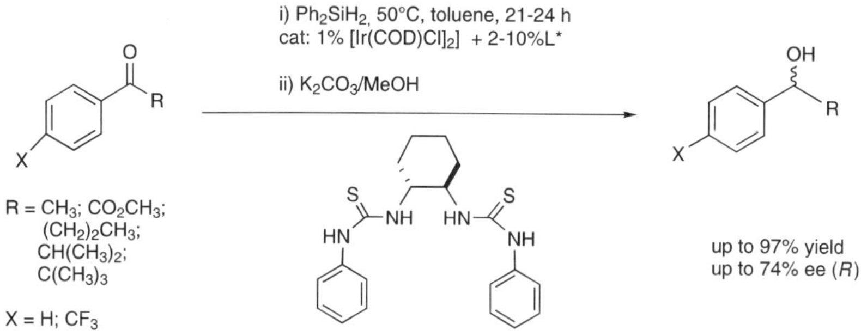

Scheme 22

4.2
Oxidation and Reaction with Epoxides

Urea hydrogen peroxide adduct (UHP) was employed in metal-catalysed asymmetric epoxidation [98] and Baeyer–Villiger oxidation [99, 100]. Since the presence of urea does not change the course of the reaction, this will not be described here. Conversion of epoxides to halohydrins with elemental halogen was accelerated in the presence of thiourea, but no chiral process has been described so far [101].

4.3 Formation of C–C Bonds

Carbon–carbon bond forming reactions via transition metal catalysts will be discussed in this section. Thiourea was also used as an organocatalyst and this topic will be discussed in the next section. Chiral polyureas were used for Pd-catalysed allylic substitution with dimethylmalonate but gave disappointing results. Chiral polyamide analogues prove to be much more efficient and selective for this reaction [75]. Thioureas were used as ligand in metal-catalysed organic synthesis. Thanks to the strong association of the sulfur atom and functional groups containing sulfur to transition metal, many examples reported the use of non-chiral thioureas. This was particularly true for Pd catalysts used for the Suzuki coupling (Scheme 23) [102], carbonylation reactions (see next section) and the Heck reaction [103]. They illustrate the potential of such types of ligand in catalysis and may inspire future developments in asymmetric catalysis.

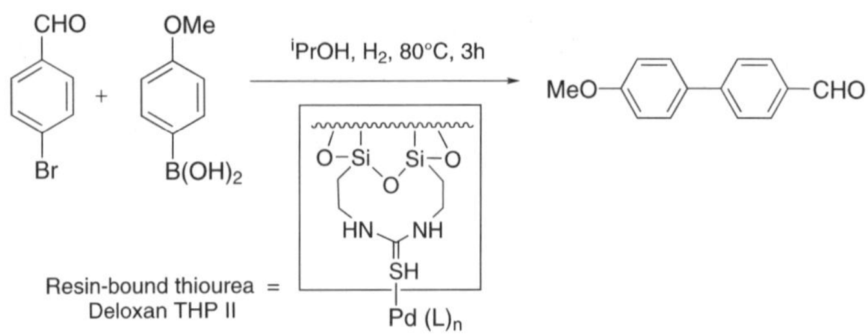

Scheme 23

The Pd-catalysed Heck reaction performed with thiourea as the ligand exhibit good activities for some catalysts. As for carbene ligands [104], steric hindrance improves catalytic results. Thus, thioureas wearing bulky substituents afford the formation of air- and moisture-stable Pd complexes [105]. For example, the catalyst obtained with 2 mol % $Pd(dba)_2$ and *N*,*N*′-dimesitylene-ethylene thiourea (Scheme 24) was still active even after 2 months in an air atmosphere.

In a very recent work, the Pd-catalysed cross-coupling reactions with arenediazonium salts under aerobic conditions in the presence of a chiral monothiourea ligand were reported (Scheme 25) [106]. Even if this ligand bears four chiral centres, no test in asymmetric Heck-type reaction has been described so far.

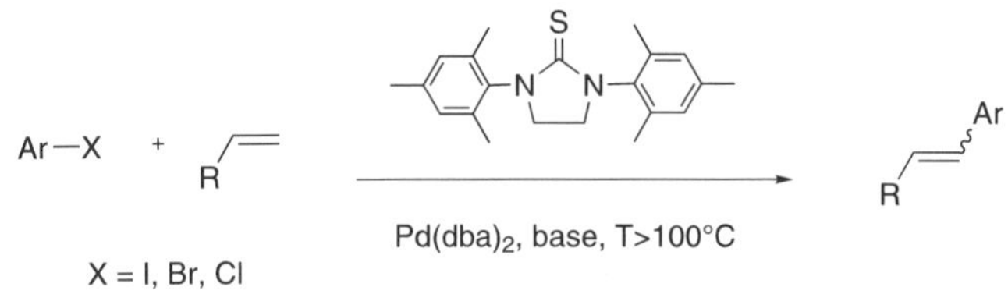

Scheme 24

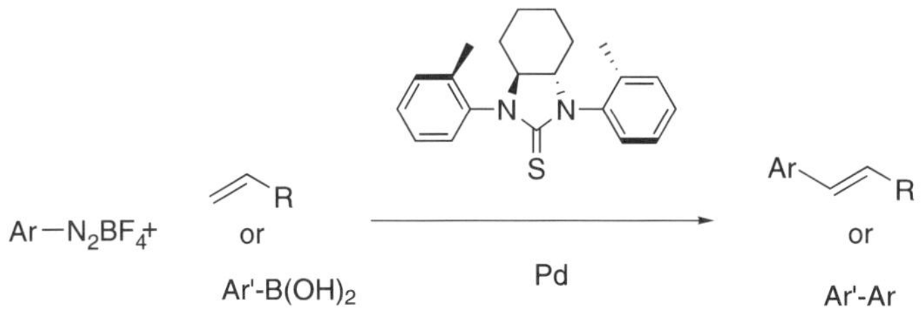

Scheme 25

4.4 Reactions Involving CO

Hydrocarbonylation of olefins, hydroformylation, hydroesterification and hydroxycarbonylation are reactions which appear to be of particular interest. Indeed, they allow the simultaneous creation of a new C – C bond as well as the introduction of a functional group (aldehyde, ester and acids). One or two new stereogenic centres can thus be formed at the same time (Scheme 26). Despite the difficulty of using high carbon monoxide pressure, the already existing industrial processes prove that such reactions can be performed on a very large scale [107].

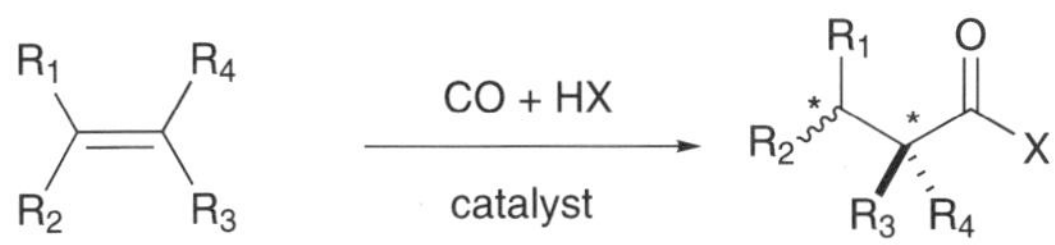

Scheme 26

4.4.1 Hydroformylation

Hydroformylation has been extensively studied since it produces optically active aldehydes which could be important precursors for pharmaceutical and fine chemical compounds. Thus, asymmetric hydroformylation of styrene (Scheme 27) is a model reaction for the synthesis of ibuprofen or naproxen. Phosphorus ligands were used for this reaction with excellent results, espe-

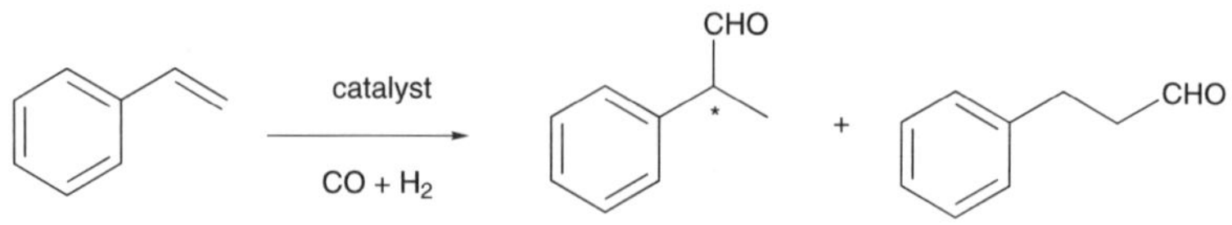

Scheme 27

cially for the Rh-Binaphos system [107]. Sulfur ligands, such as thiols and thioethers, led to low enantioselectivities although a good activity is noted for this reaction [108]. Rhodium is the preferred metal for this reaction since it forms highly active and selective catalysts, leading to the main formation of aldehydes (instead of the hydrogenation product). The neutral dimeric complex $[Rh(COD)Cl]_2$ is generally used in the presence of a base (Et_3N). More sophisticated complexes can also be used. The branched aldehyde which presents a stereogenic centre is the major product, but some linear aldehyde is also obtained (Scheme 27, *i*: hydratropaldehyde/*n*: hydrocinnamaldehyde, $i/n = 90/10$ typically) [109].

The major difficulty in this reaction consists in the formation of "naked Rh species" which are highly active and compete with the chiral species, which often have lower activity. Thus, neutral $[Rh(COD)Cl]_2$ and cationic $[Rh(COD)_2]BF_4$ precursors, without any additional ligand, are able to form active complexes for the catalytic hydroformylation of styrene [110]. It is difficult to determine if the ligand coordinates to the metal centre or not. In such cases, the stable Rh complexes act as a reservoir for the active naked species. Thus, amines or ureas are not bonded strongly enough to maintain the asymmetric catalytic cycle, and the formation of metallic Rh particles can be observed [111]. The asymmetric induction is then difficult to obtain in this reaction. Moreover, the optically active aldehyde can turn into the racemic product during the course of the reaction. Cauzzi showed that thiourea-functionalised silica xerogels containing Rh were effective and recyclable catalysts for hydroformylation (Scheme 28). In this particular case, it was not proven that the reaction occurred at the surface since leaching was observed, suggesting reaction in the homogeneous phase. However, regioselectivity towards the branched aldehyde increased for successive runs and the authors studied the Rh distribution at the surface and inside the core of the material. The spectroscopic data (IR and XPS) showed four different thiourea-Rh linkages with a decrease in the external Rh population. The Rh

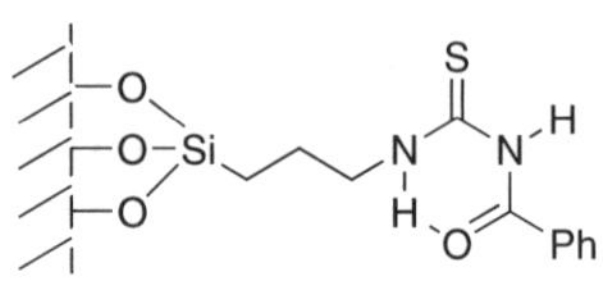

Scheme 28

species located in the core of the material were more selective towards the formation of the branched product [112, 113].

Some chiral mono-, acyl- and di-thioureas have been used as ligand for the Rh-catalysed asymmetric hydroformylation of styrene. Although thiourea ligands form inactive systems with $[Rh(COD)Cl]_2$ as the catalyst precursor, in standard conditions (40 °C, 40 bar CO + H_2: 1/1), the cationic Rh complex $[Rh(COD)_2]BF_4$ combined with monothioureas as the ligand showed moderate to good activity (Scheme 29) [114].

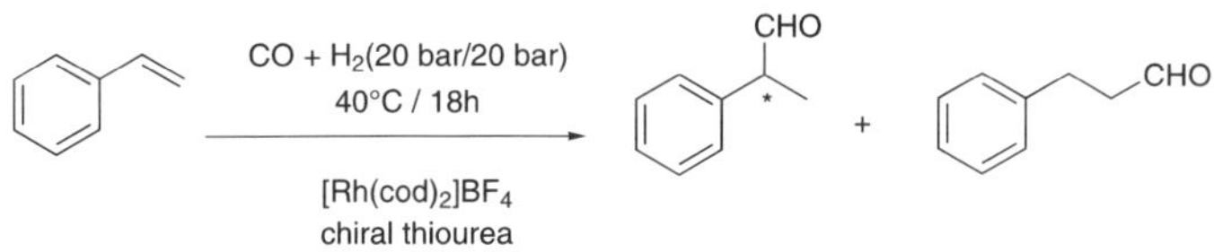

chiral thiourea	% ald. (%conv.)	b/n	% ee
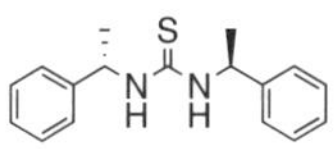	99 (18)	85/15	16 (-)
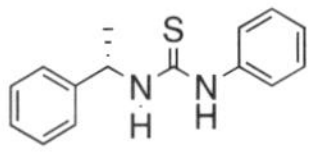	98 (18)	91/9	24 (-)

Scheme 29

Excellent chemoselectivity (> 98%) was obtained with good regioselectivity (> 85%) but generally with very low enantioselectivity, except for a few ligands with ee up to 24%. An excess of monothiourea ligand (L/Rh= 2/1 and more) or a dithiourea ligand inhibited hydroformylation, except for ligands with C_2-symmetry (ee up to 16%) [115]. Monothioureas then had to be associated to phosphorus ligands in order to give active and stable hydroformylation catalysts. Recent work of Cauzzi et al. discusses this point, but without any chiral application [116].

4.4.2 Alkoxycarbonylation

Thiourea was used as stabilising agent for zerovalent Pd species [117]. The Pd-thiourea (H_2NCSNH_2) catalysed carbonylation of terminal alkynes and allylic alcohols has been described by Chiusoli [118]. More recently, Pd-thiourea-catalysed carbonylative annulation was studied. The reaction proceeds between alkynes, iodophenol acetates and carbon monoxide, in the presence of dppp, thiourea (H_2NCSNH_2) and base at 40 °C. Flavones have been obtained in good yields (Scheme 30) [119].

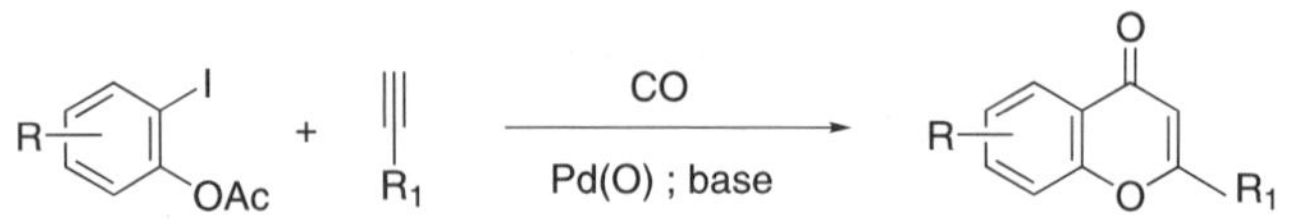

Scheme 30

2,3-Disubstituted benzo[*b*]furans were also prepared by intramolecular cyclisation in the presence of a $[Pd(thiourea)_4]I_2$ catalyst (thiourea = $H_2NCS NH_2$). No Pd precipitation occurred with this very stable thiourea complex (Scheme 31) [120].

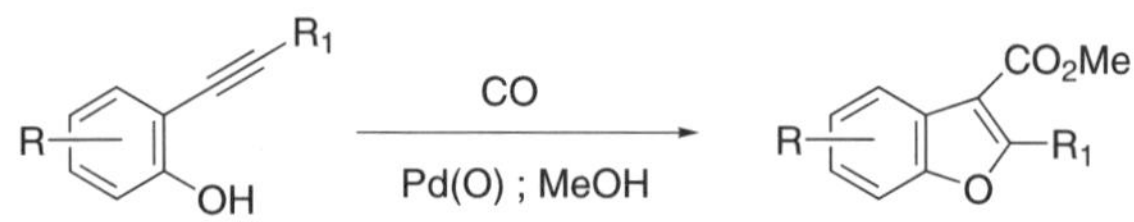

Scheme 31

Chiral lactones were also obtained by cyclocarbonylation of chiral acetylenic alcohols with Pd and thiourea (H_2NCSNH_2) (Scheme 32). No loss in chirality was observed, but large amounts of Pd and thiourea were used (10 mol %) since the catalyst deactivates by forming metal particles. The catalytic precursor (PdI_2 > $PdCl_2$) and the ratio of thiourea to Pd were very important, thiourea being necessary for this reaction. The active species was supposed to be $[Pd(thiourea)_3I]I$, which forms in situ from $[Pd(thiourea)_4]I_2$ and $[Pd(thiourea)_2]I_2$. It had to be a partially dissociated species since $[Pd(thiourea)_4](BF_4)_2$ was inactive [121].

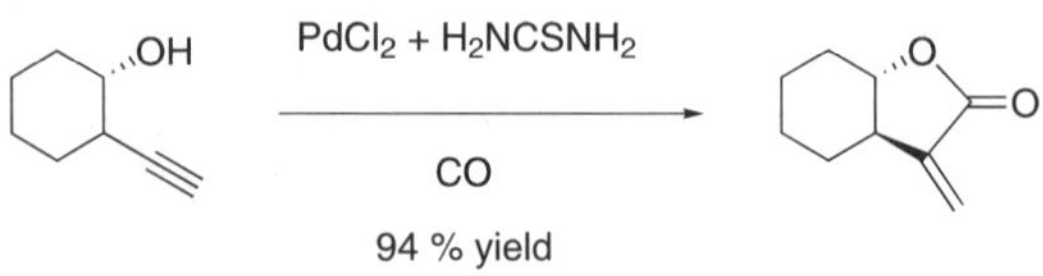

Scheme 32

Chiral lactones were also formed by cyclocarbonylation [122] with chiral catalysts, such as Pd-poly-L-leucine catalytic system. For example, but-2-en-1-ol led to the corresponding cyclic chiral lactone in the presence of Pd catalysts with chiral ligands (Scheme 33). About 10 mol % of Pd(II) chloride

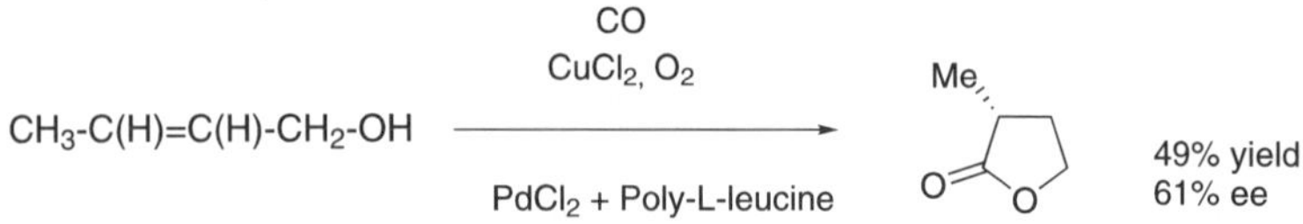

Scheme 33

and 4 – 40 mol % of the chiral ligand were needed for this reaction. The major drawback lay in the vinyl alcohol reagent, which required both CO (reducing agent) and oxygen [123].

Palladium salts are able to catalyse diyne carbonylation, so the reaction can be performed at room temperature under 1 atm of carbon monoxide. Thiourea (H_2NCSNH_2), which is added to stabilise the Pd catalyst (Scheme 34), is described as the best ligand for the efficiency of this reaction [124].

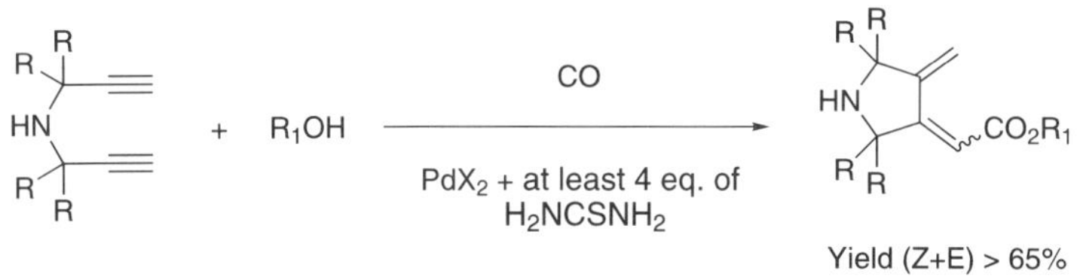

Scheme 34

Many other carbonylation reactions were carried out with thiourea [117], but no information about the use of chiral thioureas have been reported yet. Other Pd-catalysed hydrocarbomethoxylation reactions were studied with styrene and chiral phosphorus ligands, leading to low ee (52% with neomenthyldiphenylphosphine reported by Chiusoli) [125, 126]. All of these systems require large amounts of oxidising agents (e.g. $CuCl_2$ and O_2) to maintain a catalytic cycle, since the Pd(0) formed after one catalytic cycle had to be re-oxidised to Pd(II) [127, 128]. However, this could not be considered as a major drawback since the industrial Wacker process is currently using a Pd re-oxidation technology. Several metal-catalysed carbonylations were studied [129–135] with sometimes high enantioselectivities: 99% ee for the hydroesterification of styrene with the bis-diphenylphosphine derived from isosorbide [136]. Only nitrogen- and phosphorus-based auxiliaries were used [137–139] until recently. However, *N*,*N*′-disubstituted ureas can act as additives for Pd-catalysed hydrocarbonylation reactions. The accelerating effect was attributed to the interaction of the acidic urea protons with the catalysts, weakly coordinated by counter-ions [140]. A study on the use of chiral ureas or thioureas in such processes has not been performed yet.

4.4.3 Bismethoxycarbonylation

The Pd(II)-catalysed asymmetric carbonylation of olefins with a chiral thiourea as the ligand has been reported recently. Since these ligands are stable in the presence of oxidising agents, they prevent Pd precipitation and double-bond isomerisation (Scheme 35) [141].

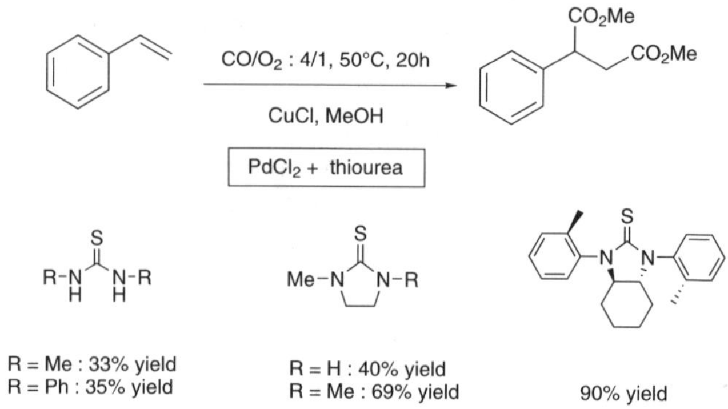

Scheme 35

The authors underline that the saturated species $[Pd(thiourea)_4]I_2$ did not lead to bis(methoxycarbonylation) of styrene under the reported reaction conditions, but to styrene polymerisation [142, 143]. With a low-coordinated Pd complex such as $[Pd(tetramethylthiourea)]Cl_2$ [144, 145], the desired product was obtained in only 30% yield. Optimisation of the reaction parameters (Pd halide, oxidising agent, solvent, temperature and CO pressure) resulted in improving the yield to over 60%. The authors also investigated the effect of the substitution of the thiourea: higher yield of carbonylated product is obtained for substituted thioureas (Scheme 35). They explain that hindered thiourea ligands were more effective for the generation of the low-coordinated active Pd complex; no Pd precipitation was observed. Double-bond isomerisation did not occur when other olefin were used. Although chiral thioureas derived from diaminocyclohexane were tested, no asymmetric induction was reported.

5
Organocatalysis: Ureas and Thioureas as Organic Catalysts

Even if organocatalysis is a common activation process in biological transformations, this concept has only recently been developed for chemical applications. During the last decade, achiral ureas and thioureas have been used in allylation reactions [146], the Baylis–Hillman reaction [147] and the Claisen rearrangement [148]. Chiral organocatalysis can be achieved with optically active ureas and thioureas for asymmetric C – C bond-forming reactions such as the Strecker reaction (Sect. 5.1), Mannich reactions (Sect. 5.2), phosphorylation reactions (Sect. 5.3), Michael reactions (Sect. 5.4) and Diels–Alder cyclisations (Sect. 5.6). Finally, deprotonated chiral thioureas were used as chiral bases (Sect. 5.7).

5.1
Strecker Reaction: CN Addition

The Strecker reaction is a condensation of an aldehyde, ammonia and cyanide source (HCN), followed by hydrolysis of the resulting amino nitrile to the corresponding amino acid [149]. In addition to metal-catalysed asymmetric cyanations, organocatalysts have been developed such as chiral guanidino-diketopiperazine [150] (Lipton) and bicyclic guanidine [151] (Corey). Imine-containing urea and thiourea derivatives were developed by Jacobsen (Scheme 36) [148, 152–155]. In these catalytic processes, asymmetric Strecker reactions were performed using preformed imines followed by quenching of the cyano adduct with trifluoroacetic anhydride (TFAA, Scheme 36).

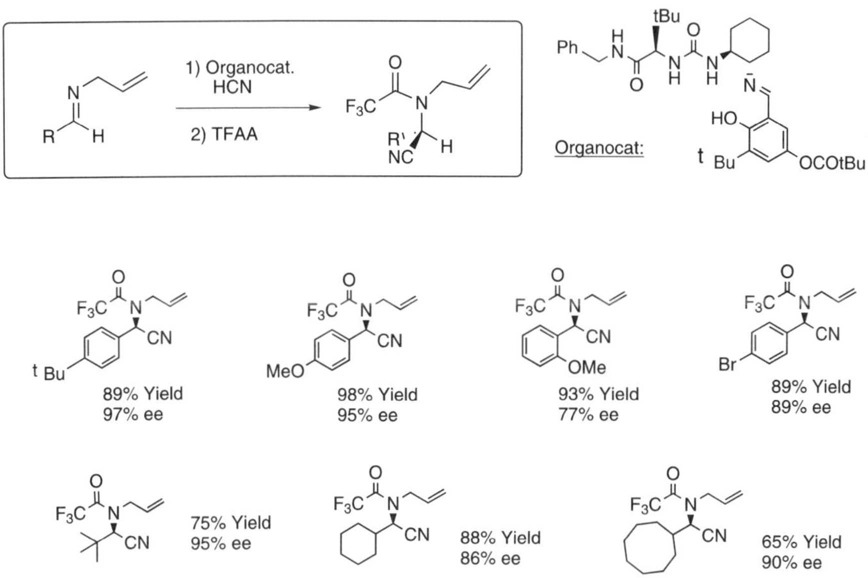

Scheme 36

5.1.1
Asymmetric Strecker Synthesis

The asymmetric Strecker-type reaction developed by the Jacobsen group is suitable for both aliphatic and aromatic imines, giving high enantiomeric excesses for a wide range of substrates. In this reaction the urea derivative also acts as the catalyst (Scheme 36).

Other ureas and thioureas were tested for this reaction. All of the molecules contain an imino bond moiety, which appeared to be beneficial for the catalysed hydrocyanation process, but urea or thiourea functional groups were of major importance.

5.1.2
Solid Phase Synthesis for High Throughput Screening

These catalysts were first tested as resin-bound derivatives via HTS, first with metals and then without. Three libraries of chiral molecules, based on three different enantiomerically pure diamines, bulky salicylidene moities and optically active *R*-amino acids were used for structure optimisation (Scheme 37 TBSCN = ${}^{t}BuMe_2SiCN$) [152].

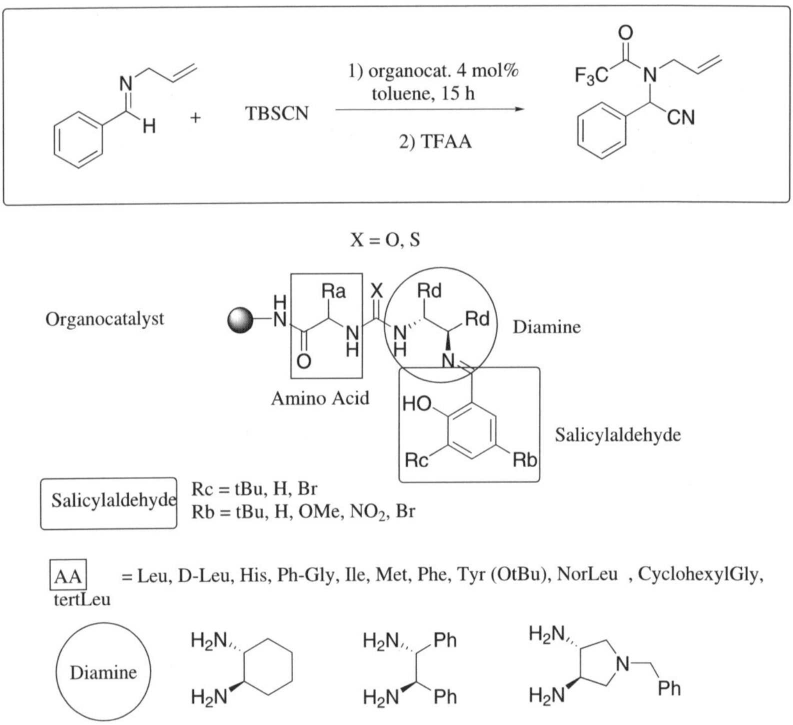

Scheme 37

The reaction was first tested with these substances as ligands but the organic molecule, in the absence of any added metal ion, proved to be the most enantioselective catalyst (library 1: 19% ee vs. less than 13% ee for the best metal catalyst). The effects of selective variations of the amino acid nature and of the salicylidene moiety on the diamine structure were investigated for urea and thiourea derivatives via HTS (library 2 : 48 urea compounds and library 3 : 132 thiourea compounds) with the model substrate *N*-allyl benzaldimine.

5.1.3
Homogeneous Strecker Synthesis

The catalysts bearing a cyclohexylamine moiety combined with a bulky salicylidene compound linked via one thiourea function to a *tert*-leucine benzylamide (Scheme 38, R_1 = Bn, R_2 = H) was the most efficient. The test was performed in solution at – 78 °C, with HCN as the cyanide source. Excellent results were obtained: 78% isolated yield with 91% ee for the optimised substrate and 70–86% ee for other imine derivatives (65–92% isolated yield) [148, 152–157].

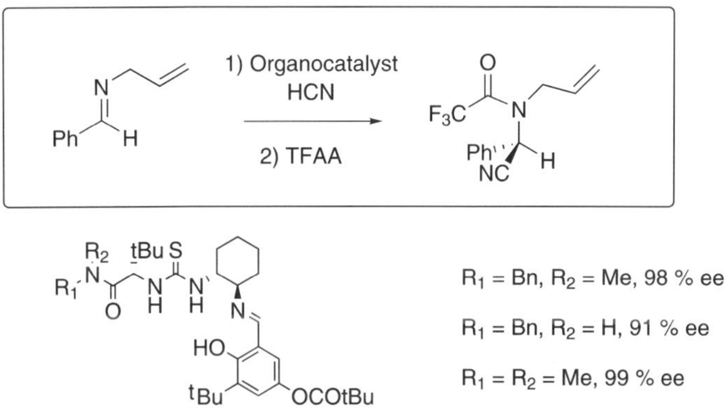

Scheme 38

5.1.4
Recent Developments: Urea vs. Thiourea Ligands

New organocatalysts prepared by the Jacobsen group showed that alkylation of the final amide bond increased the enantioselection (Scheme 38, compare R_2 = Me, 98% ee to R_2 = H, 91% ee). Thus, the reaction performed with *N*-allyl benzaldimine and with the dimethylamide-ending thiourea (Scheme 38 with R_1 = R_2 = Me) gave up to 99% ee. This compound is a structural analogue of the urea depicted in Scheme 36 [148, 152, 154].

Both the ureas and thioureas are highly suitable organocatalysts for the asymmetric Strecker synthesis. For example, the thiourea function was replaced by an urea function (note the opposite configurations). The organocatalysts thus obtained showed similar activity and slightly higher enantioselectivities with *N*-allyl benzaldimine (Scheme 39, 74% yield with 95% ee for R_1 = Bn and R_2 = H). Once again, better enantioselectivity (up to 99% ee) was at-

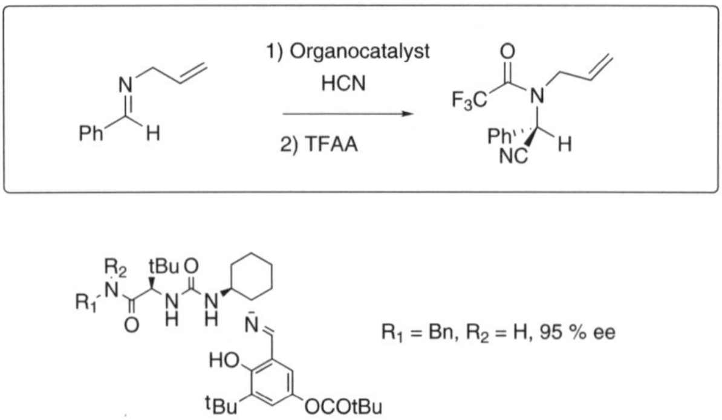

Scheme 39

tained with the dimethylamide-urea analogue (with opposite configurations, depicted in Scheme 36) [148, 152, 154].

5.1.5 Structural Characteristics and Reaction Mechanism

Other organocatalysts were prepared for the Jacobsen-type Strecker synthesis. The characteristic structural components were reviewed recently and new catalysts were used [148]. From this large number of chiral products, it appeared that the use of thiourea instead of urea gave, in general, better results. The amide bond was more efficient than the ester bond. The reaction was not sensitive to the amino acid, but sensitive to the salicylaldimine substituents: the *ortho*-substituent had little effect, while changing the *para*-substituent from ester to *tert*-butyl lowered the ee of the product. Alkylation (MeO-, $^{t}BuCO_2$-) of the phenol also resulted in lower enantioselectivities. More details concerning the mechanism of the reaction were given: it was found that the Schiff base catalyst has a well-defined secondary structure in solution (NOE NMR). The hydrocyanation reaction proceeded according to a Michaelis–Menten kinetic model, with a first-order dependence on the thiourea and HCN, and saturation kinetics with respect to the imine substrate. Reversible hydrogen bond formation between the two urea or thiourea hydrogens and the imine was observed. The binding of the imine as the *Z* isomer was noticed. A broad variety of substrates were tolerated and the asymmetric induction was independent of their steric or electronic properties. The addition of HCN was supposed to take place over the diaminocyclohexane framework (Scheme 36) [153].

5.1.6 Other Substrates

The tertiary amide containing thiourea (Scheme 38 with $R_1 = R_2$ = Me) and urea (Scheme 36) were used for a wide range of substrates as depicted in

Scheme 36. The urea derivative gave similar results to the thiourea compound. Acyclic imines (mainly *E* isomers) and *Z* cyclic imines could also be used for this process (Scheme 40, 91% ee) [148, 152, 154].

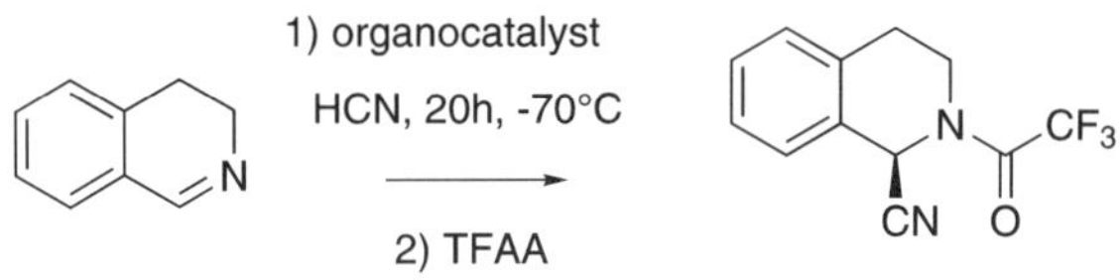

Scheme 40

5.1.7 Industrial Applications

The Jacobsen group has also shown that the recycling of the resin-bounded catalyst can be successfully performed [152, 154]. Moreover, they have developed an efficient method for the hydrolysis of the aminonitrile into the corresponding amino acid. This method was applied for the commercial production of optically active *R*-amino acids at Rhodia ChiRex (e.g. *tert*-leucine): the catalyst was immobilised on a resin support (4 mol %, 10 cycles) and the intermediate hydrocyanation adduct was trapped by simply replacing TFAA with $HCOOH/Ac_2O$, for example. Highly crystalline formamide derivatives were thus obtained in excellent yields (97–98% per cycle) with very high enantioselectivities (92–93% per cycle) [158].

Jacobsen and co-workers also described the highly enantioselective hydrocyanation of ketimines with the urea analogue. After recrystallisation of the corresponding Strecker adduct, formylation and hydrolysis, the *N*-benzyl *R*-methylphenylglycine, was obtained. The *R*-amino acid hydrochloride is obtained in 93% overall yield with > 99.9% ee on a gram scale [149].

5.1.8 Nitrone Cyanation

Finally, nitrone cyanation were performed with non-chiral urea and thiourea derivatives, the latter being more efficient for this process. No chiral compound has been described yet (Scheme 41) [159].

5.2 Mannich Reactions

Similar organocatalytic species to those successfully used for the Strecker reaction were used for the asymmetric Mannich reaction. Catalyst structure/enantioselectivity profiles for the asymmetric Strecker and Mannich reactions were compared by the Jacobsen group [160]. The efficient thiourea

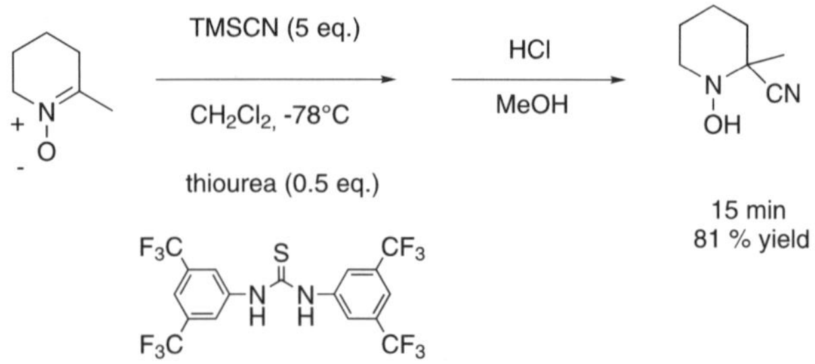

Scheme 41

ligands used for the Strecker reaction (80–99% ee) showed lower activity in the Mannich reaction (44–97% ee). However, similar enantioselectivities were obtained both for the urea- and the thiourea-containing molecules. The model reaction was the asymmetric silylketene acetal addition to *N*-Boc benzaldimine (Scheme 42).

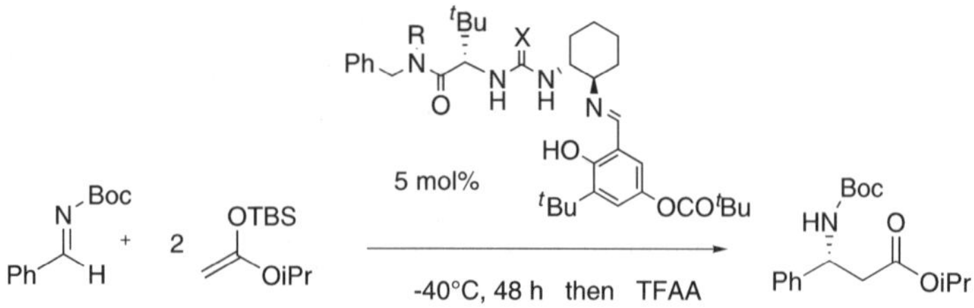

X	R	% conv.	% ee
O	H	92	81
S	H	90	91
	Me	96	97

Scheme 42

New catalysts were prepared after optimisation of the ligand structure. The most efficient organocatalyst for this reaction was an amido-thiourea derivative (Scheme 43). Interestingly, dissymmetrical ligands were more efficient and selective for this reaction.

5.3 Phosphorylation

The Jacobsen group also studied the thiourea-catalysed enantioselective hydrophosphonylation of imines (Scheme 44) [160]. Many examples were de-

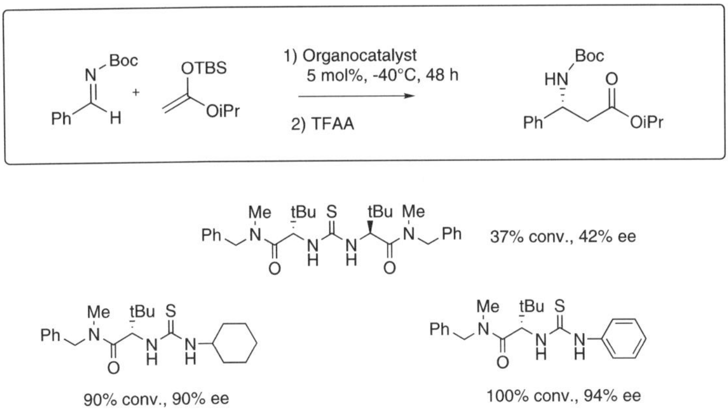

Scheme 43

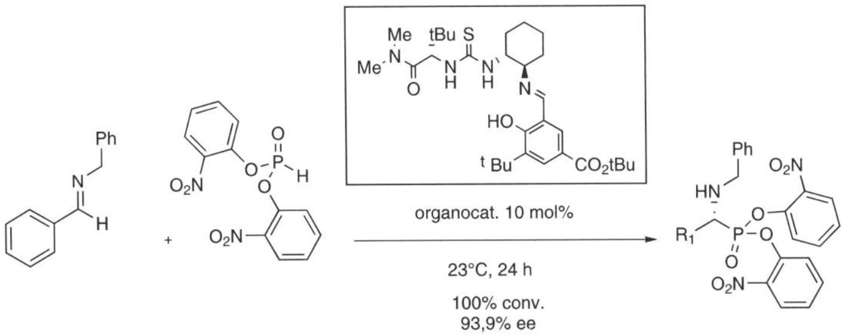

Scheme 44

scribed with excellent enantioselectivities (> 90%) for both substrates after phosphite optimisation (*o*-nitrobenzyl > Ph, 2-cyanoethyl). Aliphatic imines can also be used.

5.4 Michael Reactions

The enantioselective Michael reaction of malonates to nitroolefins catalysed by bifunctional amino-thioureas has recently been reported by Takemoto [161]. Excellent ee (75–93%) were obtained with diethylmalonate after solvent optimisation, toluene being the best solvent both for the activity and for the selectivity. Substituted malonates were then reacted with various nitroolefins under the same conditions. Excellent enantioselectivities were observed (Scheme 45).

This bifunctionnal amino-thiourea organocatalyst led to high selectivity because it was activating both the nitrone and the malonate, in its enol form, due to the acidic hydrogen atoms of the thiourea. Thus, the amino-thiourea catalyst promoted the Michael reaction of malonates to various nitroolefins

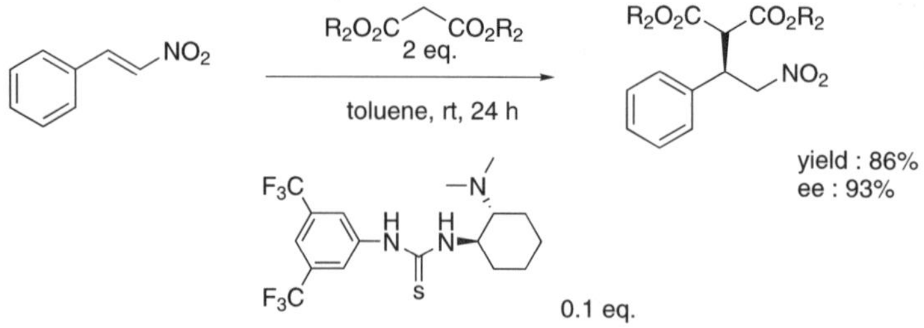

Scheme 45

with high enantioselectivities (81–93% ee). The authors reported that the reaction was also successful without solvent. Very recently, Nagasawa reported the use of chiral ureas for the hetero-Michael reaction [162]. Non-chiral ureas, guanidines and thioureas accelerated the racemic reaction of pyrrolidine with γ-crotonolactone. The asymmetric conjugate addition was then performed with chiral ureas. The most successful representatives are reported in Scheme 46.

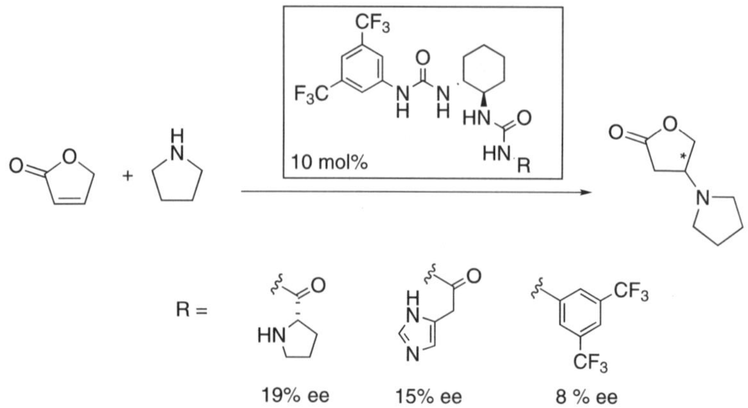

Scheme 46

The asymmetric induction depended on the solvent and on the R group. The best enantioselectivities were obtained in toluene (Scheme 46). Despite of the low ee values, those results show the importance of the R group, suggesting possible enhancement.

5.5 Aza-Henry and Nitroaldol Reactions

An enantioselective aza-Henry reaction catalysed by the same bifunctional organocatalyst was recently reported by the same group (Scheme 47) [163].

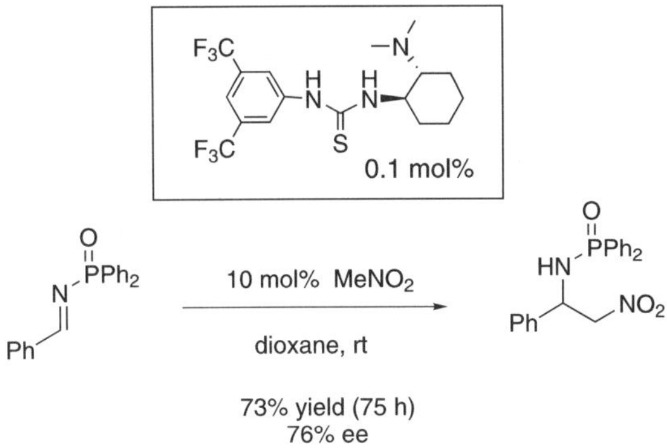

Scheme 47

The aza-Henry reaction is the nucleophilic addition of nitroalkanes to imines to give nitroamine derivatives. This reaction was also studied with metal-based catalysts [164].

Preliminary studies on the racemic reaction of protected imines with nitromethane showed that the thiourea and the amine mutually weakened their reactivities. However, the bifunctional amino-thiourea led to good results. Enantioselectivity of the adduct depended on the protecting group, $P(O)Ph_2$ affording the best results (76% ee). Then, other aromatic imines substrates were successfully phosphorylated with good to high enantioselectivities (63–76% ee).

5.6 Diels–Alder Reactions

Diels–Alder reactions [165] using thiourea as organocatalyst were recently examined [166]. Kinetic measurements showed that accelerations of the relative reaction rates were more dependent on the thiourea substituents than on the substrates or the solvent (even in highly coordinating polar solvents like wa-

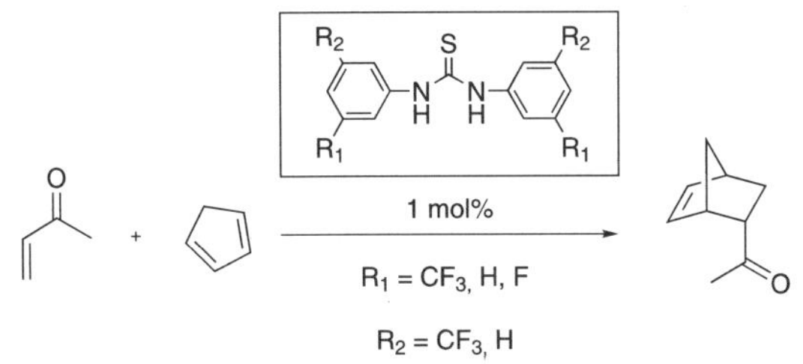

Scheme 48

ter). The catalysts increased the reaction rates and endo-selectivities of Diels–Alder reactions between cyclopentadiene and vinylketones (Scheme 48).

Aromatic thioureas were more active than alkyl (octyl, cyclohexyl) derivatives. Thioureas with trifluoromethyl substituents were even more effective. The same group also showed that these organocatalysts can act as weak Lewis acids and are thus able to alter the stereochemistry of the Diels–Alder reaction between cyclopentadiene and chiral acrylamide derivatives (Scheme 49) [167].

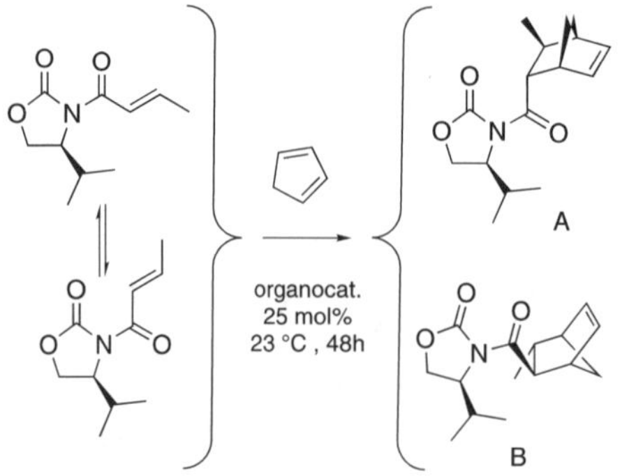

R_1	R_2	A	B	yield
CF_3	H	77	23	74%
CF_3	CF_3	81	19	78%

Scheme 49

5.7 Thioureas as Chiral Bases

Asymmetric deprotonation of prochiral cyclic ketones (Scheme 50) was performed with chiral ureas in the presence of butyllithium. Yields were good (85–88%) with high enantioselectivities (83–87%). Moderate enantioselectivity is obtained with the cyclopentyl-containing urea (Scheme 50: 37% ee with R = Ph; 7% ee with R = Me) [168, 169].

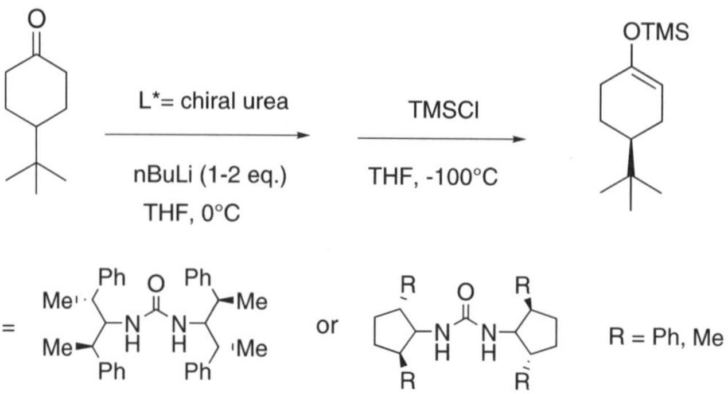

Scheme 50

6
Conclusion

Although the use of optically pure ureas and thioureas in asymmetric catalysis mainly appeared in this last decade, their potential as chiral inductors is already very important. In this relatively short time period, urea and thiourea species have found an application in numerous domains of asymmetric catalysis. This can be explained by both the easy synthesis of such compounds and the diversity of their chemical properties. Many chiral amines and diamines are available, thus enabling the straightforward synthesis of various series of optically pure ureas and thioureas in very good yields. As the formation of urea and thiourea complexes is very easy and fast, they are ideal for solid phase synthesis and high-throughput screening. On another hand, the easy polyaddition of diamines onto isocyanates or isothiocyanates opens a way to supported catalytic systems affording the recovery of the chiral active species. Various coordinating modes of urea and thiourea groups to metallic centres can be observed since they can act as L- or X-ligand types, and are thus able to present either hard or soft properties. The chemical versatility of these groups allows the formation of various complexes with several types of transition metal precursors. Some of these organometallic complexes proved to be both efficient and selective catalysts in asymmetric reductions. In some cases, similar results to those obtained with the best known phosphine ligands have been attained. In addition to their Lewis base properties, ureas and thioureas are also able to form several types of hydrogen bonding (acceptor or donor) which confer organocatalyst properties. It is therefore not surprising that these molecules are used as efficient and selective organocatalysts for important asymmetric transformations, such as the Strecker reaction. Many other enantioselective reactions also appear possible in the presence of such thiourea-based organocatalysts. Two examples have been described by the time of writing this review: the Pictet–Spengler reaction [170], which provides a highly enantioselective access to a range of substituted tetrahydro-β-carbolines, and the Baylis–Hillman reaction, for which a drastic rate increase is observed during the formation of allylic alcohols. In the later case, the bis-thiourea catalyst was quantitatively recovered by silica-gel column chromatography [171]. In this review we describe many applications of ureas and thioureas for asymmetric metal-catalysed reactions and often noted that thioureas lead to more active and enantioselective complexes than ureas. As the organocatalytic properties, including metal interactions, of chiral thioureas are widely observed, they are a very promising type of compound and their use in asymmetric catalysis will probably expand very rapidly during the next decade.

References

1. Fache F, Schulz E, Tommasino ML, Lemaire M (2000) Chem Rev 100:2159
2. Choi K, Hamilton AD (2003) Coord Chem Rev 240:101
3. Kyne GM, Light ME, Hursthouse MB, de Mendoza J, Kilburn JD (2001) J Chem Soc, Perkin Trans 1:1258
4. Chien CH, Leung MK, Su JK, Li GH, Liu YH, Wang Y (2004) J Org Chem 69:1866
5. Hamann BC, Branda NR, Rebek J Jr (1993) Tetrahedron Lett 34:6837
6. Lehn JM (1995) In: Supramolecular chemistry. VCH, Weinheim
7. Kohli P, Blanchard GJ (2000) Langmuir 16:8518
8. Kawasaki T, Ishikawa K, Sekibata H, Sato I, Soai K (2004) Tetrahedron Lett 45:7939
9. Jimenez Blanco JL, Benito JM, Mellet CO, Garcia Fernandez JM (1999) Org Lett 1:1217
10. Wu FY, Li Z, Wen ZC, Zhou N, Zhao YF, Jiang YB (2002) Org Lett 4:3203
11. Pratt MD, Beer PD (2003) Polyhedron 22:649
12. Hoveyda AH, Murphy KE (2004) In: Jacobsen EN, Pfaltz A, Yamamoto H (eds) Comprehensive asymmetric catalysis, Supplement 1, Supplement to chap 39. Springer, Berlin Heidelberg New York
13. Griffin TS, Woods TS, Klayman DL (1975) In: Katritzky AR, Boulton AJ (eds) Advances in heterocyclic chemistry, vol 18. Academic, New York, p 99
14. Strohmeier GA, Kappe CO (2004) Angew Chem Int Ed 43:621
15. Sasaki S, Hirose K, Naemura K (1997) Tetrahedron Lett 38:4791
16. Smith J, Liras JL, Schneider SE, Anslyn EV (1996) J Org Chem 61:8811
17. Yarima M, Saraça S, Ertana M, Kelicenb P, Altinokc G, Demirdamarb R (2000) Arch Pharm Pharm Med Chem 333:415
18. Yasuda N, Hsiao Y, Jensen MS, Rivera NR, Yang C, Wells KM, Yau J, Palucki M, Tan L, Dormer PG, Volante RP, Hughes DL, Reider PJ (2004) J Org Chem 69:1959
19. Hollinshead AC (1998) US Patent 5,747,526
20. Schroeder DC (1955) Chem Rev 55:181
21. Sarkis GY, Faisal ED (1985) J Heterocycl Chem 22:137
22. Chalina EG, Chakarova L (1998) Eur J Med Chem 33:975
23. Stark H, Purand K, Ligneau X, Rouleau A, Arrang JM, Garbarg M, Schwartz JC, Schunack WJ (1996) Med Chem 39:1157
24. Walpole C, Ko SY, Brown M, Beattie D, Campbell E, Dickenson F, Ewan S, Hughes GA, Lemaire M, Lerpiniere J, Patel S, Urban L (1998) J Med Chem 41:3159
25. Mallams AK, Morton JB, Reichert P (1981) J Chem Soc, Perkin Trans 1:2186
26. Goodyer CLM, Chinje EC, Jaar M, Stratford IJ, Threadgilla MD (2003) Bioorg Med Chem 11:4189
27. Venkatachalam TK, Sudbeck EA, Mao C, Uckun FM (2000) Bioorg Med Chem Lett 10:2071
28. Venkatachalam TK, Mao C, Uckun FM (2001) Antiviral Chem Chemother 12:213
29. Sharma S (1978) Synthesis 803
30. Cotarca L, Delogu P, Nardelli N, Sunjic V (1996) Synthesis 553
31. Staab HA (1962) Angew Chem, Int Ed Engl 1:351
32. Staab HA, Walther G (1962) Leibigs Ann Chem 657:98
33. Neville RG, McGee JJ (1963) Can J Chem 41:2123
34. Sridevi G, Rao J, Reddy KK (1989) Synth Commun 19:965
35. Kim S, Yi KY (1985) Tetrahedron Lett 26:1661
36. Atzrodt J, Beckert R, Darsen A (1997) Tetrahedron: Asymmetry 8:2257
37. Gamez P, Dunjic B, Lemaire M (1996) J Org Chem 61:5196

38. Touchard F, Gamez P, Fache F, Lemaire M (1997) Tetrahedron Lett 38:2275
39. Karamé I, Tommasino ML, Lemaire M (2003) J Mol Catal A: Chem 196:137
40. Katritzky AR, Ledoux S, Witek RM, Nair SK (2004) J Org Chem 69:2976
41. Patai S (1977)(ed) The chemistry of cyanates and their thio derivatives. Wiley, New York
42. Touchard F, Fache F, Lemaire M (1997) Tetrahedron: Asymmetry 8:3319
43. Bied C, Moreau JJE, Chi Man MW (2001) Tetrahedron: Asymmetry 12:329
44. Isobe T, Fukuda K, Tokunaga T, Seki H, Yamaguchi K, Ishikawa T (2000) J Org Chem 65:7774
45. Gladysz JA (ed) (2002) Recoverable catalysts and reagents. Chem Rev 102:3215
46. Dunjic B, Gamez P, Fache F, Lemaire M (1996) J Appl Polym Sci 59:1255
47. Vankelecom IFJ, Jacobs PA (2000) In: De Vos DE, Vankelecom IFJ, Jacobs PA (eds) Chiral catalyst immobilisation and recycling, chap 2. Wiley-VCH, Weinheim
48. Moreau JJE, Vellutini L, Wong Chi Man M, Bied C (2001) J Am Chem Soc 123:1509
49. Theophanides T, Harvey PD (1987) Coord Chem Rev 76:237
50. El-Bahy GMS, El-Sayed BA, Shabana AA (2003) Vibrat Spect 31:101
51. Roecker L, Akande J, Elam LN, Gauga I, Helton BW, Prewitt MC, Sargeson AM, Swango JH, Willis AC, Xin T, Xu J (1999) Inorg Chem 38:1269
52. Larsen PL, Gupta R, Powell DR, Borovik AS (2004) J Am Chem Soc 126:6522
53. MacBeth CE, Hammes BS, Young VG Jr, Borovik AS (2001) Inorg Chem 40:4733
54. Rodrigez-Fernandez E, Garcia E, Hermosa MR, Jimenez-Sanchez A, Sanchez MM, Monte E, Criado JJ (1999) J Inorg Biochem 75:181
55. Livingstone SE (1987) In: Gillard RD, McCleverty JA (eds) Comprehensive coordination chemistry, 2nd edn. Pergamon, Oxford, p 639
56. Henderson W, Nicholson BK (1996) Polyhedron 15:4015
57. Cauzzi D, Pellinghelli MA, Predieri G, Tripicchio A (1993) Gazz Chim Ital 123:713
58. Abdallah R, Breuzard JAJ, Bonnet M, Lemaire M (2001) unpublished results
59. Sacht C, Datt MS, Otto S, Roodt A (1998) J Chem Soc Dalton Trans 727
60. Koch KR, Bourne SA, Coetzee A, Miller J (1999) J Chem Soc Dalton Trans 3157
61. Kemp G, Roodt A, Purcell W, Koch KR (1997) J Chem Soc Dalton Trans 4481
62. Déchamps-Olivier I, Guillon E, Mohamadou A, Barbier JP (1996) Polyhedron 15:3617
63. Koch KR, Wang Y, Coetzee A (1999) J Chem Soc Dalton Trans 1013
64. Cauzzi D, Lanfranchi M, Marzolini G, Predieri G, Tiripicchio A, Costa M, Zanoni R (1995) J Organomet Chem 488:115
65. Hoferkamp L, Rheinwald G, Stoeckli-Evans H, Süss-Fink G (1996) Organometallics 15:1122
66. Adams RD, Huang M, Yamamoto JH, Zhang L (1996) Chem Ber 129:137
67. Piraino P, Bruno G, Tresoldi G, Faraone G, Bombieri G (1983) J Chem Soc Dalton Trans 2391
68. Sharp RP (1995). In: Puddephat MJ, Abel EW, Stone FGA, Wilkinson G (eds) Comprehensive organometallic chemistry II, vol 8. Pergamon, Oxford, p 115
69. Henderson W, Nicholson BK, Rickard CEF (2001) Inorg Chim Acta 320:101
70. Okeya S, Kameda H, Kawashima H, Shimomura H, Nishioka T, Isobe K (1995) Chem Lett 501
71. Robinson SD, Sahajpal A, Steed JW (2000) Inorg Chim Acta 306:205
72. Zassinovich G, Mestroni G, Gladiali S (1992) Chem Rev 92:1051
73. Palmer MJ, Wills M (1999) Tetrahedron: Asymmetry 10:2045
74. Wills M, Palmer MJ, Smith A, Kenny J, Walsgrove T (2000) Molecules 5:4
75. Gamez P, Dunjic B, Fache F, Lemaire M (1994) J Chem Soc, Chem Commun 1417

76. Gamez P, Dunjic B, Pinel C, Lemaire M (1995) Tetrahedron Lett 36:8779
77. Locatelli F, Gamez P, Lemaire M (1998) J Mol Catal A: Chem 135:89
78. Gamez P, Dunjic B, Lemaire M (1996) J Org Chem 61:5196
79. Touchard F, Bernard M, Fache F, Lemaire M (1999) J Mol Catal A: Chem 140:1
80. Touchard F, Bernard M, Fache F, Delbecq F, Guiral V, Sautet P, Lemaire M (1998) J Organomet Chem 567:133
81. Touchard F, Fache F, Lemaire M (2000) Eur J Org Chem 3787
82. Saluzzo C, ter Halle R, Touchard F, Fache F, Schulz E, Lemaire M (2000) J Organomet Chem 603:30
83. Saluzzo C, Lemaire M (2002) Adv Synth Catal 344:915
84. Knowles WS, Noyori R, Sharpless KB (2002) Angew Chem Int Ed 41:1998
85. Takaya H, Ohta T, Noyori R (1993) In: Ojima I (ed) Catalytic asymmetric synthesis, chap 1. VCH, New York
86. Noyori R (1994) Asymmetric catalysis in organic synthesis, chap 2. Wiley, New York
87. Brown JM, Halterman RL, Ohkuma T, Noyori R (1999) In: Jacobsen EN, Pfaltz A, Yamamoto H (eds) Comprehensive asymmetric synthesis, vol I, chaps 5 and 6. Springer, Berlin Heidelberg New York
88. Blaser HU, Malan C, Pugin B, Spindler F, Steiner H, Studer M (2003) Adv Synth Catal 345:103
89. Crépy KVL, Imamoto T (2003) Adv Synth Catal 345:79
90. Tommasino ML, Thomazeau C, Touchard F, Lemaire M (1999) Tetrahedron: Asymmetry 10:1813
91. Tommasino ML, Casalta M, Breuzard JAJ, Lemaire M (2000) 11:4835
92. Brunner H, Nishiyama H, Itich K (1993) In: Ojima I (ed) Catalytic asymmetric synthesis, chap 6. VCH, New York
93. Nishiyama H (1994) Asymmetric catalysis in organic synthesis, chap 2. Wiley, New York
94. Nishiyama H (1999) In: Jacobsen EN, Pfaltz A, Yamamoto H (eds) Comprehensive asymmetric synthesis, vol I, chap 6.3. Springer, Berlin Heidelberg New York
95. Brunner H, Kurzinger AJ (1988) J Organomet Chem 346:413
96. Brunner H, Storiko R (1988) Eur J Inorg Chem 783
97. Nishimyama H, Kondo M, Nakamura T, Itoh K (1991) Organometallics 10:500
98. Kureshy RI, Khan NH, Abdi SHR, Singh S, Ahmed I, Shukla RS, Jasra RV (2003) J Catal 219:1
99. Katsuki T, Watanabe A, Uchida T (2003) EP 1352908. To Kyushu University
100. Watanabe A, Ushida T, Katsuji I, Katsuki T (2002) Tetrahedron Lett 43:4481
101. Shargi H, Eskandari MM (2003) Tetrahedron 59:8509
102. Zhang TY, Allen MJ (1999) Tetrahedron Lett 40:5813
103. Shibasaki M, Vogl EM (1999) In: Jacobsen EN, Pfalz A, Yamamoto H (eds) Comprehensive asymmetric catalysis, vol I. Springer, Berlin Heidelberg New York, p 458
104. Dötz KH (2004) In: Dötz KH (ed) Metal carbenes in organic synthesis. Springer, Berlin Heidelberg New York
105. Yang D, Chen YC, Zhu NY (2004) Org Lett 6:1577
106. Dai M, Liang B, Wang C, Chen J, Yang Z (2004) Org Lett 6:221
107. Nozaki K (1999) In: Jacobsen EN, Pfaltz A, Yamamoto H (eds) Comprehensive asymmetric synthesis, vol I. Springer, Berlin Heidelberg New York, p 382
108. Bayon JC, Claver C, Masdeu-Bulto AM (1999) Coord Chem Rev 193:73
109. Claver C, van Leeuwen PWNM (2001) In: van Leeuwen PWNM, Claver C (eds) Rhodium catalysed hydroformylation, chap 5. Kluwer, Dordrecht
110. Feng J, Garland M (1999) Organometallics 18:417

111. Breuzard JAJ, Bonnet M, Tommasino ML, Lemaire M (1999) unpublished results
112. Cauzzi D, Lanfranchi M, Marzolini G, Predieri G, Tiripicchio A, Costa M, Zanoni R (1995) J Organomet Chem 488:115
113. Cauzzi D, Costa M, Gonsalvi L, Pellinghelli MA, Predieri G, Tiripiccio A, Zanoni R (1997) J Organomet Chem 541:377
114. Breuzard JAJ, Tommasino ML, Touchard F, Lemaire M, Bonnet MC (2000) J Mol Catal A: Chem 156:223
115. Breuzard JAJ, Tommasino ML, Bonnet MC, Lemaire M (2000) CR Acad Sci Paris, Série IIc, Chimie 557
116. Cauzzi D, Costa M, Cucci N, Graiff C, Grandi F, Predieri G, Tiripicchio A, Zanoni R (2000) J Organomet Chem 593:431
117. Tsuji J (1995) Palladium reagents and catalysts. Wiley, Chichester, p 19
118. Gabriele B, Salerno G, Costa M, Chiusoli GP (1995) J Organomet Chem 503:21
119. Miao H, Yang Z (2000) Org Lett 2:1765
120. Nan Y, Miao H, Yang Z (2000) Org Lett 2:297
121. Norton JR, Shenton KE, Schwartz J (1975) Tetrahedron Lett 16:51
122. El Ali B, Alper H (2000) Synlett 161
123. Alper H, Hamel N (1990) J Chem Soc, Chem Commun 135
124. Chiusoli GP, Costa M, Masarati E, Salerno G (1983) J Organomet Chem 255:C35
125. Feringa BL (1998) In: Beller M, Bolm C (eds) Transition metals for organic synthesis, vol 2. Wiley-VCH, Weinheim, p 307
126. Lautens M, Klute W, Tam W (1996) Chem Rev 96:49
127. Hosakawa T, Ueno T, Inui S, Murahashi SI (1981) J Am Chem Soc 103:2318
128. Backvall JE (1983) Acc Chem Res 16:335
129. Tamaru Y, Hijo M, Yoshida ZI (1991) J Org Chem 56:1099
130. Suzuki T, Uozumi Y, Shibasaki M (1991) J Chem Soc, Chem Commun 1593
131. Ukaji Y, Miyamoto M, Mikuni M, Takeuchi S, Inomata K (1996) Bull Chem Soc Jpn 69:735
132. Ogawa A, Kawabe K, Kawakami J, Mihara M, Hirao T (1998) Organometallics 17:3111
133. Aoyagi S, Hasegawa Y, Hirashima S, Kibayashi C (1998) Tetrahedron Lett 39:2149
134. Cao P, Zhang X (1999) J Am Chem Soc 121:7708
135. Kreimerman S, Ryu I, Minakata S, Komatsu M (2000) Org Lett 2:389
136. Zhou H, Hou J, Chen J, Lu S, Fu H, Wang H (1997) J Organomet Chem 543:227
137. Nefkens SCA, Sperrle M, Consiglio G (1993) Angew Chem Int Ed Engl 32:1719
138. Hayashi M, Takezaki H, Hashimoto Y, Takaoki K, Saigo K (1998) Tetrahedron Lett 39:7529
139. Takeuchi S, Ukaji Y, Inomata K (2001) Bull Chem Soc Jpn 74:955
140. Scheele J, Timmerman P, Reinhoudt DN (1998) Chem Commun 2613
141. Dai M, Wang C, Dong G, Xiang J, Luo T, Liang B, Chen J, Yang Z (2003) Eur J Org Chem 4346
142. Drent E, Budzelaar PHM (1996) Chem Rev 96:663
143. Reddy KR, Chen CL, Liu YH, Peng SM, Chen JT, Liu SY (1999) Organometallics 18:2574
144. Gosavi RK, Agarwaia U, Rao CNR (1967) J Am Chem Soc 89:235
145. Gosavi RK, Rao CNR (1967) J Inorg Nucl Chem 29:1937
146. Curran DP, Kuo LH (1994) J Org Chem 59:3259
147. Maher DJ, Connon SJ (2004) Tetrahedron Lett 45:1301
148. Curran DP, Kuo LH (1995) Tetrahedron Lett 36:6647
149. Gröger H (2003) Chem Rev 103:2795

150. Iyer MS, Gigstad KM, Namdev ND, Lipton M (1996) J Am Chem Soc 118:4910
151. Corey EJ, Grogan MJ (1999) Org Lett 1:157
152. Sigman MS, Jacobsen EN (1998) J Am Chem Soc 120:4901
153. Jacobsen EN, Sigman MS (1999) PCT Int Appl WO9951546. To Harvard College
154. Sigman MS, Vachal P, Jacobsen EN (2000) Angew Chem, Int Ed Engl 39:1279
155. Vachal P, Jacobsen EN (2000) Org Lett 2:867
156. Vachal P, Jacobsen EN (2002) J Am Chem Soc 124:10 012
157. Wenzel AG, Lalonde MP, Jacobsen EN (2003) Synlett 1919
158. Rhodia ChiRex Novel Technologies/Amino Acid Technology http://www.rhodiachirex.com/default.htm
159. Okino T, Hoashi Y, Takemoto Y (2003) Tetrahedron Lett 44:2817
160. Joly GD, Jacobsen EN (2004) J Am Chem Soc 126:4102
161. Okino T, Hoashi Y, Takemoto Y (2003) J Am Chem Soc 125:12 672
162. Sohtome Y, Tanatani A, Hashimoto Y, Nagasawa K (2004) Chem Pharm Bull 52:477
163. Okino T, Nakamura S, Furukawa T, Takemoto Y (2004) Org Lett 6:625
164. Shibasaki M, Gröger H (1999) In: Jacobsen EN, Pfaltz A, Yamamoto H (eds) Comprehensive asymmetric synthesis, vol III. Springer, Berlin Heidelberg New York, p 1076
165. Evans DA, Johnson JS (1999) In: Jacobsen EN, Pfaltz A, Yamamoto H (eds) Comprehensive asymmetric synthesis, vol III. Springer, Berlin Heidelberg New York, p 1178
166. Wittkopp A, Schreiner PR (2003) Chem Eur J 9:407
167. Schreiner PR, Wittkopp A (2002) Org Lett 4:217
168. Graf CD, Malan C, Knochel P (1998) Angew Chem Int Ed 37:3014
169. Graf CD, Malan C, Harms K, Knochel P (1999) J Org Chem 64:5581
170. Taylor MS, Jacobsen EN (2004) J Am Chem Soc 126:10 558
171. Sohtome Y, Tanatani A, Hashimoto Y, Nagasawa K (2004) Tetrahedron Lett 45:5589

Top Organomet Chem (2005) 15: 271–288
DOI 10.1007/b136353
© Springer-Verlag Berlin Heidelberg 2005
Published online: 10 September 2005

Chiral-at-Metal Complexes as Asymmetric Catalysts

Marc Fontecave (✉) · Olivier Hamelin · Stéphane Ménage

Laboratoire de Chimie et Biochimie des Centres Rédox Biologiques,
UMR Université J. Fourier-CEA-CNRS no 5047, CEA-Grenoble, DRDC/CB,
17 rue des Martyrs, 38054 Grenoble Cedex 9, France
mfontecave@cea.fr

Abstract In general, asymmetric catalysts are based on the combination of a chiral organic ligand and a metal ion. Here we show that future research should also focus on complexes in which the chirality resides only at the metal center, as the result of a given topology of coordination of achiral ligands to the metal ion. Here we make a brief presentation of the methods available for preparing such compounds as well as the very few examples of enantioselective reactions catalyzed by chiral-at-metal complexes.

Keywords Asymmetric transformation · Bipyridine · Catalysis · Chiral-at-metal · Phenanthroline · Ruthenium

1 Introduction

Metal-assisted enantioselective catalytic reactions are one of the most important areas in organic chemistry [1–3]. They require the appropriate design and the preparation of chiral transition metal complexes, a field also of major importance in modern synthetic chemistry. These complexes are selected on both their ability to catalyze a given reaction and their potential as asymmetric inducers. To fulfill the first function, it is absolutely required that the catalysts display accessible metal coordination sites where reactants can bind since activation would result from a direct interaction between the metal ion

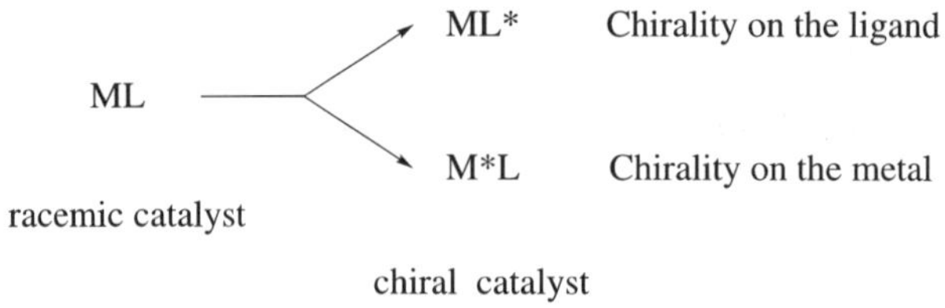

Fig. 1

and the reactants. Asymmetric induction results from the chiral environment provided by the metal complex and thus enjoyed by the reactants as a consequence of the formation of the catalyst–reactants association.

The almost unique strategy developed so far for the preparation of chiral catalysts is based on the association of a chiral organic ligand with a metal ion (Fig. 1) [1–4]. This strategy proved to be in many cases extremely efficient showing that localization of a stereogenic center in the coordination sphere of a metal ion may be enough to generate a strong asymmetric effect and to obtain large enantiomeric excesses. However, the chirality of the catalyst might also be located at the metal center itself and not at the ligands (Fig. 1). Until only very recently, the potential of chiral metal complexes, with the chirality only at the metal center resulting from a given topology of coordination of the achiral ligands, as asymmetric catalysts has not been explored.

In this contribution we want to raise this possibility by discussing the very few examples of enantioselective reactions catalyzed by such chiral-at-metal catalysts. On purpose we have selected only those catalysts containing achiral ligands. Indeed, there are some examples of enantioselective reactions catalyzed by complexes containing both a chiral metal site and a chiral ligand and thus in most cases it is not possible to identify the source of the inducing effect. This paper will also include a brief presentation of the methods available for the preparation of these compounds, which is a topic of increasing importance in coordination chemistry. The requirement for both exchangeable coordination sites and high configurational stability, in the absence of any other source of chiral information, makes the preparation of enantiomerically pure chiral-at-metal catalysts a very challenging problem and is probably one of the reasons why there are still very few examples of such catalysts.

2 Chirality at Metal Centers

As in organic chemistry, there are several sources of chirality at a metal center. As for an asymmetric carbon atom in an organic molecule, the coordination of the metal ion by four different monodentate ligands in a tetrahedral configuration generates a chiral complex with two possible enantiomers, *R* and *S*. Such a chiral rhenium complex of tetrahedral configuration [5] is shown in Fig. 2. In practice examples of this type are rare.

There are more examples of a second type in which the chirality of the metal center is the result of the coordination of polydentate ligands. The easiest case is that of octahedral complexes with at least two achiral bidentate ligands coordinated to the metal ion. The prototype complex with chirality exclusively at the metal site is the octahedral tris-diimine ruthenium complex $[Ru(diimine)_3]^{2+}$ with diimine = bipyridine or phenanthroline. As shown in Fig. 2 such a complex can exist in two enantiomeric forms named Δ and Λ [6, 7]. The bidentate ligands are achiral and the stereoisomery results from the helical chirality of the coordination and the propeller shape of the complex. The absolute configuration is related to the handness of the helix formed by the ligands: when rotated

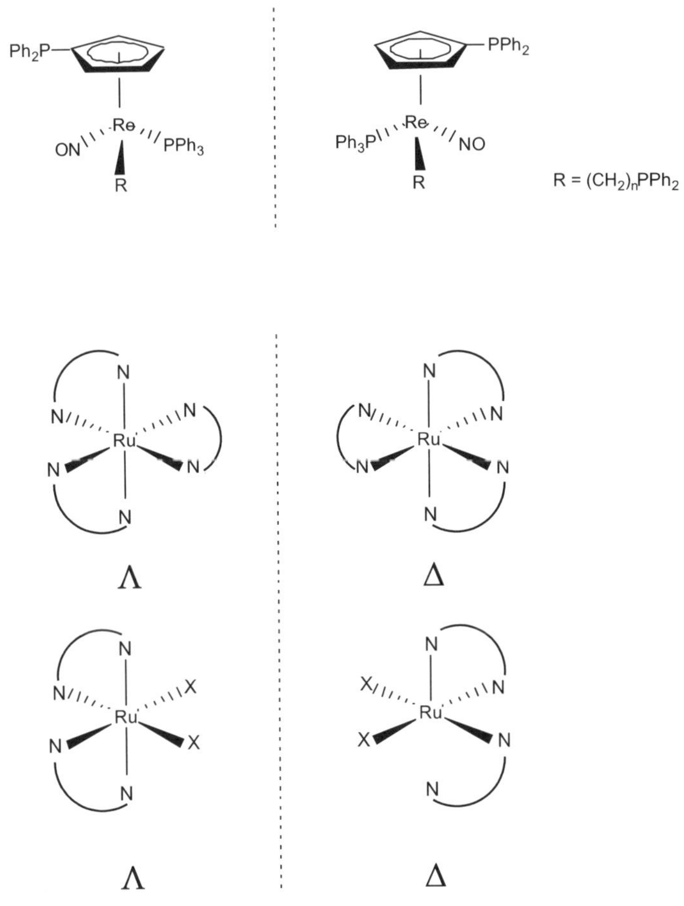

Fig. 2

clockwise the Δ isomer appears to screw into the plane while with a left rotation the Λ isomer appears to screw out of the plane. Clearly such a complex has no potential as a catalyst since due to its relatively large stability and low kinetic lability it does not contain exchangeable coordination sites where substrates would be able to bind for activation. However, as shown in Fig. 2, bis-diimine complexes $[RuX_2(\text{diimine})_2]^{2+}$, where X is a solvent molecule for example, also exist in two enantiomeric forms, Δ and Λ [8].

3 Preparation of Chiral-at-Metal Complexes

There are three ways, borrowed from organic chemistry, of preparation of an optically pure chiral-at-metal complex with achiral ligands. The most extensively used strategy involves an initial racemic synthesis, using the standard methods of coordination chemistry, followed by a resolution procedure. The second method implies a stereoselective synthesis involving the use of a chiral auxiliary to induce stereoselectivity at the metal center during preparation of the complex. To date the number of successful preparations of enantiomerically enriched complexes via an asymmetric synthesis is still very small. The most well-known and prototypic example is the highly diastereoselective synthesis of ruthenium bis(bipyridine) complexes by von Zelewsky using the tetradentate "Chiragen" ligand which will be discussed below [9]. Finally, crystallization-induced asymmetric transformation, a rarely observed reaction in coordination chemistry, proved to be also efficient to afford chiral-at-metal complexes in optically pure (or enriched) form and in high yield.

3.1 Resolution Methods

The first resolution of an octahedral complex into its enantiomers was achieved in 1911 by A. Werner, who got the Nobel Prize in 1913, with the complex $[Co(\text{ethylenediamine})(Cl)(NH_3)]^{2+}$ [10]. Obviously, resolution is to be considered only in the case of kinetically inert complexes whose enantiomers do not racemize quickly after separation. This is a very important remark since, as noted above, the interesting complexes are those containing exchangeable sites required for catalytic activity and thus more sensitive to racemization. We will not discuss here the very rare cases of spontaneous resolution during which a racemic mixture of complexes forms a conglomerate (the Δ and Λ enantiomers crystallize in separate crystals) [11, 12].

Most methods for the resolution of enantiomers contained in a reaction mixture consist in the conversion of the compounds into stable or transient diastereoisomers and separation of the latter on the basis of their different physico-chemical properties.

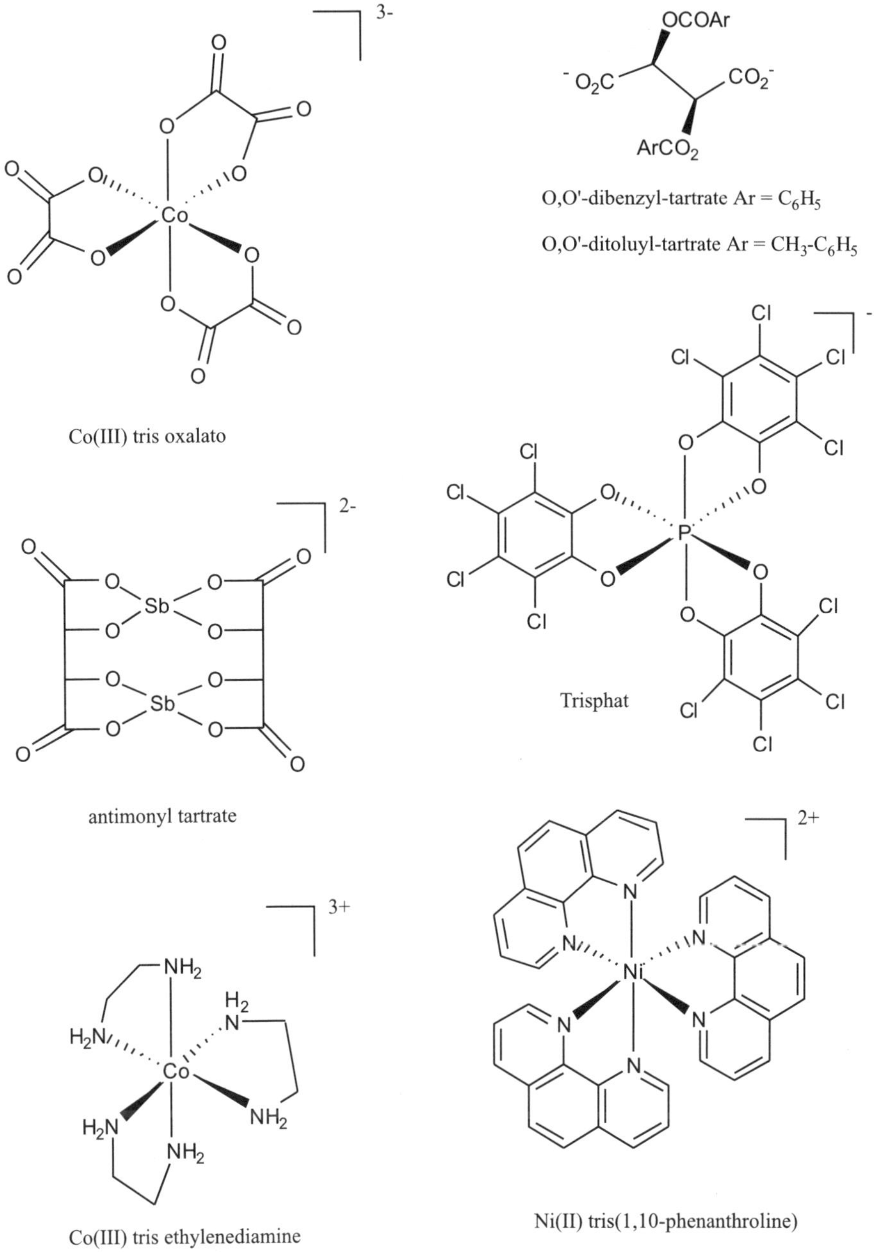

Co(III) tris oxalato

O,O'-dibenzyl-tartrate Ar = C_6H_5

O,O'-ditoluyl-tartrate Ar = CH_3-C_6H_5

antimonyl tartrate

Trisphat

Co(III) tris ethylenediamine

Ni(II) tris(1,10-phenanthroline)

Fig. 3

The most direct method is the separation of the enantiomers by HPLC using chiral columns. It has the advantage that there is no risk of con tamination from chiral resolving agents. The formation of diastereoisomers transiently occurs during association of the compounds with the chiral components of the solid support, which allows elution of the enantiomers at different rates. This has been extensively used with Ru(II) complexes. Examples of such columns are:

- SP Sephadex C-25 cation exchangers whose dextran support, itself composed of propane-sulfonate-functionalized cross-linked α-D-glucopyranoside units, provides the chiral environment. Thus chiral eluents are not always necessary and achiral solvents can be used for separation. However, in some cases, separation is improved using an eluent containing a chiral ion [13].
- Commercially available chiral columns such as CHIRACEL OD-R or Crest-Pak [14–16]. Such columns have, for example, been successfully used for the separation of the enantiomers of the bis-diimine $[Ru[(N–N)_2(dmso)Cl]^+]$ complexes [14, 17, 18].

In general the metal complexes are charged. It is thus possible to convert the racemic mixture of such a complex into a pair of diastereoisomeric species with different physico-chemical properties, in particular solubility, by association with an enantiomerically pure chiral counterion [19]. Examples of frequently used such ions are shown in Fig. 3. Then the separation can be achieved by:

- Selective crystallization: Practically, the choice of the counterion should be dictated not only by the largest difference in solubility between the two resulting diastereoisomers, but also by the need for complete elimination of the resolving agent at the end of the process. There are many examples of octahedral complexes resolved by selective crystallization. Obviously this was shown to be possible with kinetically inert complexes, based for example on chromium [20], cobalt [21], or ruthenium [8, 9, 22–24]. More recently, it has also been shown with kinetically labile complexes, such as high-spin Fe(III) (d5) [25] and Ga(III) (d10) [26] complexes which have no ligand field stabilization. The method has thus been widely used in the case of Ru(II) complexes such as $Ru(N-N)_3^{2+}$, $Ru(N-N)_2(X)_2^{2+}$ (where N – N stands for bipyridine and substituted derivatives or phenanthroline and derivatives and X stands for CO, pyridine, phenylpyridine, or acetonitrile) [8, 9, 22–24]. Most of these complexes can serve as chiral building blocks for the synthesis of a number of chiral-at-metal complexes since substitution of the monodentate pyridine or CO ligands is possible with retention of configuration [8, 9, 23, 27, 28]. It is remarkable that resolution is also possible in the case of cationic complexes such as $Ru(dmp)_2(CH_3CN)_2^{2+}$, where dmp is 2,9-dimethyl 1,10-phenanthroline, in which two exchangeable solvent ligands are in *cis* position [24]. These complexes are catalytically active during oxidation because of these exchangeable sites and can be resolved into enantiomers upon association with the chiral Trisphat anion as the resolving agent and selective crystallization of the resulting diastereoisomers.

- Chromatography: Separation of the two enantiomers, via the formation of diastereomerics ion pairs, can be achieved on polar chromatographic phases (SiO_2, Al_2O_3) and on preparative thin-layer chromatographic plates using achiral organic eluents. For example, a ferrous complex associated with Trisphat can be resolved on thin plate of silica gel (eluent CH_2Cl_2) [19, 29].
- Selective asymmetric extraction: This can be achieved using a chiral solvent, in particular when the racemic mixture to be resolved is not ionic, or using a solution containing a chiral counterion in the case of charged complexes. As an example, Lacour achieved an excellent resolution of tris-diimine Ru(II) complexes, in an aqueous solution, by combination with an organic phase containing the enantiomerically pure Trisphat salt [30]. Upon vigorous stirring of the biphasic mixture, selective transfer (35 : 1 ratio) of one enantiomeric cation from water to the organic layer occurred. It is interesting to mention the case reported recently of the dynamic resolution of a tetrahedral Ru complex (de = 70%) using encapsulation into a chiral Gd-based chiral cavity [31].

3.2 Stereoselective Synthesis

To date, direct asymmetric synthesis of optically active chiral-at-metal complexes, which by definition leads to a mixture of enantiomers in unequal amounts thanks to an "external" chiral auxiliary, has never been achieved. The most studied strategy is currently indirect asymmetric synthesis, which involves (i) the stereoselective formation of the chiral-at-metal complex thanks to a chiral inductor located either on the ligand or on the counterion and then (ii) removal of this "internal" chiral auxiliary (Fig. 4). Indeed, when the isomerization of the stereogenic metal center is possible in solution, in-

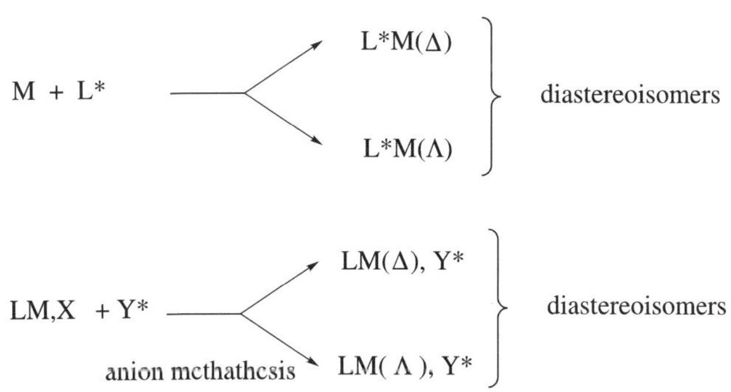

Fig. 4

troduction of a stereogenic element to their ligands generates *intramolecular diastereoselective interactions* which can control the configuration around the metal center and thus favor one of the equilibrating diastereoisomers. If the complexes are charged, an alternative strategy to control their configuration during synthesis is to consider their asymmetric ion pairing with a chiral counterion; *non-covalent intermolecular diastereoselective interactions* then control the stereoselectivity of the reaction. Again it is important to note that this can be a major problem if those chiral auxiliaries cannot be removed after the synthesis in order to generate a complex in which the only chiral center is the metal itself. When the auxiliary is a counterion, substitution by an achiral ion is often possible by a methathesis reaction even though so far we have failed to do it with Trisphat as the counteranion. When the auxiliary is located on the ligand itself, it is much more difficult to remove.

3.2.1 Induction by a Chiral Ligand

Stereoselective methods of synthesis in inorganic chemistry using chiral ligands have been described and discussed in excellent recent review articles by Von Zelewsky [7, 32–34]. We will thus just indicate the two recent prototypic examples leading to stereoselective synthesis with very large diastereoselectivities.

Probably the most well-known example is that based on the Chiragen ligands (named from "chiral generator" and consisting of bridged chiral bis(bipyridine) ligands) (Fig. 5) introduced in 1993 by A. Von Zelewsky for the preparation of chiral-at-metal bis-diimine and tris-diimine Ru(II) complexes [35]. It was shown, in particular by X-ray crystallography, that Chiragen, a chiral tetradentate ligand, can occupy four non-planar coordination

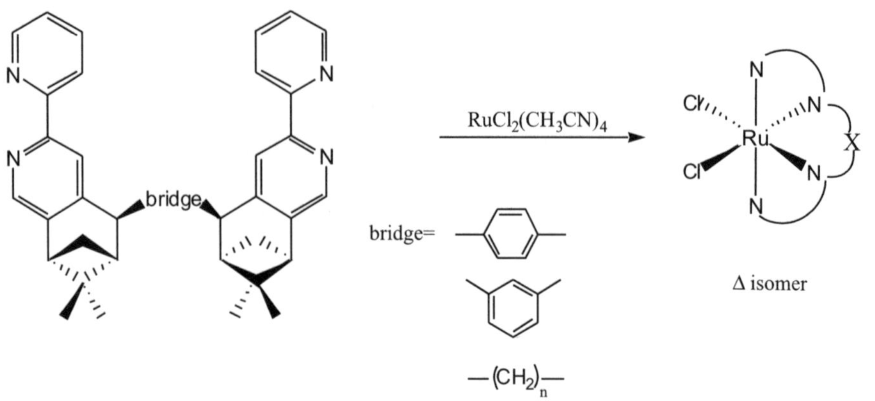

Fig. 5

sites in an octahedral complex, leaving two *cis* quite labile positions. The chirality of the rigid ligand predetermines the Δ or Λ chirality of the complex. As a result, only one stereochemically highly stable diastereoisomer was generated, with no possibility for conversion into the diastereoisomer with the other configuration at the metal center, mainly for sterical reasons. The products of the synthetic process can be further used as chiral building blocks for the synthesis of a variety of stereochemically well-defined mononuclear Ru complexes because of the possibility of substitution of the two *cis* monodentate ligands with retention of configuration, as well as for the synthesis of optically pure polynuclear(binuclear) species. These ligands can also be used for the synthesis of optically pure osmium and rhodium complexes.

A chiral sulfoxide, *p*-tolyl sulfoxide, has also been used for the stereoselective synthesis of chiral-at-metal bis(bipyridine) or bis(phenanthroline) Ru complexes containing exchangeable sites (Fig. 6) [18, 36, 37]. The diastereoselectivity of the reactions, which nevertheless does not reach the level obtained with the Chiragen ligands, is also only governed by the chirality of the sulfoxide (de = 76%) [36]. Control by the metal-bound sulfoxide is based on its ability to generate a hydrogen bond between its oxygen atom and a proton of one of the diimine ligand, as well as a π–π interaction between the tolyl moiety and one of the diimine ligands, only in the major isomer. The advantage of this method is in the possibility to replace both the chiral sulfoxide and the chloro ligand by a bidentate nitrogen-based ligand with retention of configuration, which lead to chiral-at-metal complexes with achiral ligands (Fig. 6) [37].

We would like to indicate a very interesting and efficient strategy for synthesis of chiral-at-metal tris-diimine-Co(III) complexes with achiral ligands [38]. The ligands used are 2,2′-bipyridine-4-monoboronic or -4,4′-diboronic acids which allow intermolecular interactions with saccharides through the boronate functions. The resulting chiral ligand proved to be very useful for asymmetric synthesis of optically active cobalt complexes (Fig. 7). The chiral information provided by the saccharide is relieved to the octahedral metal center via the bipyridinyl moieties which adopt a given configuration. Thus the chirality of the sugar determines the configuration at the

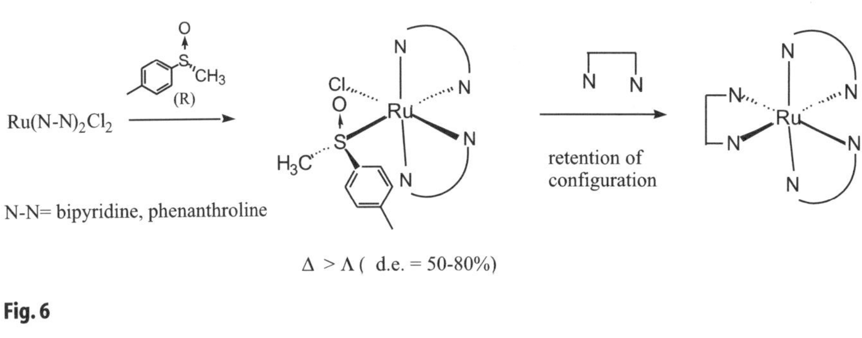

Fig. 6

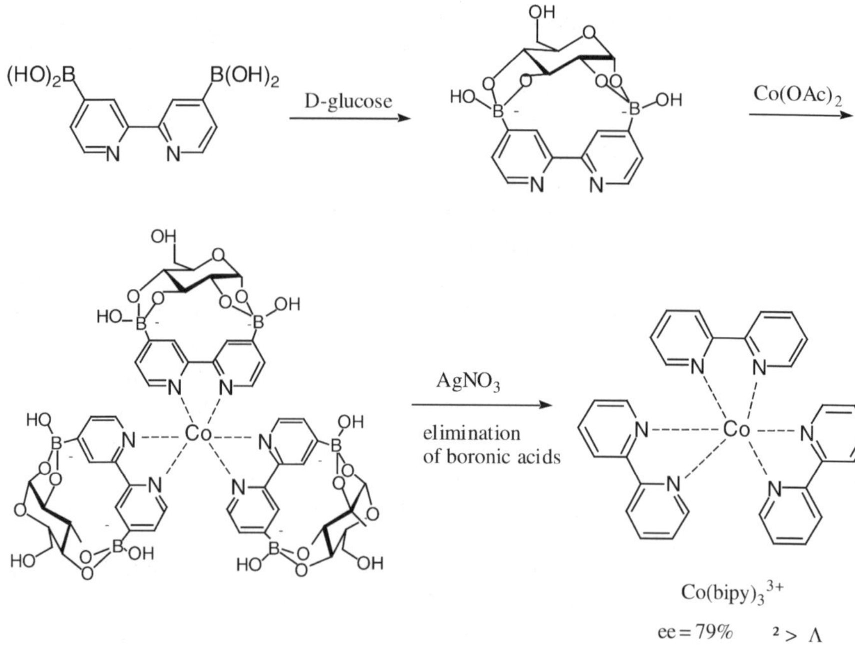

Fig. 7

metal center. For example D-glucose allows the preferential formation of the Δ enantiomer whereas the L-glucose allows that of the Λ enantiomer. One more important point is that the chiral sugar auxiliary located on the ligand can be easily removed, affording chiral-at-metal complexes with only achiral ligands with large enantiomeric excesses (up to 79% ee).

Other chiral ligands such as BINAP (where BINAP is bis(diarylphosphino)-1,1 binaphthyl) or aminophosphines are also efficient for stereoselective synthesis of chiral-at-metal Ru complexes [39–41].

Stereoselective synthesis can also be carried out in the case of tetrahedral complexes. An interesting example was described in 2001 (Fig. 8) [42].

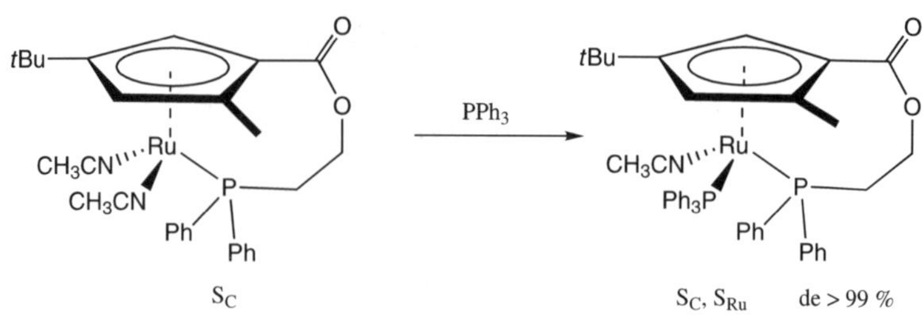

Fig. 8

It was shown that planar-chiral cyclopentadienyl-phosphine ligands were excellent chirality inducers during the synthesis of chiral-at-metal tetrahedral Ru-phosphine or -phosphite complexes (99% de).

3.2.2
Induction by Chiral Ions

There are very few examples of asymmetric synthesis using optically pure ions as chiral-inducing agents for the control of the configuration at the metal center. Chiral anions for such an application have recently been reviewed by Lacour [19]. For example, the chiral enantiomerically pure Trisphat anion was successfully used for the stereoselective synthesis of tris-diimine-Fe(II) complex, made configurationally stable because of the presence of a tetradentate bis(1,10-phenanthroline) ligand (Fig. 9) [29]. Excellent diastereoselectivity ($> 20:1$) was demonstrated as a consequence of the preferred homochiral association of the anion and the iron(II) complex and evidence for a thermodynamic control of the selectivity was obtained. The two diastereoisomers can be efficiently separated by ion-pair chromatography on silica gel plates with excellent yields.

Chiral-at-metal cations can themselves serve as chirality inducers. For example, optically pure $Ru[(bipy)_3]^{2+}$ proved to be an excellent chiral auxiliary for the stereoselective preparation of optically active 3D anionic networks $\{[M(II)Cr(III)(oxalate)_3]\text{-}\}_n$ (with M = Mn, Ni), which display interesting magnetic properties. In these networks all of the metallic centers have the same configuration, Δ or Λ, as the template cation, as shown by CD spectroscopy and X-ray crystallography [43].

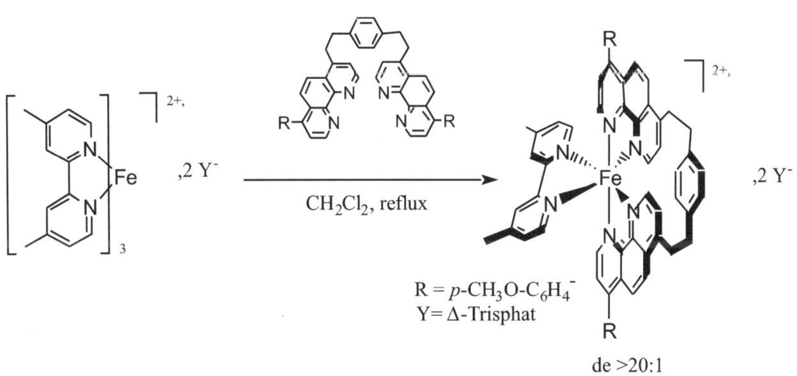

Fig. 9

3.3 Asymmetric Transformation

All the resolution methods indicated above obviously only generate a given enantiomer from a racemic mixture with a 50% yield. It is often desirable to obtain the targeted complex or catalyst in large amounts and one might need to improve that yield. Indeed it is sometimes possible to obtain more than 50% of one component of diasteroisomeric mixtures if these components are equilibrated. The spontaneous equilibration of stereoisomers in solution is called "an asymmetric transformation of the first kind". When equilibration is rapid, as a consequence of epimerization of both diastereoisomers, mixtures are obtained that reflect the relative thermodynamic stability of the diastereosiomers. Racemization can be spontaneous or instead induced, for example by increased temperature, light, or pH changes. When a single diastereoisomer crystallizes from a solution in which reagents epimerize quickly, the yield can thus exceed that given by its solution concentration and theoretically the diastereoisomer can be prepared with 100% yield and 100%

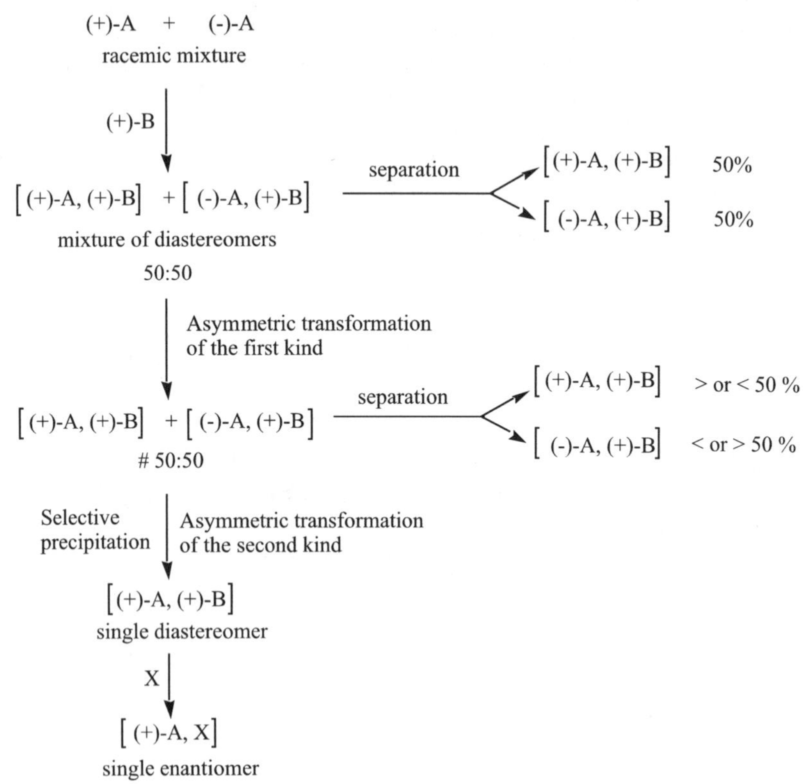

Fig. 10

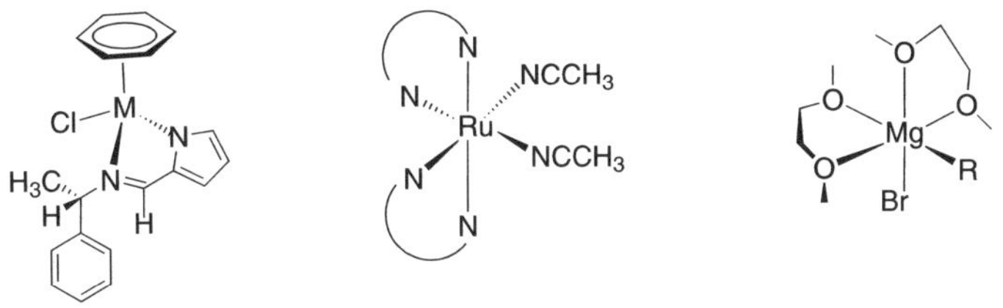

Fig. 11

optical purity. Such a process is termed "asymmetric transformation of the second kind", also named "crystrallization-induced asymmetric transformation". The various resolution processes are represented in Fig. 10. Once the diastereoisomers are separated, the enantiomers are obtained by substitution of the resolving agent (counterion or ligand) by an achiral equivalent and this has to be done without racemization. It should be noted that there is no such reported case of removal of the resolving agent after the process.

Whereas many examples of such reactions have been reported in organic chemistry, very few asymmetric transformations of coordination complexes have been observed and all of them are crystallization-induced asymmetric transformations. Brunner et al. reported the preparation of chiral-at-metal pentamethylcyclopentadienyl-rhodium(III) and iridium(III) complexes, with the optically active bidentate Schiff base ligand [(+)-2-*N*-(5-*S*)-1-phenylethyl]-pyrrolocarbaldimine or -salicylaldimine (Fig. 11) [44]. Only one of the two diastereoisomers crystallized and in solution the labile compounds spontaneously epimerized via a change of the metal configuration. We recently showed for the first time an example of asymmetric transformation leading to a chiral-at-metal Ru(II) complex with achiral ligands, $Ru[(dmp)_2(CH_3CN)_2]^{2+}$ (Fig. 11) [45]. The combination of the selective precipitation of the heterochiral pair using enantiopure Trisphat as the resolving agent and the light-induced racemization in solution resulted in the quantitative conversion of the initial racemate into only one of the enantiomers associated with Trisphat, with excellent diastereoselectivity $>98:2$. We may cite a recent work regarding the isolation of an enantiopure six-coordinate chiral-at-metal Grignard reagent thanks to an absolute crystallization-induced asymmetric transformation issued from the combination of the selective crystallization of one enantiomer (spontaneous resolution) and racemization of the metal center in solution [11].

4
Enantioselective Catalysis with Chiral-at-Metal Complexes

Enantioselective catalysis requires that the distance between the inductor and the substrate is not too long. Basically, the smaller the distance, the better the chirality transfer and optical induction. As indicated above, in general the catalysts are based on optically active ligands in which the chiral centers are located in the substituents. Therefore, the inducing chirality is rather far away from the metal atom at which catalysis takes place.

On the other hand, if the metal atom of the catalyst itself were the inducing chirality, this would be the shortest possible distance. One would expect the largest stereoselectivities during reactions catalyzed by such complexes, thus taking place within the coordination sphere of these chiral metal ions. So far this possibility has been very little exploited even though, as discussed extensively above, methods are available to prepare optically pure chiral-at-metal complexes. In fact, one should again recall that an active catalyst should not be too stable and should contain exchangeable ligands so that the reactants can penetrate the coordination sphere and bind to the metal ion where they are activated and where they enjoy the chiral environment which stereoselectively orientates the reaction. So far, resolution methods and stereoselective synthesis of metal complexes have been mainly applied to complexes lacking such exchangeable coordination sites.

We will present only the very few examples that we are aware of of enantioselective catalysis using a chiral catalyst with achiral ligands and with the chirality exclusively at the metal center. We exclude from that list the large number of reported chemical systems in which the catalyst is a diastereoisomer with two stereogenic centers, each with a given configuration, one at the metal ion itself and one in the ligand. In almost all these cases it is not known which center is contributing to the enantioselectivity of the catalytic reaction and to which extent each one is contributing.

It is noteworthy that, as early as 1929, Shibata and Tsuchida reported a kinetic resolution of rac-3,4-dihydroxyphenylalanine by selective oxidation of one enantiomer using a chiral cobalt complex $[Co(en)_3NH_3Cl]Br_2$ as a catalyst [46, 47]. Figure 12 shows a highly enantioselective addition of diisopropylzinc to 2-(*tert*-butylethynyl)pyrimidine-5-carbaldehyde via an autocatalytic process in the presence of a chiral octahedral cobalt complex with ethylenedi-

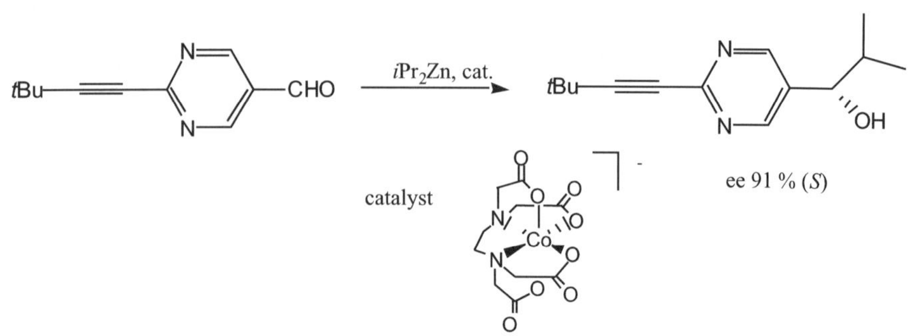

Fig. 12

aminetetraacetate (edta) [48]. Under these conditions the reaction affords the pyrimidyl alkanol with large yields and high enantiomeric excesses (85–94%) and the reaction proved to be stereospecific since the absolute configuration of the chiral complex determines the absolute configuration of the product.

Figure 13 illustrates our recent investigation of chiral-at-metal bis-diimine Ru(II) complexes as catalysts for the oxidation of sulfides to sulfoxides by hydrogen peroxide [24]. This class of Ru complexes was selected on the basis of their reported ability to catalyze oxidation reactions using peroxides or molecular oxygen through metal-based pathways, rather than autoxidation radical-based mechanisms [49], and of the observation that an optically enriched (60% optical purity of the Λ configuration) bis-diimine Ru-oxo complex, $[Ru(bipy)_2(py)O]^{2+}$, was able to stereoselectively transfer its oxygen atom to a prochiral sulfide (20% ee for the isolated sulfoxide) [27]. The complex $[Ru(dmp)_2(CH_3CN)_2]^{2+}$ was easily obtained under separated enantiomerically pure forms by selective precipitation during association with the chiral anion Trisphat. One diastereosiomer could also be obtained quantitatively from the racemic mixture by a light- and crystallization-induced asymmetric transformation after addition of Trisphat of a given configuration [45]. The optically pure complex was found to be a catalyst for the enantioselective oxidation of sulfides to sulfoxides. No sulfones were formed during the reaction. Even though the enantiomeric excesses were not large enough for a synthetic application (5–18% ee), this is the first demonstration that reactions as complex as oxidations can be made enantioselective with rather simple chiral-at-metal catalysts with standard commercially available ligands such as 1,10-phenanthrolines. Certainly, improvement of the stereoselectivity is possible. Even though the chirality is as close as possible to the reactants, if one assumes that the reaction takes place between the peroxide (or the oxo derivative) and the sulfide, one reactant or both bound to the Ru center, the electronic environment and the steric hindrance provided by the achiral ligands at the active site might play important roles in the relative orientation of those reactants and thus in the stereoselectivity of the oxy-

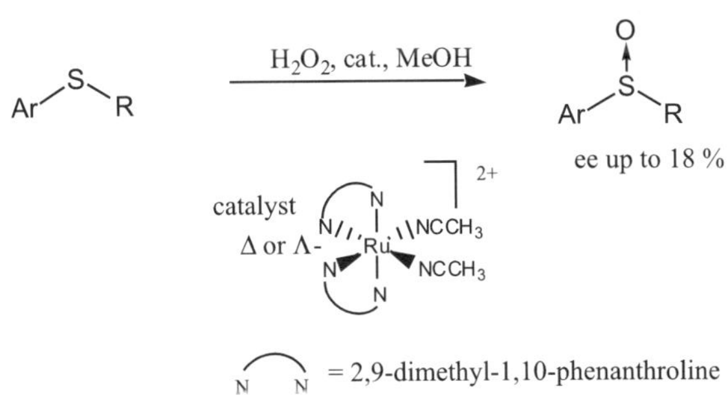

Fig. 13

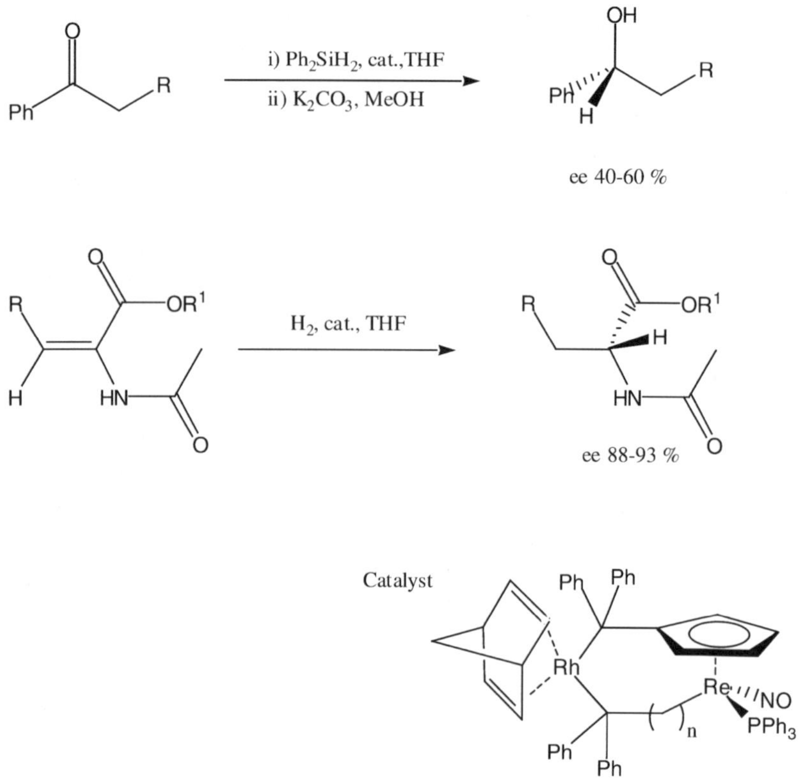

Fig. 14

gen atom transfer. We thus believe that such a system will provide larger ees simply by manipulation of the diimine ligands.

We would also like to indicate that chiral-at-metal complexes might have applications in enantioselective catalysis, not only as catalysts per se but also as ligands for catalysts. An interesting example of such a strategy using a chiral transition-metal-containing chelating ligand with the chirality at the metal center [50] is shown in Fig. 14. In this case, the ligand is a tetrahedral chiral rhenium complex containing phosphine moieties used to chelate the catalytic rhodium ion. The rhodium/rhenium complex shown in Fig. 14 was found to be an excellent enantioselective catalyst for hydrosilylation of ketones and hydrogenation of olefins giving large enantiomeric excesses. Configurations in the reaction products are determined by the rhenium configuration of the catalyst.

Considering that:

1. It is greatly advantageous to have active chiral catalysts with easily accessible, commercially available, achiral ligands

2. In a diastereoisomeric complex the two types of chirality inducers might have opposite rather than additional effects, a feature so far impossible to predict
3. A complex with the metal atom as the inducing chirality can behave as an enantioselective catalyst

we are fully convinced that use of chirality at the metal ion in enantioselective catalysts is an exciting and new challenge for the future.

References

1. Jacobsen EN, Pflatz A, Yamamoto H (1999) Comprehensive asymmetric catalysis. Springer, Berlin Heidelberg New York
2. Ojima I (2000) Catalytic asymmetric synthesis. Wiley-VCH, New York
3. Trost BM (2004) PNAS 101:5348
4. Thummel RP, Chelucci G (2002) Chem Rev 109:3129
5. Brunner H (1999) Angew Chem int Ed 38:1194
6. IUPAC (1970) IUPAC nomenclature of inorganic chemistry, definitive rules, 2nd edn. Butterworth, London
7. von Zelewsky A, Mamula O (2000) J Chem Soc Dalton Trans 219
8. Rutherford TJ, Quagliotto MG, Keene FR (1995) Inorg Chem 34:3857
9. Hua X, von Zelewsky A (1995) Inorg Chem 34:5791
10. Werner A (1911) Chem Ber 44:1887
11. Verstregen M, Eriksson J, Hakansson M (2003) Chem Eur J 9:4678
12. Breu J, Seidl W, Huttner D, Kraus F (2002) Chem Eur J 8:4454
13. Fletcher NC, Junk PC, Reitsma DA, Keene FR (1998) J Chem Soc Dalton Trans 133
14. Hesek D, Inoue Y, Everitt SRL, Ishida H, Kunieda M, Drew MGB (1999) Chem Comm 403
15. Hesek D, Inoue Y, Everitt SRL, Ishida H, Kunieda M, Drew MGB (1999) J Chem Soc Dalton Trans 3701
16. Hasegawa T, Matsuura K, Kobayashi K (2000) Chem Letters 466
17. Hesek D, Hembury GA, Drew MGB, Taniguchi S, Inoue Y (2001) Inorg Chem 40:2478
18. Hesek D, Inoue Y, Everitt SRL, Ishida H, Kunieda M, Drew MGB (2000) Inorg Chem 39:317
19. Lacour J, Hebbe-Viton V (2003) Chem Soc Rev 32:373
20. Liu CF, Doyle J (1967) Chem Comm 412
21. Worrel JH (1972) Inorg Synth 13:195
22. Kolp B, Viebrock H, von Zelewsky A, Abeln D (2001) Inorg Chem 40:1196
23. Brissard M, Convert O, Gruselle M, Guyard-Duhayon C, Thouvenot R (2003) Inorg Chem 42:1378
24. Chavarot M, Ménage S, Hamelin O, Charnay F, Pécaut J, Fontecave M (2003) Inorg Chem 42:4810
25. Abu-Dari K, Raymond KN (1977) J Am Chem Soc 99:2003
26. Brumaghim JL, Raymond KN (2003) J Am Chem Soc 125:12066
27. Hua X, Lappin A G (1995) Inorg Chem 34:992
28. Patterson BT, Keene FR (1998) Inorg Chem 37:645
29. Monchaud D, Jodry JJ, Pomeranc D, Heitz V, Chambron J-C, Sauvage, Lacour JJ-P (2002) Angew Chem Int Ed 41:2317

30. Lacour J, Goujon-Ginglinger C, Torche-Haldimann, Jodry JJ (2000) Angew Chem Int Ed 39:3695
31. Fiedler D, Leung DH, Bergman RG, Raymond KN (2004) J Am Chem Soc 126:3674
32. Knof U, von Zelewsky A (1999) Angew Chem Int Ed 38:302
33. von Zelewsky A (1999) Cood Chem Rev 811
34. Pierre JL (1998) Coord Chem Rev 178–180:1183
35. Hayoz P, von Zelewsky A, Stoeckli-Evans H (1993) J Am Chem Soc 115:5111
36. Pezet F, Daran JC, Sasaki I, Aït-Addou H, Balavoine GGA (2000) Organometallics 19:4008
37. Hesek D, Inoue Y, Ishida H, Everitt SRL, Drew MGB (2000) Tetrahedron Lett 41:2617
38. Mizuno T, Takeuchi M, Hamachi I, Nakashima K, Shinkai S (1998) J Chem Soc Perkin Trans 2:2281
39. Ohta T, Takaya H, Noyori R (1988) Inorg Chem 27:566
40. Kuznetsov VF, Yap GPA, Alper H (2001) Organometallics 20:1300
41. Kuznetsov VF, Jefferson GR, Yap GPA, Alper H (2002) Organometallics 21:4241
42. Onitsuka K, Dodo N, Matsushima Y, Takahashi S (2001) Chem Comm 521
43. Andres R, Brissard M, Gruselle M, Train C, Vaissermann J, Malézieux B, Jamet J-P, Verdaguer M (2001) Inorg Chem 40:4633
44. Brunner H, Köllnberger A, Burgemeister T, Zabel M (2000) Polyhedron 19:1519
45. Hamelin O, Pécaut J, Fontecave M (2004) Chem Eur J 10:2548
46. Shibata Y, Tsuchida R (1929) Bull Chem Soc Jpn 4:142
47. Shibata Y, Tanaka Y, Goda S (1931) Bull Chem Soc Jpn 6:210
48. Sato I, Kadowaki K, Ohgo Y, Soai K, Ogino H (2001) Chem Comm 1022
49. Goldstein AS, Beer RH, Drago RS (1994) J Am Chem Soc 116:2424
50. Kromm K, Osburn PL, Gladysz JA (2002) Organometallics 21:4275

Author Index Volumes 1–15

The volume numbers are printed in italics

Subject Index

Printing: Krips bv, Meppel
Binding: Stürtz, Würzburg